Macromolecular Systems – Materials Approach

Springer
Berlin
Heidelberg
New York
Barcelona
Budapest
Hong Kong
London
Milan
Paris
Santa Clara
Singapore
Tokyo

V.P. Shibaev (Ed.)

Polymers as Electrooptical and Photooptical Active Media

With 117 Figures and 30 Tables

Springer

Prof. Valery P. Shibaev
Chemistry Department
Moscow State University
119899 Moscow B-234, Russia

ISBN 3-540-59486-8 Springer-Verlag Berlin Heidelberg New York

Library of Congress Cataloging-in-Publication Data, Polymers as electrooptical and photooptical active media/ edited by V.P. Shibaev, p. cm. Includes bibliographical references. ISBN 3-540-59486-8 (hardcover: alk. paper) 1. Electrooptical devices – Materials. 2. Photopolymers. 3. Polymer liquid crystals. 4. Thin films – Materials. I. Shibaev, V.P. (Valerij Petrovich) II. Series. TA1750.
P65 1996 620. 1' 9204295 – dc20

Cover design: de'blik, Berlin

Typesetting: Scientific Publishing Services (P) Ltd, Madras

SPIN: 10135768 2/3020/SPS – 5 4 3 2 1 0 – Printed on acid-free paper

Preface

The rapid development of various systems for the recording, storage and retrieval of information, telecommunication transmission and visualization of images requires not only the fundamental elaboration of the physicotechnical principles of these methods, but also the creation of new materials which are capable of carrying out rapidly and reliably the necessary "commands", as a rule given by means of optical, electric and magnetic signals.

Polymeric materials, which successfully combine a lot of advantages, occupy a special position among a large number of such organic and inorganic materials. One of the major advantages of macromolecular systems is their ability to produce thin and stable films as well as coatings for large areas which may be deposited on a variety of substrates using sufficiently simple methods such as spin coating, dipping, lamination, etc.

There are, at least two principal methods which allow us to obtain highly sensitive macromolecular systems.

The first is the simplest and consists of obtaining mixtures of polymers with special compounds containing photo- or electroresponsive groups. In this case a polymer acts as the matrix ("host"), while the introduced additive plays the role of the "guest". Such mixtures allow us to control significantly the properties of the "guest" by means of the external field without disturbing the matrix itself. In such systems, a polymer serves, to a certain extent, only as a passive "host" giving "cart blanche" to "guest" molecules.

The second method seems more complicated but is more effective. In this case electrosensitive or photoactive molecules or their fragments are chemically "incorporated" into macromolecules. Now the polymer matrix, as a rule, is no longer indifferent to the "guest" and may exert an influence on its behavior or, in contrast, may "feel" a certain "perturbing" effect due to the "guest".

In both cases, two principal questions arise – what happens to "guest" molecules under the action of the external field on the one hand and how the polymer matrix responds to the changes caused by the action of the excited "guest" molecules on the other.

These questions are comprehensively considered in all the chapters of this book in which the authors, although in different ways, are trying to decide how "to force" the polymer systems to change directly (and preferably reversibly) their structural organization (and become oriented) under the action of external factors such as light or an electric field.

We are aware of the fact that the field of electrooptical and photooptical phenomena observed in polymers is exceedingly wide and although it has been dealt with in many separate reviews and chapters of some books (see, for example, [1–11]). However, it seemed expedient to discuss the most essential scientific research into obtaining new polymer systems which could be used as photooptical and electrooptical active media as well as their function.

Our attention is focused on polymer systems and liquid-crystalline (LC) polymers which could be used as new electro- and photooptical media on the basis of some of the newly-discovered physico-chemical effects and phenomena [12–17]. From the comparatively large amount of information published in the literature concerning photo- and electrooptical phenomena, only those lines of investigation have been chosen for this book which could soon result in the creation of new recording media, the design of passive and controllable optical elements, optically nonlinear materials and other optically active media easily regulated by means of the electric field and light illumination.

Chapter 1 is written by the Japanese authors (K. Horie and Sh. Machida) and is devoted to photochemical hole burning and the photochemical properties of the doped dye molecules in linear polymers.

In Chapter 2 (V. Shibaev, S. Kostromin, S. Ivanov), the electrooptical and photooptical behavior of LC side-chain mesogenic polymers (so-called comb-shaped polymers) in the amorphous and LC state is discussed. This chapter provides the reader with the introduction to the field of liquid crystal polymers containing electrosensitive or photosensitive molecules chemically-linked to the backbone via a spacer. Such polymers can be considered as a new class of high-quality laser addressed thin layer materials for reversible optical data storage and holography, the creation of optical discs and integrated optics-field wave guides as well as new modulators and optical switches.

In Chapter 3 (F. Kreuzer, Ch. Bräuchle, A. Miller and A. Petri), a relatively new class of optical recording materials is considered. These materials are cyclic-siloxanes which might be classified as side-chain oligomers/polymers, forming different types of mesophases. The authors of this chapter demonstrate interesting possibilities of reversible thermal and photooptical recording on cholesteric films of photochromic and nonphotochromic LC cyclic siloxanes, using simultaneously the polarized light and electric field that opens up very wide possibilities for holography and the creation of nonlinear optical wave guides.

Chapter 4 (K. Ichimura) reviews the results of investigations devoted to very interesting and unusual polymeric photoregulated systems – the so-called "command surfaces". Using these modified surfaces or thin photochromic films, one can regulate the alignment of low molar mass liquid crystals deposited on such "command surfaces" simply by means of light illumination. It provides researchers with a powerful and very sophisticated method of photoregulation of liquid crystal alignment by photochromic "command molecules" incorporated in polymeric materials.

And finally Chapter 5 (C.P. van der Vorst and S.J. Picken) includes the consideration of nonlinear optical properties in polymers with polar hyperpolarizable groups attached as side chains to the backbone. The electric field induced ordering of initially isotropic and LC systems is actively discussed and different theoretical models for the calculation of the ordering degree are considered. The experimental results relating to isotropic and LC polymers are compared with each other and with theoretical models. The examples of a positive and negative influence of liquid crystallinity on the polar order in polymeric materials are comprehensively discussed.

Each chapter of the book is written by well-known and acknowledged experts in the appropriate scientific research field providing a high scientific level of this up-to-date edition. Electrooptics and photooptics of polymers are attracting a great deal of attention of scientists and researchers in connection with the possibility of potential applications of these materials.

Unfortunately, it is impossible to embrace in one book all the interesting aspects of electrooptical and photooptical phenomena observed in polymers. These and related problems form a specific sort of interface between polymer science (including synthesis, study and processing techniques of polymeric materials) and photochemistry, optics, the physics of solids and information techniques in both industrial and academic research. This creates certain difficulties for finding the necessary information in the "sea of publications" taking into account the very rapid growth of the number of articles, collections, reviews and books covering very wide aspects of problems connected with electrooptical and photooptical phenomena in polymers.

We hope that this book will help the reader find the necessary information in the most concentrated and prepared form. The editor and the authors would consider their job successful if their reader became interested in these new polymer materials and found some new stimulating ideas in this book. Other aspects close to the problems considered here can be found in recent publications and books indicated in the references.

This book is intended for a wide range of scientists and researchers including polymer and material scientists, applied scientists and technologists in chemistry, physics, electronics and optics. It is my hope that this book will stimulate the scientific activity of graduate students in this new and fascinating interdisciplinary field.

I would like to take this opportunity to express my deepest thanks to the authors for their efforts and to the publisher and the staff of Springer-Verlag, especially to Dr. M. Hertel, for such efficient cooperation.

Finally I warmly acknowledge the encouragement given by my wife Marina.

Moscow, June 1996 Valery P. Shibaev

References

1. Nonlinear optical properties of organic molecules and crystals. vol 1 and 2 (ed by D Chemla and J Zyss). Academic Press, New York, 1987
2. Properties of photosensitive materials and their application in holography (ed by VA Barachevsky), (in Russian), Nauka, Leningrad, 1987
3. Nonlinear optical and electroactive polymers (ed by P Prasad and D Ulrich), Plenum, New York, 1988
4. Polymers for lightwave and integrated optics. Technology and applications (ed by L Hornak), Marcel Dekker Inc., New York, 1992

5. Macromolecular host-guest complexes: optical, optoelectronic and photorefractive properties and applications (ed by S Jenekhe), Materials Research Society, Pittsburg, 1992
6. Kumar GS. Azo Functional Polymers. Technomic Publishing AG, Basel, 1992
7. Applied photochromic polymer systems (ed by C McArdle), Blackie, Glasgow-London, 1994
8. Polymeric materials for microelectronic applications: science and technology (ed by H Ito, S Tagawa and K Horie), ACS Symposium Series No 579, Clothbound, 1995
9. Special polymers for electronics and optoelectronics (ed by J Chilton and M Goosey). Chapman & Hall, London, 1995
10. Applications of electroactive polymers (ed by B Scrosati). Chapman & Hall, London, 1995
11. Processes in photoreactive polymers (ed by V Krongauz and A Trifunac). Chapman & Hall, London, 1995
12. Plate NA, Shibaev VP. Comb-shaped polymers and liquid crystals. Chemistry, Moscow, 1980 (in Russian). New edition (in English), Plenum, New York, 1987
13. Side chain liquid crystal polymers (ed by C McArdle), Blackie, Glasgow, London, 1989
14. Applied liquid crystal polymers (ed by M Takeda, K Iimura, N Koide and N Plate). Mol Cryst Liq Cryst 189, p 1–192, 1989
15. Shibaev VP, Belyaev SV. Prospects of applications of functional liquid-crystalline polymers and compositions. Vysokomolek. Soedin. A32, No 12, 2266(1990) (in Russian); Polymer Sci USSR 32, 2361(1991) (English translation)
16. Liquid-crystal polymers (ed by N Plate), Plenum, New York, 1993
17. Liquid-crystalline and mesomorphic polymers (ed by V Shibaev and Lui Lam), Springer-Verlag, New York, 1994
18. Shibaev VP. Some new physico-chemical aspects of side chain liquid crystal polymers, Mol Cryst Liq Cryst 243, 201(1994)

Contents

CHAPTER 4

Photoregulation of Liquid Crystal Alignment by Photochromic Molecules and Polymeric Thin Films

CHAPTER 5

Electric Field Poling of Nonlinear Optical Side Chain Polymers

Contributors

Bräuchle, Ch.
Institute of Physical Chemistry, University of Munich, Sophienstraße 11,
80333 Munich, Germany

Horie, K.
Department of Chemistry and Biotechnology, Faculty of Engineering,
The University of Tokyo, 7-3-1 Hongo, Bunkyo-ku, Tokyo 113, Japan

Ichimura, K.
Research Laboratory of Resources Utilization, Tokyo Institute of Technology,
4259 Nagatusta, Midori-ku, Yokohama 227, Japan

Ivanov, S.A.
Physics Department, Moscow State University, 119899 Moscow B–234, Russia

Kostromin, S.G.
Chemical Department, Moscow State University, 119899 Moscow B–234, Russia

Kreuzer, F.H.
Consortium für Electrochemische Industrie GmbH, Central Research Company
of Wacker-Chemie GmbH, Zielstattstraße 20, 81379 Munich, Germany

Machida, S.
Department of Chemistry and Biotechnology, Faculty of Engineering,
The University of Tokyo, 7-3-1 Hongo, Bunkyo-ku, Tokyo 113, Japan

Miller, A.
Consortium für Electrochemische Industrie GmbH, Central Research Company
of Wacker-Chemie GmbH, Zielstattstraße 20, 81379 Munich, Germany

Petri, A.
Institute of Physical Chemistry, University of Munich, Sophienstraße 11,
80333 Munich, Germany

Picken, S.J.
Akzo Nobel Central Research, Physical Chemistry Department, P.O. Box 9300,
6800 SB Arnhem, The Netherlands

Shibaev, V.P.
Chemical Department, Moscow State University, 119899 Moscow B-234, Russia

van der Vorst, C.P.J.M.
Akzo Nobel Central Research, Applied Physics Department, P.O. Box 9300, 6800 SB Arnhem, The Netherlands

Symbols and Abbreviations

CHAPTER 1

PHB	Photochemical hole burning
DAQ	1,4-dihydroxyanthraquinone
NPHB	Nonphotochemical hole burning
TLS	Two level systems
TPPS	Tetraphenylporphine sulfonate
PVA	Poly(vinyl alcohol)
LIHF	Laser-induced hole filling
ZnTTBP	Zinc tetratolyltetrabenzoporphine
PMMA	Poly(methyl methacrylate)
GAP	Glicydil azide polymer
SMD	Single molecule detection
SMS	Single molecule spectroscopy

CHAPTER 5

α	molecular linear polarizability
$\Delta\alpha$	anisotropy of α
(α_i, β_i)	polar and azimuthal angles of dipoles i=1, 2 (Fig. 5.5)
β	molecular hyperpolarizability
Γ_{12}	angle between dipoles 1 and 2
δn	EO modulation of refractive index
Δn	anosotropy of refractive index
$\Delta\epsilon$	anisotropy of permittivity
ϵ	(1) permittivity (2) "strength" of Maier-Saupe potential U_0
θ	angle between permanent dipole moment and poling field
μ	molecular permanent electric dipole moment
π	(1) 3.1416 (angle of π radials is 180°) (2) dipole pair states defined in Section 5.3.6
πa	anti-parallel π dipole pair state
πp	parallel π dipole pair state
Π	pi electron cq. conjugated pi electron system
σ	co-linear dipole pair state (Section 5.3.6)
σa	anti-parallel σ dipole pair state

σp	parallel σ dipole pair state
$\chi^{(1)}$	macroscopic linear susceptibility
$\chi^{(2)}$	macroscopic second-order NLO susceptibility
ω	angular frequency
a	reduced dipole energy ($a=\mu E/kT$)
A	(1) general angle-dependent function $A(\theta)$ (2) pi electron Acceptor group
A∏D	Acceptor-Donor substituted conjugated pi electron system
b	reduced dipole-dipole energy defined in Section 5.3.7 ($b=\mu^2/r^3kT$)
B	magnetic field
C_1, C_2, C_3	constants in Landau type expression Eq. (14)
DC-SHG (= EFISH)	DC-induced Second-Harmonic-Generation (also called Electric Field Induced Second Harmonic generation)
D	pi electron Donor group
E	(1) general electric field (2) short notation for local poling field strength
E′	local electric field
E_{mod}	modulation field strength
E_{pol}	poling field strength
EO	Electro-Optic (abbr.)
i, j, k=x, y, z	molecular coordinates (z is direction of μ)
I, J, K=X, Y, Z or 1, 2, 3 resp.	macroscopic coordinates (3≡Z is poling direction)
f	(1) local field factor (2) free energy term
F	combination of local field factors
$G(\theta)$	normalized orientational distribution function
h	temperature coefficient of C_1 in Eq. (14)
k	Boltzmann's constant
n	refractive index
n_e	extraordinary refractive index (3 direction)
n_o	ordinary refractive index (1 and 2 directions)
ñ	director axis (± Z direction)
N	hyperpolarizability density
O_{Ii}, O_{Jj} O_{Kk}	projections of molecular unit vectors on corresponding macroscopic unit vectors
p	(1) induced molecular dipole moment (2) parallel polarization direction
P	induced macroscopic polarization density
$P_1(\cos\theta)$	Legendre polynomial of order 1 in $\cos\theta$
$\langle P_1 \rangle$	average of corresponding Legendre polynomial (orientational order parameter, polar order for odd 1, axial order for even 1)
Q	tensor order parameter in Landau type expression Eq. (14)
r	(1) Pockels coefficient (2) separation between dipoles 1 and 2

s	perpendicular polarization direction ("senkrecht")
T_{cl}	zero field clearing temperature
$T_{cl}(E)$	field-dependent clearing temperature
T_{crit}	critical end temperature
TSC (=TSP)	Thermally Stimulated Current (Thermally Stimulated Poling)
TSD	Thermally Stimulated Discharge
U	angle dependent mean-field potential (Section 5.3.1)
U_0, U_1, U_2	energy terms of U defined in Table 5.1
X_D	dimer fraction (Section 5.3.8)
Z	partition function

CHAPTER 1

Photochemical Hole Burning and Photooptical Properties of Doped Dye Molecules in Linear Polymers

K. Horie and S. Machida

1.1 Introduction

Photochemical hole burning (PHB) is a phenomenon where a stable dip or a hole is created in the absorption spectrum of a chromophore by site-selective excitation with a narrow-band light such as a laser. The phenomenon of PHB was discovered by two Russian groups in 1974 [1, 2]. The site-selective excitation is possible when the absorption line width of each chromophore (this is called, the "homogeneous width", Γ_h) is smaller than the width of the whole absorption band profile (called the "inhomogeneous width", Γ_i). In such a case, the whole absorption line is called, the "inhomogeneous band". This situation is often realized for doped dye molecules in amorphous matrices like polymers at low temperatures because the optical linewidth of each chromophore decreases drastically as temperature becomes lower and because the resonant frequency of each chrompohore in amorphous solids has an inhomogeneous distribution due to the difference in its microenvironment (Fig. 1.1). The sharp absorption line of individual chromophore at low temperatures is called the "zero-phonon line".

The width of a zero-phonon line or a homogeneous width, Γ_h, can be expressed as follows [3, 4]:

$$\Gamma_h = 1/2T_1 + 1/T_2 \tag{1.1}$$

where T_1 is the population lifetime or the so-called lifetime of an upper level, T_2 is pure dephasing time or phase relaxation time. The pure dephasing time, T_2 is determined by the extent of interaction between electronic levels and phonon or other excitations. Since T_2 more strongly depends on temperature than T_1, the temperature dependence of Γ_h is determined by T_2.

The production of a photochemical hole occurs when the excited chromophore changes its resonant frequencies after the relaxation from the excited states by a photochemical reaction (Fig. 1.2). A spectral hole can also be created by a photophysical process such as a conformation change in the matrix, which is called "nonphotochemical hole burning" (NPHB). However, the boundary between PHB and NPHB is not so clear. Sometimes the term of "PHB" is used to represent the hole production phenomena via both photochemical and nonphotochemical processes. Here we use "PHB" as a general term for the hole production phenomena. The term "persistent spectral hole burning (PSHB)" is also used as a general term for hole burning phenomena.

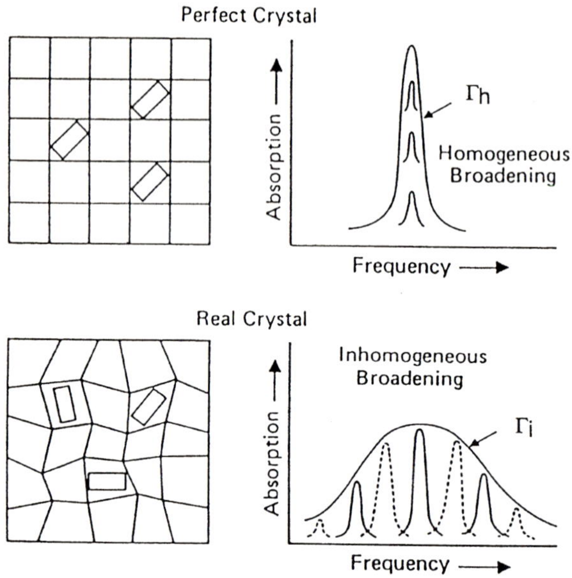

Fig. 1.1. Schematic representation of homogeneous and inhomogeneous broadening [4]

PHB has been of interest both for pure science and for applications [3–5]. The first proposed and the most well-known application of the PHB phenomenon is its use as a frequency-domain optical storage system [4, 6]. The recording density of the present optical memory systems is limited by the diffraction-limit of the laser spot. If the PHB phenomenon is applied to an optical memory system, the frequency-domain can be utilized for data storage by using the presence or absence of a hole for 1 bit of digital data (Fig. 1.3). Then, the recording density may increase by more than 10^3 times because in many PHB

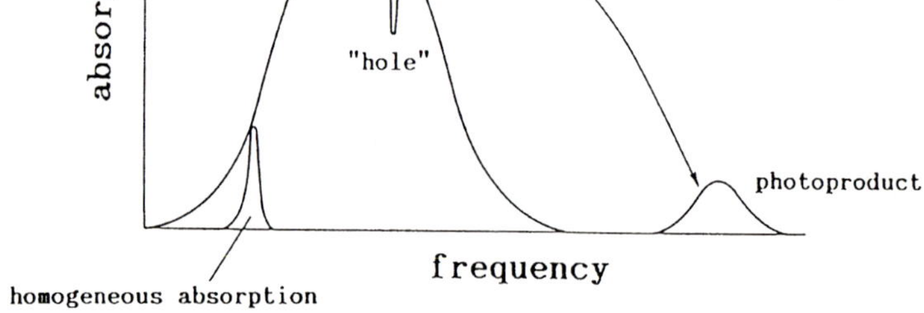

Fig. 1.2. Schematic representation of hole formation [5]

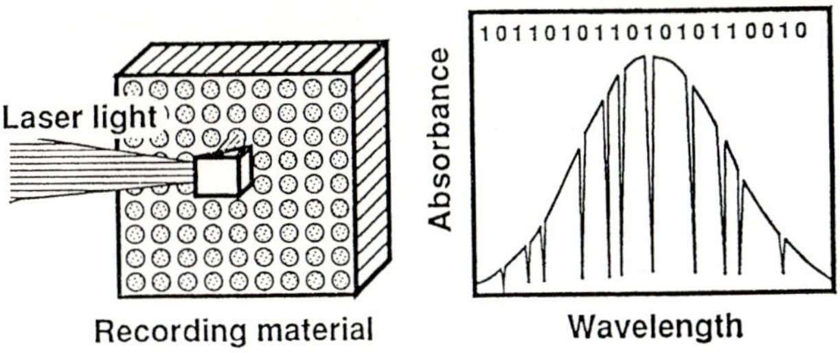

Fig. 1.3. Schematic representation of PHB memory [13]

systems the ratio of the inhomogeneous width to the homogeneous width is more than 10^3 at the temperature of liquid helium.

In addition to an optical digital memory system, possible applications of PHB have grown to cover a variety of optical devices. PHB materials can be used for laser pulse shaping based on Fourier syntheis [7] or voltage modulation [8]. Frequency-multiplexed optical spatial filters can be produced using PHB materials [9]. By using holographic techniques, PHB materials can also be applied to a frequency-domain hologram media [10]. The electric field effects provide us with the possibility of using PHB materials as an electric-field-domain optical memory [11] and as optical operating devices [12].

For the application of PHB to an optical memory system, the following properties are required for PHB materials: 1) high-temperature hole burning and persistence, 2) highly efficient photoreaction, 3) nondestructive readout, 4) formation of deep holes with high multiplicity. The second of the required properties is needed to achieve high-speed writing, and the third and fourth properties are necessary for high-speed readout. It was pointed out that the fourth one is indispensible for realizing a high-density optical storage system with PHB materials [13].

The PHB phenomenon is quite attractive not only as a means of obtaining new optical devices but also as a spectroscopic method of low-temperature solid-state photophysics and photochemistry. PHB is a very sensitive detection method of photoinduced processes because we can observe the photoreactions site-selectively and detect the change of the extremely small number of chromophores compared to conventional-absorption spectroscopy.

Table 1.1 lists some of the phenomena which can be probed by PHB spectroscopy and the PHB properties affecting these phenomena. Using PHB spectroscopy we can obtain the homogeneous width of a chromophore which offers information about the excited-state lifetime (T_1) [14] or pure dephasing lifetime (T_2) [15]. The phonon side hole and pseudo-phonon side hole, which are produced by the electron-phonon coupling, inform us of the low-energy excitation modes of host polymers [16]. By the measurement of the hole shapes at various temperatures we can also get information about thermally induced relaxation process of the matrices [17] and the temperature dependence of the Debye-Waller factor [18]. The measurement of frequency dependence of growth rates of the resonant and satellite holes makes it possible to analyze the vibronic components of an inhomogeneous band [19, 20].

Table 1.1. Phenomena Probed by PHB spectroscopy

PHB property		Phenomena probed
Hole width	Energy dependence	Homogeneous width (pure dephasing lifetime and excited state lifetime)
	Temperature dependence	Dephasing mechanism (electron-phonon coupling, energy transfer etc.)
	Cycle temperature dependence	Thermally induced spectral diffusion (structural relaxation, intrinsic TLS)
	Time dependence	Spectral diffusion (intrinsic TLS, structural relaxation)
Hole area	Energy dependence	Quantum efficiencies
	Intensity dependence	Bottleneck lifetimes
	Temperature dependence	(Temperature depedence of) Debye-Waller factors (DWF)
	Cycle temperature depedence	Thermally induced hole filling (barrier heights to reverse reaction, structural relaxation which induces site redistribution)
	Time depedence	Spontaneous hole filling (barrier heights to reverse reaction)
Energy difference from Zero-phonon hole to pseudo-phonon side hole or phonon-side hole		Low-energy excitation modes of host matrices
Satellite hole	Positions	Vibronic and other excited state splittings
	Growth rate	Vibronic overlaps of inhomogeneous band (real values of photoreaction, absorption cross sections and DWF for vibronic transition)
Antihole positions		Nature and number of product states, ground state splittings
Laser-induced hole filling		Laser-induced back reaction, non-site-selective excitation
External field effects	E-field	Stark effect, site symmetry
	H-field	Zeeman effect, degeneracy of transition
	Strain field	Stress coupling coefficients, symmetry
Gating light effects		Photoreactions from highly excited state (electron transfer, energy transfer, etc.)

A typical experimental apparatus for PHB is shown in Fig. 1.4 [21]. The holes are usually detected through the change in transmittance in the absorption spectrum of the sample. If the sample is opaque and cannot be measured with transmission, the holes are observed through the decrease in emission intensity in the fluorescence or phosphorescence excitation spectrum [22]. For extremely high resolution and fast detection, the frequency modulation (FM) spectroscopy method was invented [23].

The quantum efficiency for hole formation, Φ, is one of the most essential parameters for PHB and is defined as the ratio of the number of the chromophores which contribute to hole production to the number of the absorbed photons in the system. We evaluate Φ by the following equation [24, 25]:

$$\Phi = -\{dA/dt\}_{t=0}/\{10^3 I_0 \ (1 - 10^{-A_0})\varepsilon R\} \tag{1.2}$$

where A is the time-varying absorbance at the burning wavelength, A_0 is the absorbance at the burning wavelength before irradiation, I_0 is the incident laser

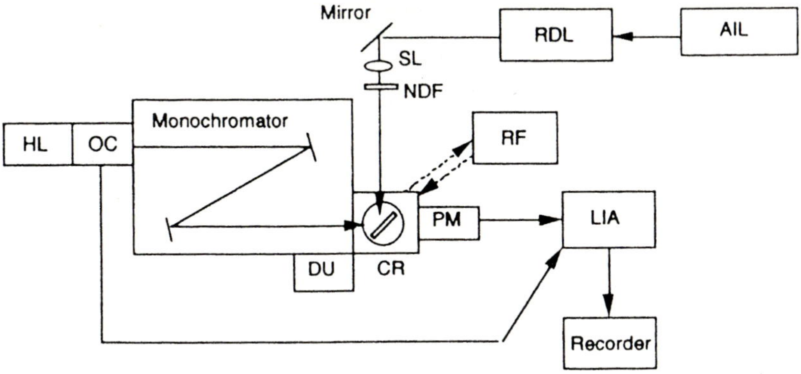

Fig. 1.4. A typical hole burning and detection system [21]. *HL*, halogen lamp; *OC*, optical chopper; *DU*, driving unit for the grating; *CR*, cryostat; *AIL*, argon ion laser; *RDL*, ring dye laser; *RF*, refrigerator; *LIA*, lock-in amplifier; *NDF*, neutral density filter; *SL*, simple lens; *PM*, photomultiplier

intensity given in einstein $cm^{-2}s^{-1}$, ε is the molar extinction coefficient for the peak absorption at room temperature, and $R - \Gamma_i/\Gamma_{hole}$ is the reciprocal initial ratio of the hole width, Γ_{hole}, to an inhomogeneous width, Γ_i. The Φ calculated by Eq. (1.2) should agree with a real value of quantum efficiency of photoreaction if the whole observed inhomogeneous band consists only of homogeneous purely electronic zero-phonon lines. However, in general dye/polymer systems phonon-side band and vibronic components are included in the observed inhomogeneous band. So the quantum efficiency of hole burning for dye/polymer systems has a wavelength dependence and is much lower than that of photoreaction [19, 20].

1.2 Photochemical Hole Burning of Dye/Polymer Systems

1.2.1 Hole Burning Mechanisms

Since the first observation of the PHB phenomenon [1, 2], many (probably more than 100) materials have been proved to show PHB. The reported PHB materials can be classified into several groups based on the mechanism of hole formation. Table 1.2 lists the hole formation mechanisms and typical materials. The mechanisms are; proton tautomerization of free-base porphyrins and phthalocyanines [26], hydrogen-bond rearrangement from intra- to intermolecular bonding [27], ionization [28], donor-acceptor electron transfer [29], decomposition [30], isomerization [31], sensitization of photoreactive matrix [32], and non-photochemical processes [33].

Table 1.2. PHB mechanisms and typical materials

	Guest Molecule or Color Center	Matrix
Ionization	Sm^{2+}	CaF_2, BaClF, $BaFCl_{0.5}Br_{0.5}$, $SrFCl_{0.5}Br_{0.5}$, $Sr_{0.5}Mg_{0.5}F$ $Cl_{0.5}Br_{0.5}$, Fluorohafnate glasses
	Co^{2+}	$LiGa_5O_8$
	Ti^{3+}	$Y_3Al_5O_{12}$
	Cr^{3+}	$SrTiO_3$
	F_2 color center	Neutron-irradiated sapphire
	Carbazole	Boric acid glass
Decomposition	Anthracene-tetracene photoaduct	PMMA
	s-tetrazine derivaties	Benzene, Terphenyl
	5,10-dihydrophenazine	Fluorene
Donor-Acceptor Electron Transfer	Zn (or Mg) tetrabenzoporphine derivatives	Organic halogenated compounds-PMMA
		Organic cyanide compounds -PMMA
	Zn porphyrin	Pyridine dimer anion -Ether
	Tetraphenylporphine	p-Benzoquinone-PMMA
	Zn porphyrin	Sulfonium Salt-EDTA-PVA
Isomerization	1,3,5,7 -octatetraene	n-Hexane
Hydrogen Tautomerization	Free-base porphyrins and Free-base phthalocyanines	n-Alkanes, polymers, and organic and inorganic glasses etc.
Inter-to Intra-Hydrogen-bond rearrangement	Quinizarine and Hydroxy-naphthoquinone	Polymers and inorganic glasses
Sensitization	Zn tetratoyltetrabenzoporphyrin (ZnTTBP)	Glycidyl azide polymer
		Acyloxyimino polymers
		Norbornadiene-PMMA
Nonphotochemical processes	Porphyrins, Ionic dyes, Ions, and many photostable molecules	Polymers and glasses

The free-base porphyrin system is one of the most well-investigated systems for the PHB reaction from the early stage of PHB investigation. Figure 1.5 illustrates the scheme of the tautomerization reaction of free-base porphyrine [26]. In the free-base porphyrin system, the chemical structure (or electronic structure) itself does not change during the tautomerization reaction. However, there is the difference in guest-host interaction which induces the change in resonant frequency of the porphyrin molecules. The tautomerization reaction can proceed in various types of host matrices such as polymers and organic or inorganic glasses, which promises the hole burning ability of a system with a free-base porphyrin. For this reason the free-base porphyrin is used for a guest chromophore for studying the low-temperature behavior of polymer matrices [16, 18].

Hydrogen-bond rearrangement from intra-to intermolecular bonding is also a well-investigated PHB mechanism. This mechanism can proceed in the case of 1,4-dihydroxyanthraquinone (DAQ) and its derivatives when a host matrix undergoes hyrdogen-bonding [27]. A hole burnt in the DAQ system is thermally less stable than free-base porphyrin systems [34].

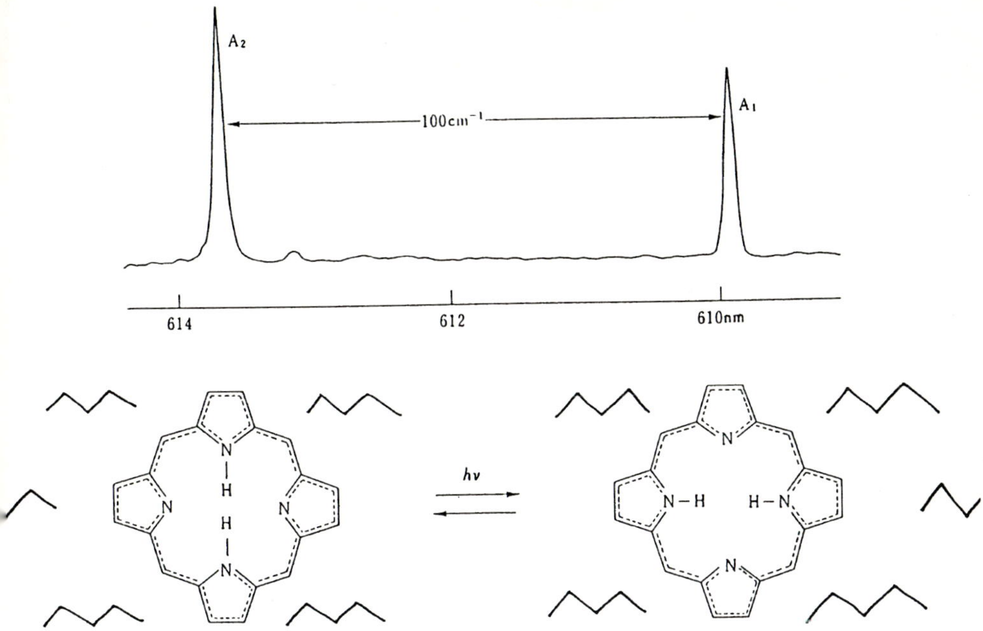

Fig. 1.5. Reaction scheme of tautomerization of free-base porphyrin [26]

Site-selective electron transfer reaction is one of the recently investigated mechanisms for PHB. The first observation of PHB by the electron transfer mechanism was achieved for a pyridine dimer anion as a donor and metaloporphyrin as a PHB chromophore and acceptor [35]. In other systems a porphyrin derivative is commonly used as a donor PHB chromophore and an organic electron acceptor is doped together in a polymer matrix. In particular, organic halogenated compounds are used for various systems as electron acceptors [29]. Recently, enhancement of hole burning by the use of sacrificially consumed molecules as electron donors was reported [36].

Ionization of an excited chromophore is also one of the popular PHB mechanisms. All the systems of this mechanisms consist of an inorganic color center as a guest and an inorganic matrix as a host except carbazole/boric acid [28] and perylene/boric acid [37] systems.

The photodecompositions of dimethyl-s-tetrazine [30], anthracenetetracene photoadduct [38], and dihydrophenazine [39] are known to cause the PHB phenomena. The isomerization of 1,3,5,7-octatetraene is also reported to be a PHB mechanism [31]. In these systems organic molecular crystals are often used as host matrices.

The above-mentioned photochemical reactions for PHB proceed in the guest chromophore itself, but a photochemical hole is also created by the chemical change in host matrices. The sensitization of a photoreactive matrix via triplet–triplet energy transfer is a recently found PHB mechanism and belongs to this type [32]. Since triplet sensitization is a very popular process in organic photochemistry, this type of mechanism can be applied to various photoreactive matrices such as photosensitive polymers. The sensitization mechanism promises an efficient hole production of a dye with zero-phonon line even if the dye itself does not react by photoexcitation.

PHB systems with the mechanisms of electron transfer, ionization, decomposition, and sensitization are usually utilized in photon-gated PHB systems, as mentioned in Sect. 1.2.2.

The persistent spectral holes can also be produced by a photophysical change in host matrices or guest chromophores. This process is called NPHB (non-photochemical hole burning). The NPHB phenomenon is treated as tunneling transitions in asymmetric double-well potentials, so-called two-level systems (TLS), and used as a probe for host dynamics in amorphous solid and biological systems [33]. In general, the quantum efficiency of NPHB processes is lower than that for the case of photochemical mechanisms. In the matrices which form hydrogen-bonding, NPHB proceeds with relatively high efficiency due to the rearrangement of hydrogen bonds [40]. Typical materials which show NPHB are ionic dyes doped in poly(vinyl alcohol), but NPHB is observed for a lot of polymer systems doped with photostable dyes. Since the activation energy of backward process of the photophysical change is not so high, a hole made by the NPHB mechanism is thermally less stable than that by photochemical mechanisms.

The phenomenon of persistent spectral hole burning is not only possible for an electronic transition but also for an vibronic transition in the infrared spectra [41]. The hole production in the infrared spectra is caused by conformer interconversion or photoinduced reorientation of guest molecules.

1.2.2
High-Temperature Hole Burning and Thermal Stability of Holes

Generally, the PHB phenomenon is difficult to observe at higher temperatures. This is mainly because the homogeneous width of each chromophore becomes drastically large at higher temperatures. The decrease in the Debye-Waller factor at higher temperatures also prevents high-temperature PHB.

The Debye-Waller factor, DW (T), represents a fraction of a zero-phonon line in a transition of a dye molecule, and is defined by [4]:

$$DW(T) = S_0(T)/[S_0(T) + S_p(T)] \tag{1.3}$$

Where $S_0(T)$ is the integrated intensity of a zero-phonon line and $S_P(T)$ is the integrated intensity of a phonon-side band.

The ability to store holes at higher temperatures is also an important thermal property of PHB systems as well as the capability of high-temperature PHB. The cyclic annealing experiment is a good way to observe the changes in a hole profile induced by the thermal excursion which is performed as follows: A hole is burnt at low temperature and is measured on the hole profile, then the temperature of the system is elevated and the system is annealed. In some cases the hole profile is measured at the elevated temperature. The system is then cooled to the initial hole-burning temperature, and a hole profile is measured again. This tempera-

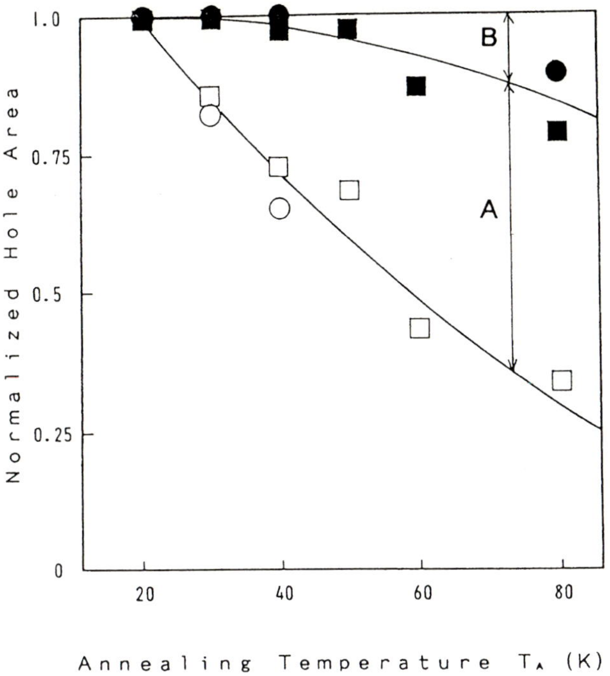

Fig. 1.6. Normalized hole area as a function of annealing temperature for ZnTTBP/GAP [52]. Holes were burnt at 20 K and measured at annealing temperatures (□, ○) and at 20 K after cooling from the annealing temperatures (■, ●)

ture cycle is repeated with increasing elevated temperatures. Figure 1.6 and Fig. 1.7 show the hole area and the hole width, respectively, for dye/polymer systems during cyclic annealing. The reversible change in the hole area (region A in Fig. 1.6) represents the temperature dependence of the Debye-Waller factor. The irreversible decrease in hole area (region B in Fig. 1.6) reflects structural relaxation of the matrix which induces site redistribution, or a thermally induced

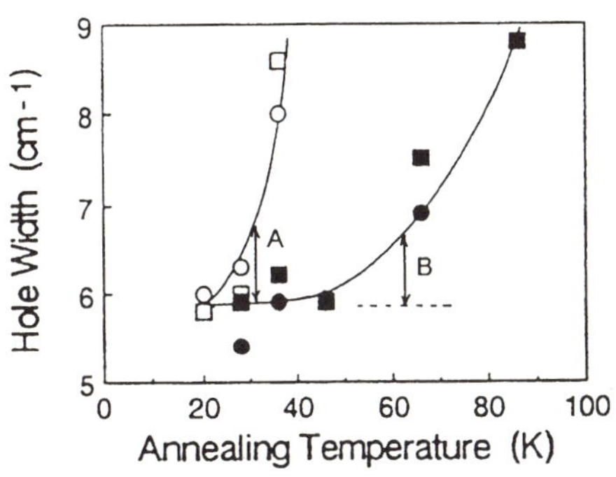

Fig. 1.7. Hole width as a function of annealing temperature for ZnTTBP/PVHP [50]. Holes were burnt at 20 K and measured at annealing temepratures (□, ○) and at 20 K after cooling from the annealing temperatures (■, ●)

Table 1.3. Factors affecting thermal properties of PHB

Factors	Phenomena
Homogeneous width, Γ_h	Reversible broadening of hole width
Debye-Waller factor, DW(T)	Reversible change in hole area
Thermally activated spectral diffusion	Irreversible broadening of hole width after temperature cycle (no change in hole area)
Laser-induced spectral diffusion	Irreversible broadening of hole width caused by laser irradiation (no change in hole area)
Thermally activated hole filling	Irreversible change in hole area after temperature cycle due to structural relaxation and/or backward reaction

back reaction which is usually observed for NPHB systems. The reversible change in hole width (region A in Fig. 1.7) is due to temperature dependence of a homogeneous width, and the irreversible increase in hole width (region B in Fig. 1.7) is called "spectral diffusion" which is caused by thermally induced structural relaxation. In the case of spectral diffusion, the correlation distance of the relaxation would be larger and the scale of the relaxing motion would be smaller than those leading to hole filling. Table 1.3 lists the factors affecting the thermal properties of PHB.

For the realization of high-temperature PHB for a system, the temperature dependence of the Debye-Waller factor should be small as well as the inhomogeneous width should be considerably higher than the homogeneous width.

Table 1.4. Highest hole-burning and hole-preserving temperatures for typical PHB materials

Materials	Highest Preserving Temperatures (K)	Highest Burning Temperatures (K)
Tetraphenylporphine (TPP)/Poly (methyl methacrylate)	100	50
Tetraphenylporphine sulfonate/Poly (vinyl alcohol)	150	120
TPP/Phenoxy resin	120	100
TPP/Epoxy resin		80
Octaethylporphine/$\gamma - Al_2O_3$ (adsorbed)		90
Quinizarine/SiO_2	65	
Quinizarine/$\gamma - Al_2O_3$ (adsorbed)		[illegible]
ZnTTBP-$CHCl_3$/PMMA		90
ZnTTBP/Glycidyl azide polymer	> 80	80
ZnTTBP/TiO_2		140
5,10-Dihydrophenazine/Fluorene	320	
Neutron-irradiated sapphire	520	
Sm^{2+}/BaClF	300	
Sm^{2+}/$SrFCl_{0.5}Br_{0.5}$		296
Sm^{2+}/$Sr_{0.5}Mg_{0.5}FCl_{0.5}Br_{0.5}$		R.T.
Sm^{2+}/$BaFCl_{0.5}Br_{0.5}$		R.T.
Sm^{2+}/Fluorohafnate glasses		R.T.
Sm^{2+}/Borate glass		R.T.
Neutron-irradiated diamond		R.T.

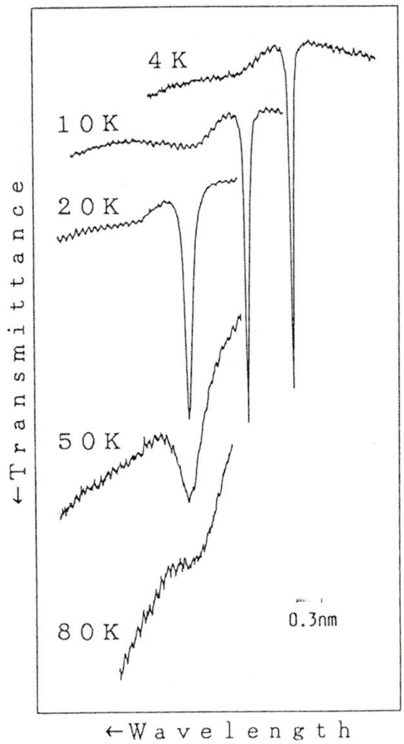

Fig. 1.8. Typical hole profiles burnt at 4.2 to 80 K for TPPS/PhR [25]

The small temperature dependence of the Debye-Waller factor means a weak electron-phonon coupling of the system. Suppression of the structural relaxation and the back reaction is also important for high-temperature hole burning and preserving. The extent of structural relaxation depends on the rigidity of the matrix.

The highest temperatures at which a hole can be burnt or preserved for typical PHB systems are listed in Table 1.4. In some inorganic materials, PHB can be performed even at room temperature [42, 43]. Preservation of a hole at room temperature, i.e., observing again at 2 K of a hole burnt at 2 K after its excursion at room temperature, was first observed for Sm^{2+} in BaClF system [44], and was then also observed for dihydrophenazine doped in fluorene molecular crystal [39].

In polymer systems, photochemical holes have been observed at 120 K at the highest and preserved at 150 K at the highest as shown in Table 1.4. Tetraphenylporphine sulfonate (TPPS) doped in poly(vinyl alcohol) (PVA) is the first material which shows PHB above the liquid nitrogen temperature [45]. Most of the polymer matrices in which hole burning can be realized above 77 K have hydroxyl groups [45–47]. Figure 1.8 shows a series of hole profiles burnt at 4.2-80 K for tetraphenylporphine doped in phenoxy resin (PhR) which has hydroxyl groups. The hydrogen bond of hydroxyl groups is considered to suppress the structural relaxation and the electron-phonon interaction. Hydrogen-bonding polymer systems have a small extent of laser-induced hole filling (LIHF) and large values of low-energy excitation modes, which should be important factors for high-temperature PHB [18].

The energy of low-energy excitation modes of host polymers are estimated with PHB measurements from the energy difference between zero-phonon holes and pseudo-phonon side holes [16]. Heat capacity measurements and neutron inelastic measurements also gives the low-energy excitation modes of host polymers. The low energy excitation modes are supposed to reflect the microscopic ordered structure of polymer matrices and to contribute to the excess heat capacity of amorphous polymers [5].

Persistent spectral hole burning at room temperature is also reported by using morphology dependent resonances of polymer microparticles which have a random distribution of sizes [48]. This phenomenon cannot be applied to an ultra-high-density optical memory, but may be utilized for a frequency-domain optical storage in an optical fiber system.

1.2.3 Photon-Gated Hole Burning Systems

If a PHB material with a single-photon mechanism is applied to an optical memory, high-speed writing and non-destructive readout are contradictory to each other, because, as the efficiency for hole formation becomes higher, the readout of the memory becomes more destructive. The concept of photon-gated PHB was proposed by an IBM research group to solve this problem [49]. Figure 1.9 illustrates the principle of photon-gated PHB. In the photon-gated materials, a photochemical hole is produced efficiently only when a gating light, λ_2, of another wavelength from a burning light, λ_1, is irradiated simultaneously with λ_1. The readout of the memory is possible by scanning the wavelength of λ_1 only, so a non-destructive readout is achieved. When a hole is produced via two-photon excitation with a one-color of λ_1, PHB systems are also classified as photon-gated materials [30]. In this case the hole depth becomes proportional to the square of irradiation intensity, and so the intensity of reading light should be much lower than that of writing light.

Table 1.5 lists the reported photon-gated PHB materials and their properties. For photon-gating the photoreaction must have a threshold of excitation energy. The mechanisms for photon-gated PHB are ionization [28], decomposition [38], donor-acceptor electron transfer [29], sensitization of photoreactive matrix [32], and a non-photochemical process in the presence of carboxylic acid [50]. The gating ratio, G, represents the ratio of the quantum efficiency for hole formation in the case of two-color excitation, Φ_2, to that without λ_2, Φ_1.

Donor-acceptor electron transfer is a most popular mechanism of photon-gated PHB for dye/polymer systems. Figure 1.10a shows an energy diagram of photon-gated PHB by the electron transfer reaction with zinc tetratolyltetrabenzoporphine (ZnTTBP) with chloroform ($CHCl_3$) doped in poly (methyl methacrylate) (PMMA) [28]. In the ZnTTB-$CHCl_3$/PMMA system, the ZnTTBP molecules are excited to a higher triplet state, T_n, via T-T absorption, and eject electrons to nearby $CHCl_3$ molecules. The fast detachment of chlorine anion prevents backward electron transfer, which leads to efficient formation of

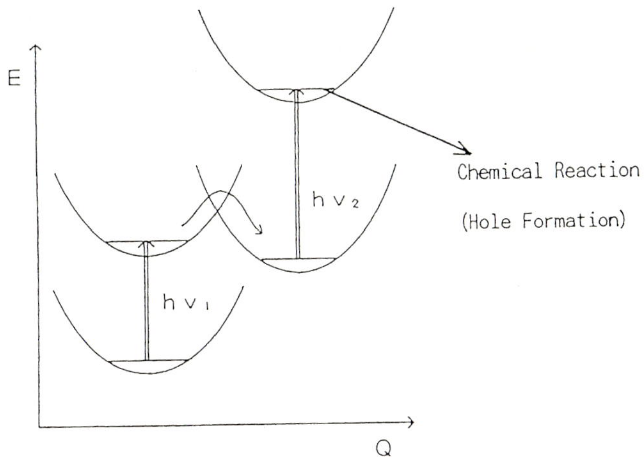

Fig. 1.9. Schematic representation of photon-gated PHB

Table 1.5. Photon-gated PHB mechanisms and typical materials

Mechanism	Material	λ_1(nm)	λ_2(nm)	Gating ratio
Ionization	Sm^{2+}/BaClF	688	454–630	10^6
	Sm^{2+}/SrClF	691	350	
	Sm^{2+}/CaF_2	696	515	
	Co^{2+}/$LiGa_5O_8$	660	680	20
	Cr^{3+}/$SrTiO_3$	794	780–1060	
	Carbazole/Boric acid glass	335	407–515	400
	Perylene/Boric acid glass	440	440	
Electron transfer	ZnTTBP-$CHCl_3$/PMMA	630	350–500	10^2–10^4
	ZnTTBP-ArCN/PMMA	630	488–515	10
	ZnTTBP-*p*-hydroxybenzaldehyde	630	488–515	
	TPP/ArX	643	515	
	ZnTSPP-Sulfonium salt-EDTA/PVA	600	488	8
Decomposition	Anthracene-tetracene aduct/PMMA	327	442	~2
	s-tetrazine derivatives/Benzene	580	580	
	5,10-dihydrophenazine/Fluorene	360	500	
	Carbazole/PMMA	337	488	
Sensitization	ZnTTBP/Glycidyl azide polymer	630	488–515	2500
	ZnTTBP/Acyloxyimino polymer	630	488–515	90
	ZnTTBP-norbornadiene/PMMA	630	488–515	36
Nonphoto chemical process	ZnTTBP/Poly(vinyl hydrogen-phthalate)	630	488–515	3

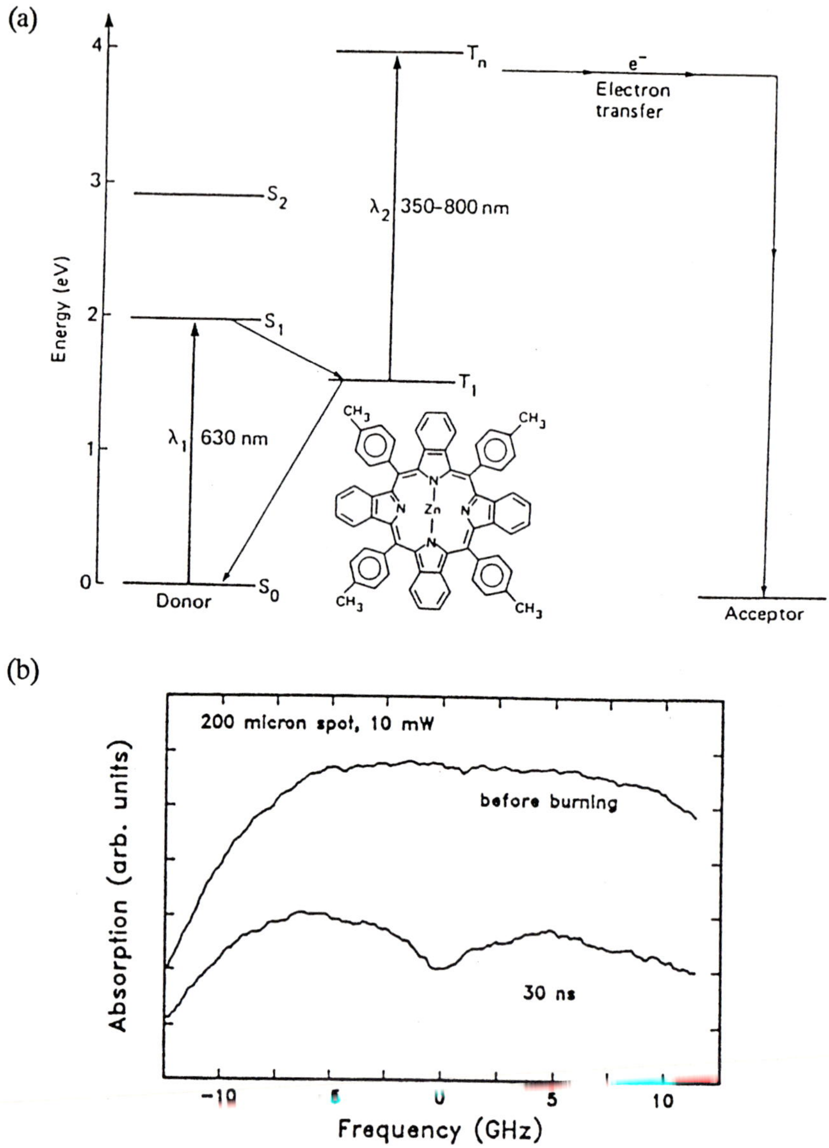

Fig. 1.10a,b. Photon-gated PHB for ZnTTBP-CHCL$_3$/PMMA [29]. **a** Level diagram via donor-acceptor electron transfer meachanism. **b** Fast burning in 200 μm spots. Upper trace: Baseline before burning. Lower trace: after burning with a 10 mW beam at λ_1 for 30 ns, followed immediately with a 200 ms gating beam at 488 nm of 17 mW

the gated holes. Fast burning of a gated hole is also observed for the ZnTTBP – $CHCl_3$/PMMA system. In Fig. 1.10b a hole is produced by 30 ns irradiation of λ_1 and 200 ms irradiation of λ_2. Photon-gated materials have the advantage that the chromophore can be excited with λ_2 for more than once within the lifetime of the lowest excited state. In PHB systems with single-photon mechanisms, a hole cannot be formed in a short time (<1–10 ns) with larger relative depth than the quantum efficiency because the molecules which can contribute to the hole production cannot be excited more than once. Subnanosecond (0.5 ns) hole burning is observed for a one-photon electron transfer system [51], suggesting the high quantum efficiency of the material.

Figure 1.11 shows another example of a photon-gated PHB system using ZnTTBP as a guest chromophore and glycidyl azide polymer (GAP) as a photoreactive polymer matrix [52]. In the ZnTTBP/GAP system, the energy of the T_n of ZnTTBP is transferred to the T_1 state of a nearby azide group leading to the decomposition of azide and the hole production.

Photon-gated PHB systems are also attractive for purely scientific studies on low-temperature photochemistry and photophysics because by using photon-gated PHB measurement, the photoreactions from highly excited states can be detected and distinguished from the photoreactions from the lowest excited state. Energy gap dependence of electron transfer rates [53] and hydrogen-bond rearrangement of carboxylic acid [50] that proceed from highly excited triplet states are measured and discussed. Photoreactions from the highly excited states

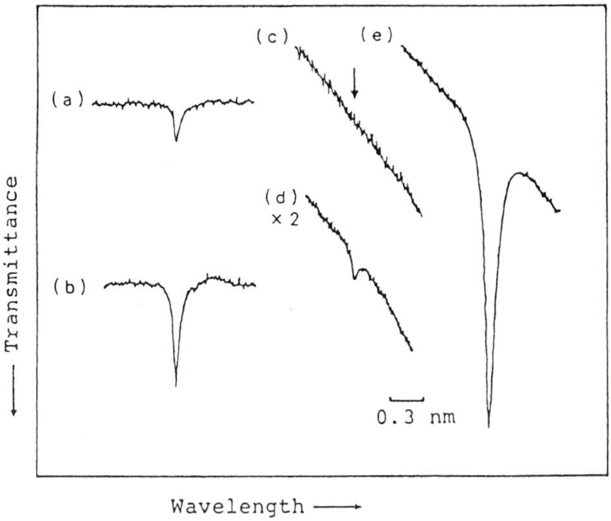

Fig. 1.11. Typical hole profiles in photon-gated PHB for ZnTTBP/GAP at 20 K [52]. Traces (*a*) and (*b*) show the two-color holes burnt by 1 min irradiation of 19 $\mu W/cm^2 \lambda_1$ for both holes and of 8 and 40 $mW/cm^2 \lambda_2$, respectively. Trace (*c*) shows the transmission spectrum after one-color irradiation (66 μ W/cm^2, 60 min). Trace (*d*) shows a one-color hole (380 $\mu W/cm^2$, 10 min) with relative hole depth of 2.2%. The addition of 8 $mW/cm^2 \lambda_2$ to λ_1 of the same irradiation conditions as trace (*d*) produced trace (*e*) (relative hole depth: 34%)

are one of the recent topics in photochemistry [54], so the photon-gated PHB materials are good systems for studying this field.

1.2.4
Required Properties for PHB Materials for Practical Application

Moerner et al. [55] investigated the required properties for PHB materials to realize a frequency-domain optical storage system first in 1985. According to their results, it is difficult to realize a PHB memory by using single-photon materials due to the destructive readout. In photon-gated materials, they concluded that there is an allowed region in the values of the absorption cross section and the concentration per unit area of guest chromophore [56]. Figure 1.12 shows the constraints for photon-gated PHB materials to achieve practical S/N ratios. In Fig. 1.12, the allowed region becomes large as the effective hole burning yield (or the relative absorbance change in the case of short burning time) increases.

The burning conditions for getting the maximum recording density was investigated by Murase et al. [57]. They pointed out that there is an optimized value of the recording spot diameter which gives the maximum recording density when the values of absorption cross section of chromophore, its concentration,

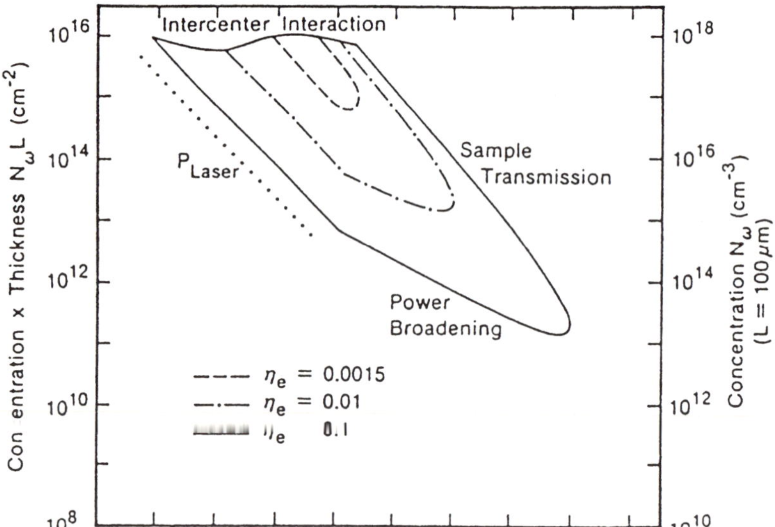

Fig. 1.12. Materials constraints for gated PHB systems in order to achieve practical S/N ratios [56]. Effective hole burning yield, η_e, is defined as the relative absorbance change produced by the best-case two-color hole burning during the writing time of 30 ns. Materials within the shown boundaries for particular values of η_e permit data readout with SNR equal to or greater than the required value with a maximum of 10 mW reading power

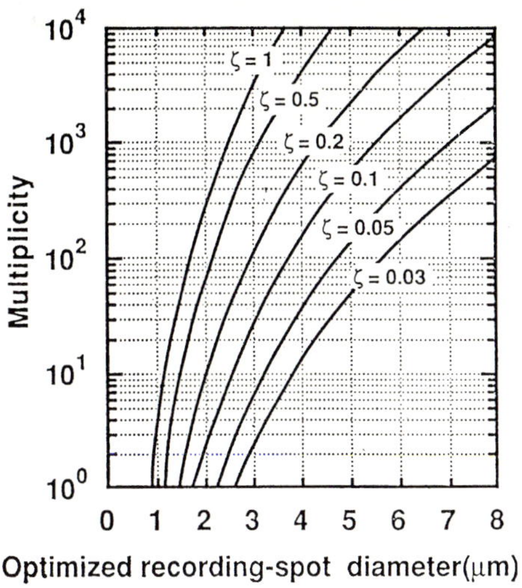

Fig. 1.13. Optimized recording spot diameter as a function of multiplicity and hole depth ζ in case of TPPS/PVA [57]

hole depth, and wavelength multiplicity are given. The experimental results of multiple hole burning with TPPS/PVA system indicated the maximum achievable recording density of this system to be a little bit less than 20 Gbit/cm^2 at 4 K (Fig. 1.13) [13]. A kind of LIHF, which is observed for the holes in the shorter wavelength region than a newly burnt hole and becomes prominent as the hole interval becomes small, is concluded to be the main factor for determining the achievable recording density.

Free-base porphyrin systems such as TPPS/PVA and chlorine/poly(vinyl butyral) show a low extent of LIHF compared to other dye/polymer systems. The mechanisms of LIHF for the TPPS/PVA system was also studied by Murase et al. [58]. The non-site-selective excitations due to the thermal fluctuation of the chromophore in the ground state was found to be the cause of LIHF. The LIHF observed for the holes in a longer wavelength region than that in a newly burnt hole is very conspicuous and is attributed to the non-site-selective excitation due to the overlapping of phonon-side bands and vibronic components. The LIHF phenomena due to the non-site-selective excitations were suppressed by using mesoporphine as a guest dye which is a flatter, lighter, and more rigid molecule than TPPS [59].

The extent of LIHF observed for NPHB systems is larger than PHB systems with photochemical mechanisms. The mechanism of the LIHF for NPHB systems is a back reaction of the photoproducts formed by the previous hole burning [60].

1.2.5 Electric-Field Readout and Holographic Detection

PHB measurements are very sensitive to the microenvironments of guest chromophores, so one can observe the influence of external fields (e.g., electric field [61], magnetic field, or pressure [62], etc.) with very high resolution. The external

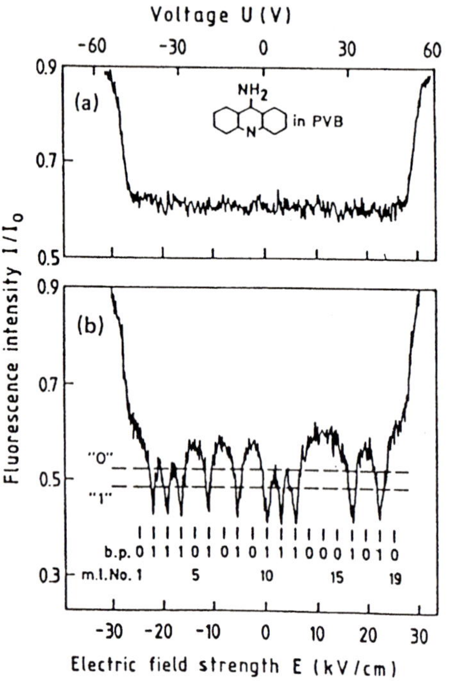

Fig. 1.14a,b. Electric-field readout of spectral holes for 9-aminoacridine in poly(vinyl butyral) films using a fixed frequency He-Cd laser [63]. (**a**) Background fluorescence signal before burning. (**b**) Fluorescence signal after burning spectral holes at various values of the applied field

field expands the dimension of memory for the PHB memory system, in addition to the two-dimentional space and frequency domain. Figure 1.14 shows an example of writing spectral holes in the electric-field domain [63]. One hole is written under certain electrical field conditions and another is burnt under different electrical field conditions. The Stark effect caused by the external electric field induces inhomogeneous frequency shifts to the guests due to the difference in microenvironments. The redistribution of resonant frequencies leads to the disappearance and appearance of holes with the change in the electrical field.

One of the ways to detect photochemical holes with high sensitivity is to use a holographic technique. Figure 1.15a shows the experimental setup for burning and detecting photochemical holes using a holographic technique [64]. A spectral hole is recorded as a diffraction grating in space domain by irradiating with a burning light from two directions. The grating is formed by the change in (real part of) refractive index as well as the change in absorption (or imaginary part of refractive index). The hole is detected as the increase in diffraction intensity by irradiating reference beam only. This method promises intrinsically zero-background measurement and thus a hole can be measured with high sensitivity and high signal-to-noise ratio as shown in Fig. 1.15b,c.

Fig. 1.15a-c. Holographic detection of spectral holes [64]. (**a**) Experimental arrangement. (**b**) and (**c**) Comparison of transmission and holographic detection. (**b**) transmission signal for a very shallow hole, (**c**) Holographic detection for the same hole

(a)

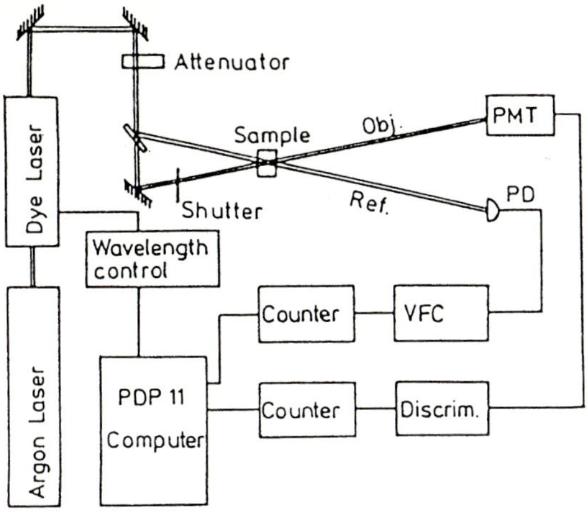
Attenuator
Dye Laser
Sample
Obj.
PMT
Shutter
Ref.
PD
Wavelength control
Argon Laser
Counter
VFC
PDP 11 Computer
Counter
Discrim.

(b)

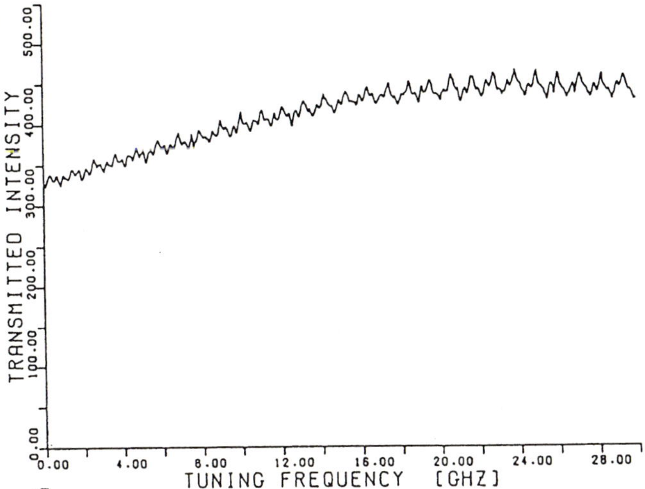
TRANSMITTED INTENSITY
0.00
100.00
200.00
300.00
400.00
500.00
0.00
4.00
8.00
12.00
16.00
20.00
24.00
28.00
TUNING FREQUENCY [GHZ]

(c)

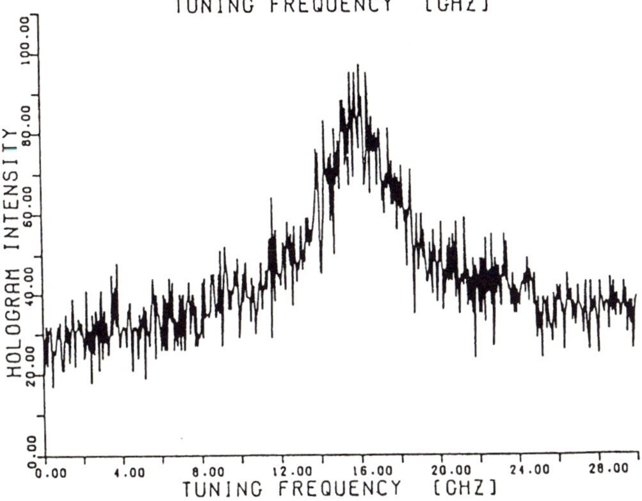
HOLOGRAM INTENSITY
0.00
20.00
40.00
60.00
80.00
100.00
0.00
4.00
8.00
12.00
16.00
20.00
24.00
28.00
TUNING FREQUENCY [GHZ]

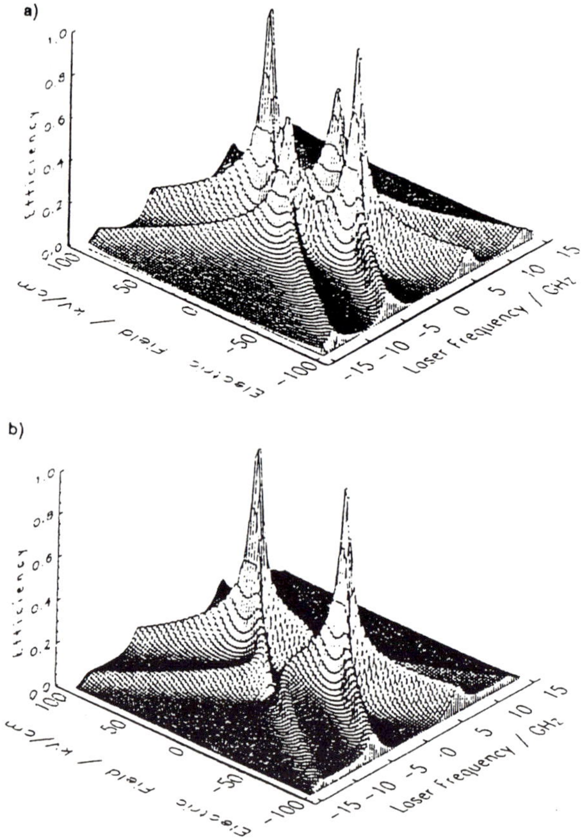

Fig. 1.16a,b. "Molecular computing" using diffraction of two holograms in the frequency-electric field plane [65]. The holograms were stored at the same frequency, separated by different field strengths. In the region where the Stark components overlap, the coherent superposition of the holograms leads to constructive interference (phase differnce 0 (**a**)), or destructive interference (phase difference π (**b**)). It could correspond to AND or XOR operation for parallel computing

The holographic recording technique can also be applied for the electric-field domain. A concept of "molecular computing" has been proposed and demonstrated by Wild et al. (Fig. 1.16) [65]. Two holograms are stored with different electric fields, E_1 and E_2 , and restored via the electric field $(E_1 + E_2)/2$. The interference of the two holograms makes the diffraction efficiency strong or weak depending on the phase shift between the two holograms. It could correspond to the AND or XOR operation for parallel computing. Recently, they prepared a photo-bistable device using the non-linear response of the diffraction efficiency of a hologram to an electric field [66].

1.2.6 Single Molecule Spectroscopy

PHB measurements can be regarded as the selection of a certain number of dye molecules in host matrices in the frequency domain. If the number of the selected molecules is reduced to unity, we can observe the absorption spectrum of a single molecule. The first attempt at single molecule detection (SMD) was done by using a frequency modulation technique [67]. Orrit et al. utilized a fluorescence excitation method for SMD of pentacene doped as an impurity in terphenyl molecular crystal (Fig. 1.17) [68]. For the pentacene/terphenyl system, power broadening and temperature-dependent dephasing [69], spectral diffusion [70], Stark effects [71], photon bunching [72], quantum antibunching [73], dispersed fluorescence spectra [74], and fluorescence spectra with polarized light [75] have been reported. These measurements are called "single molecule spectroscopy" (SMS) as a general term.

The SMS measurements were extended also to polymer systems. The systems of perylene [76] and terylene (a higher homologue of perylene) [77] in poly-

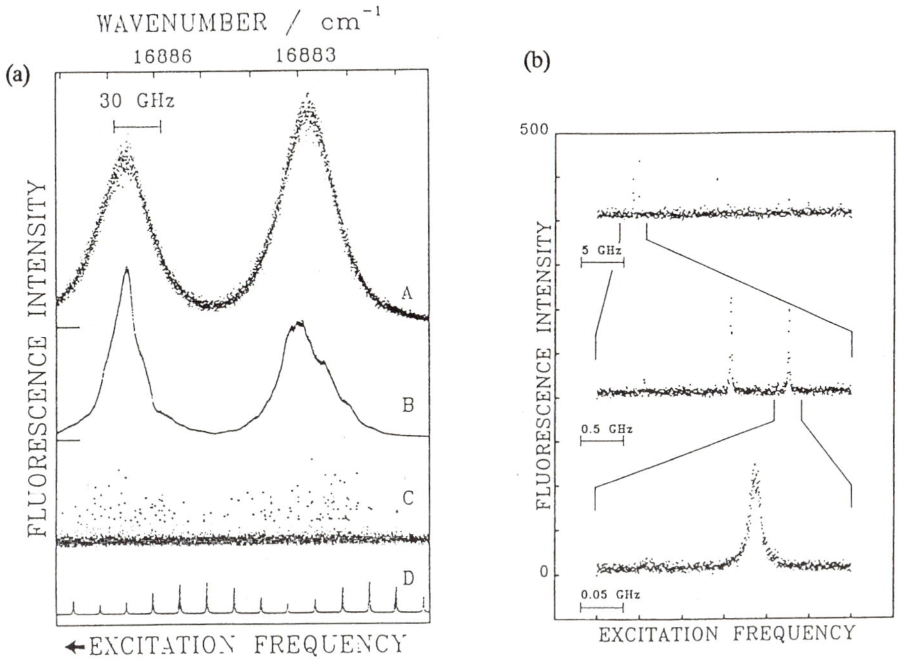

Fig. 1.17a,b. Detection of single pentacene molecules in p-terphenyl crystal by fluorescence excitation spectra [17]. (**a**) Fluorescence excitation spectra of pentacene in different *p*-terphenyl cyrstals. (*Curve A*), thick melt-grown crystal showing the Gaussian inhomogeneous bands. *B*, sublimation flake presenting narrower bands and substructure presumably due to cooling-induced defects. *C*, spectrum of a very small volume of a sublimation flake. *D*, calibration spectrum of an etalon. (**b**) Shape of a single molecule's excitation peak at different frequency scales. The bottom spectrum is approximately Lorentzian with FWHM about 12 MHz

ethylene are used for SMS. In polymer systems, the stochastic kinetics of a single molecule interacting with the matrix can be measured [78].

SMS seems to be one of the most attractive topics in current photophysics. These techniques may correspond to tunneling spectroscopy in the frequency domain.

1.3 Photon Echo of Dye/Polymer Systems

1.3.1 Principle of Photon Echo

Photon echo is a typical coherent transient phenomenon in quantum optics which has been studied for a long time [79]. The idea of application of the photon echo phenomenon to a time-domain optical storage was proposed by Mossberg in 1982 [80]. Figure 1.18 illustrates the scheme of a stimulated photon echo which is the basis of the photon echo phenomenon [81]. Three laser pulses which are resonant to the absorption of a medium are irradiated to the medium at the times of t_1, t_2, and t_3 ($t_1 < t_2 < t_3$). The stimulated photon echo is emitted as a light pulse from the medium at the time of $t_3 + t_2 - t_1$ (Fig. 1.18a). If the second pulse (at the time t_2) has a time-dependent signal-profile, the photon echo is observed as a replica of the second pulse (Fig. 1.18b). In this case, the pulses at t_1, t_2, and t_3 are called "write pulse", "data pulse", and "read pulse", respectively, and operate as a time-domain optical memory.

The mechanism of operation of the photon echo memory is illustrated in Fig. 1.19 [81]. The write pulse is irradiated at the time of t_1 to a medium which has an inhomogeneous absorption band, and excites the whole inhomogeneous band of the chromophore. The polarization of the excited chromophores begins

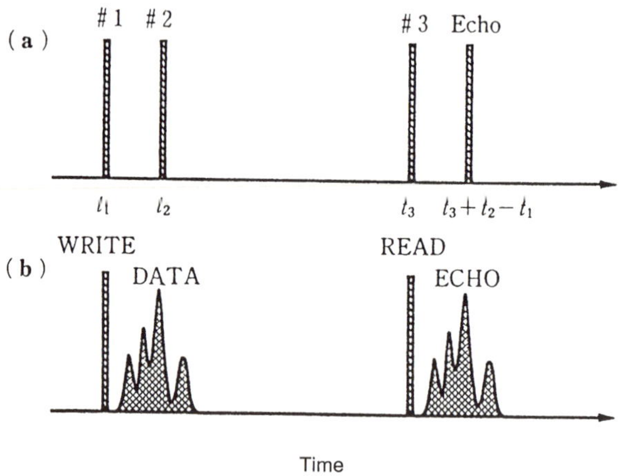

Fig. 1.18a,b. Schematic representation of stimulated photon (**a**) and photon echo memory (**b**) [81]

to oscillate with the same phase and lasts after the first pulse is finished, with the lifetime of homogeneous phase relaxation time (or dephasing time), T_2. However, the oscillation phase of each polarization relaxes within T_2 due to the inhomogeneous distribution of resonant frequencies. Then, the data pulse is irradiated at the time of t_2, (Here we suppose the data pulse to have a single-pulse profile as the write pulse for simplification). As illustrated in Fig. 1.19a, the molecule A with the same oscillation phase as the data pulse is excited to have a larger amplitude, while the polarization oscillation of the molecule B with π-radian-different phase from the data pulse is forced to decay and disappear. In other words, it can be assumed that there were two laser pulses for the molecule A but no pulse irradiation for the molecule B. If the irradiated medium undergoes PHB, the molecules which have their polarization oscillation induced, react and change their resonant frequency. Consequently, a population grating is produced in the absorption spectrum of the medium, which corresponds to a one-bit data of the photon echo memory (Fig. 1.19b). The population difference in the inhomogeneous band, $w(\Delta)$, can be expressed as [81]:

$$w(\Delta) = \varepsilon_1(t_1)\ \varepsilon_2(t_2)\ \cos(\Delta(t_2 - t_1)) \tag{1.4}$$

where ε_1 and ε_2 are the amplitude of electric fields of the irradiated laser pulses at the times of t_1 and t_2, respectively, and Δ is the difference between the resonant frequency of each chromophore, ω, and the center frequency of the laser pulse, ω_L. (Here we ignore the quantum efficiency for hole formation and the hole lifetime for simplification.) When the write and data pulses are irradiated repeatedly to make the population difference larger, the photon echo is called the "accumulated photon echo".

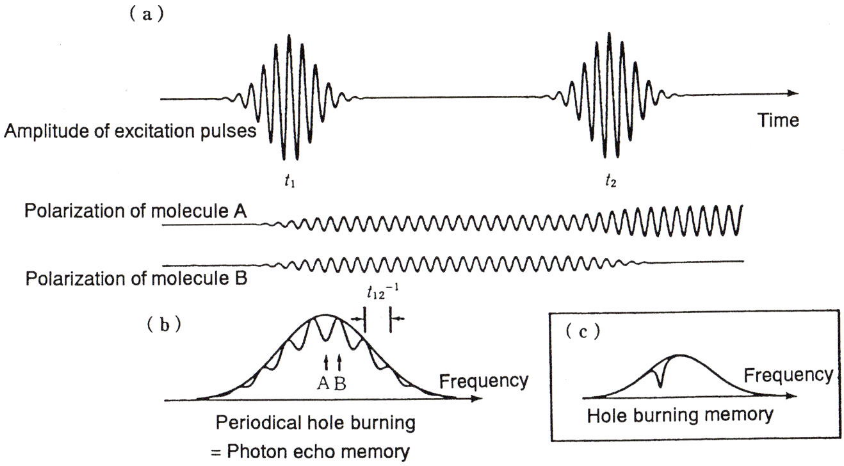

Fig. 1.19a-c. Schematic representation of the mechanism of operation of photon echo memory [81]. (**a**) Polarization oscillation of molecule A and B, which have different resonant frequencies , induced by the write and data pulses. (**b**) Periodical hole burning or 1 bit of photon echo memory, which is clearly different from 1 bit of PHB memory (**c**)

The read pulse at the time of t_3 excites the whole inhomogeneous band as the write pulse, but the population is modulated in this case as is represented in the equation (1.4). The polarizations of the molecules begin to oscillate at the time t_3 and thus the observed signal intensity, S(t), can be obtained by integrating the polarizations on Δ [81]:

$$S(t) = \varepsilon_3(t_3) \int d\Delta w(\Delta) \cos(\Delta(t - t_3)) \tag{1.5}$$

where ε_3 is the amplitude of electric field of the read pulse. By inserting Eq. (1.4) into Eq. (1.5), the following equation Eq. (1.6) is obtained [81]:

$$S(t) = \varepsilon_1(t_1)\, \varepsilon_2(t_2)\, \varepsilon_3(t_3)\, \delta(t_3 + t_2 - t_1 - t) \tag{1.6}$$

where δ is a delta function. Equation (1.6) means that the echo signal can be observed only at the time $t_3 + t_2 - t_1$, and represents the stimulated photon echo itself (Fig. 1.18a).

If the data pulse, $\varepsilon_2(t_2)$, has information in the time domain, it can be regarded as an assembly of short pulses at various times of t_2. Thus the population difference, $w(\Delta)$, can be expressed as [81]:

$$w(\Delta) = \varepsilon_1(t_1) \int dt_2\, \varepsilon_2(t_2) \cos(\Delta(t_2 - t_1)) \tag{1.7}$$

Equation (1.7) means that the data is stored in the frequency domain of the medium as the Fourier transform. The echo signal can be obtained by integrating Eq. (1.6) at t_2 [81]:

$$S(t) = \varepsilon_1(t_1)\, \varepsilon_3(t_3)\, \varepsilon_2(t - t_3 + t_1) \tag{1.8}$$

Consequently, the echo signal repeats the data pulse after $t_3 - t_1$ of the data pulse.

Furthermore, if both the write pulse and data pulse have time-domain information, the correlation between the two pulses is stored in the medium, and the convolution of the read pulse and the correlation is observed as the echo signal. Therefore, the photon echo can be used for time-domain signal processing [83] or pattern recognition. This characteristic of the photon echo is quite similar to that of holography which uses a spatial grating instead of a frequency grating in the case of the photon echo. Therefore the photon echo phenomenon is sometimes called "time-domain holography" [84].

The photon echo phenomenon can be used for studying the relaxation processes in amorphous systems as well as the PHB. The direct measurement of the pure dephasing time, T_2, and the Debye-Waller factor, DW(T), is possible using the photon echo technique. The echo signal is regarded as a Fourier transform of a hole profile [85] and reflects the fractal structure of the host polymers [86]. Ultrafast relaxation processes for dye/polymer systems have been studied using incoherent light photon echo and are interpreted based on the multiple energy levels of the systems [87].

1.3.2
Kinds of Photon Echo Materials

For the application of the photon echo to a time-domain optical memory, the following points are required for materials [81]: 1) small homogeneous width, $\Gamma_h (= T_2^{-1})$, 2) large inhomogeneous width, $\Gamma_i (= (T_2^*)^{-1})$, 3) long lifetime of memories (or holes), 4) no addition of modulation to echo signals, 5) high signal intensity, and 6) good thermal properties. In the photon echo memory the storage capacity is determined as T_2/T_2^* which corresponds to that of PHB memory.

There are three kinds of photon echo materials reported so far, namely, metal vapor [88], rare earth metals in crystals [89], and dye molecules in polymers [84]. Table 1.6 lists typical materials for these kinds and their properties.

Metal vapors were the earliest materials used for the experiments of photon echo memory. This type of material does not need a low temperature, but the storage time is determined by the excited state lifetime and thus the metal vapor can be used only for short-time storage systems.

Because rare earth metal ions in solids are known to have long T_2, a real-time measurement of photon echo is possible with these materials. Eu^{3+} is one of the most popular metal ions in this class and provides several hours of storage time. In the case of rare earth ions, there is a problem of echo modulation caused by the sublevel structure of the metal atom. So it is important to choose a material with a small degree of modulation. Figure 1.20 shows a demonstration of time- and frequency-domain hybrid optical memory using $Eu^{3+}:Y_2SiO_5$ [90]. In this system the ASCII codes of periodic table of elements are recorded in 103 addresses in the frequency domain each of which has 16 bits in the time domain. Image processing using the photon echo technique [91] and spectrally programmed coherent transient [90] are also reported for systems of rare earth metal ions.

The above mentioned materials are not usually classified as being PHB materials because the holes do not persist but decay with the lifetime of an excited state or a metastable state. The doped dye molecules in polymer systems are most preferable from the aspect of storage time. The memory can be stored almost permanently in dye/polymer systems when they are kept at a low temperature. The large inhomogeneous width of the dye/polymer systems allows ultrafast writing and reading. However, it is quite difficult to make and detect picosecond

Table 1.6. Kinds of photon echo materials and their properties

	Metal Vapor	Dye/Polymer	Rare earth metal ions in crystals
Typical material	Yb Vapor	OEP/polystyrene	$Eu^{3+}: YAlO_3$
T_2	1 μs	1 ns	10 μs
T_2^*	1 ns	100 fs	100 ps
Recording capacity (T_2/T_2^*)	10^3	10^4	10^5
Memory lifetime	1 μs	∞	5 hr
Working temperature	500°C	<10 K	<10 K

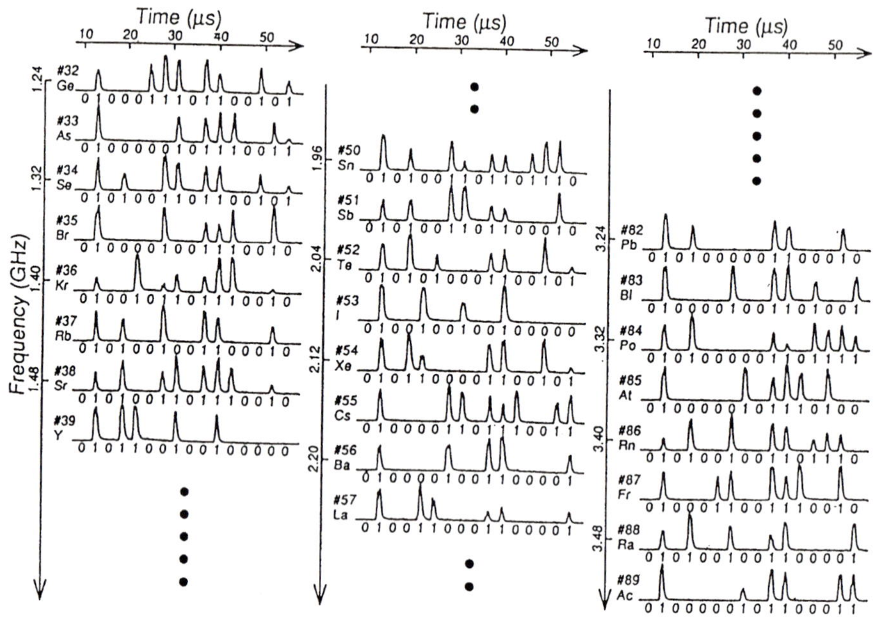

Fig. 1.20. Time- and frequency-domain hybrid optical memory using Eu^{3+}: Y_2SiO_3 [87]. Readout signal wave forms represent the ASCII-coded periodic table of elements. Each address in the frequency domain has 16 bits in the time domain. Only 24 out of 103 addresses are shown

data pulses in real time with the present optical technique. Thus photon echo experiments for dye/polymer systems are performed by writing and reading per bit.

An example of photon echo experiments for dye/polymer systems is shown in Fig. 1.21 [92]. The material is resorufin doped in PVA, and the experiment was performed at 10 K. The data bits were written by changing the glass plates with different thickness (or optical path) to make the time delay of picoseconds between the write pulse and the data pulse (Fig. 1.21a). The photon echo was detected as a heterodyne signal with a reference pulse which was irradiated at each delay time after the readout pulse. Figure 1.21 b shows the output signal as a function of delay time.

Recently, fluorescence detection of femtosecond accumulated photon echo [93], electric-field effect on accumulated photon echoes [94], and temporally multiplexed Fourier holography of femtosecond-duration images [95] for dye/polymer systems have been reported. The research of photon echo memory will involve both techniques and materials.

(a)

(b)

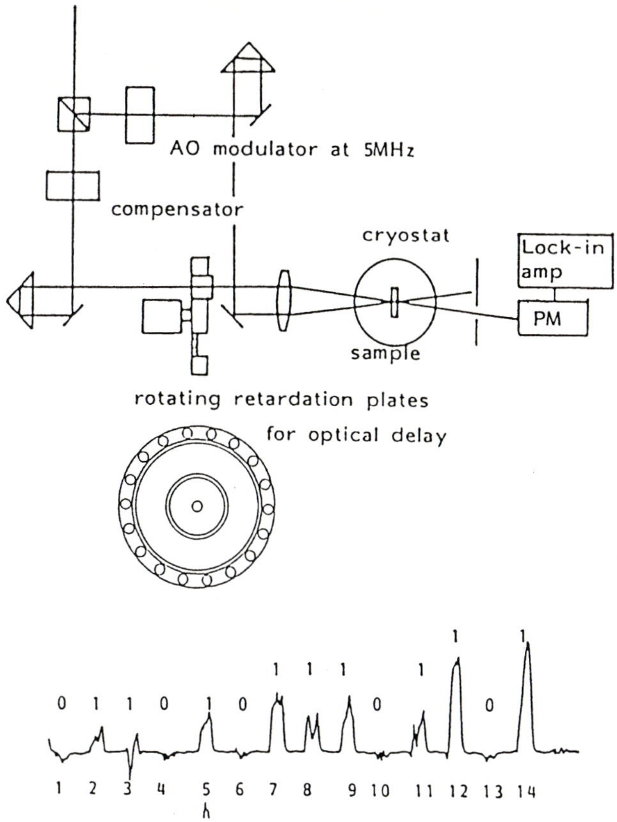

Fig. 1.21a,b. Photon echo for resorufin doped in PVA [89]. (**a**) Experimental set-up based on heterodyne detection. (**b**) A sequence of 14 data bits obtained by rotating the disk

1.4 Photoinduced Refractive Index Change and Birefringence of Dye/Polymer Systems

1.4.1 Photooptical Effects Induced by Photochromic Reactions

For the control of optical properties of materials by photoirradiation, nonlinear optical effects are commonly used. However, there are also possibilities of controlling optical properties (i.e., absorption, refractive index, birefringence, optical rotation, etc.), by photochemical or photophysical processes. The phenomena of changes in optical properties by photochemical or photophysical processes can be called "photooptical effects" generally irrespective of their origin.

Photochromism is a well-known phenomenon that a profile of absorption spectrum is changed by photoirradiation. Since absorption and refractive index are connected by the Kramers-Kronig relationship, photochromic reactions can induce refractive index change as well as absorbance change.

The *trans-cis* photoisomerization of azobenzene derivatives is one of the most popular photochromic reactions. Todorov et al. [95–97] have studied photoinduced formation of holographic gratings and birefringence using azo dye compounds doped in polymer systems. The isomerization reaction caused by

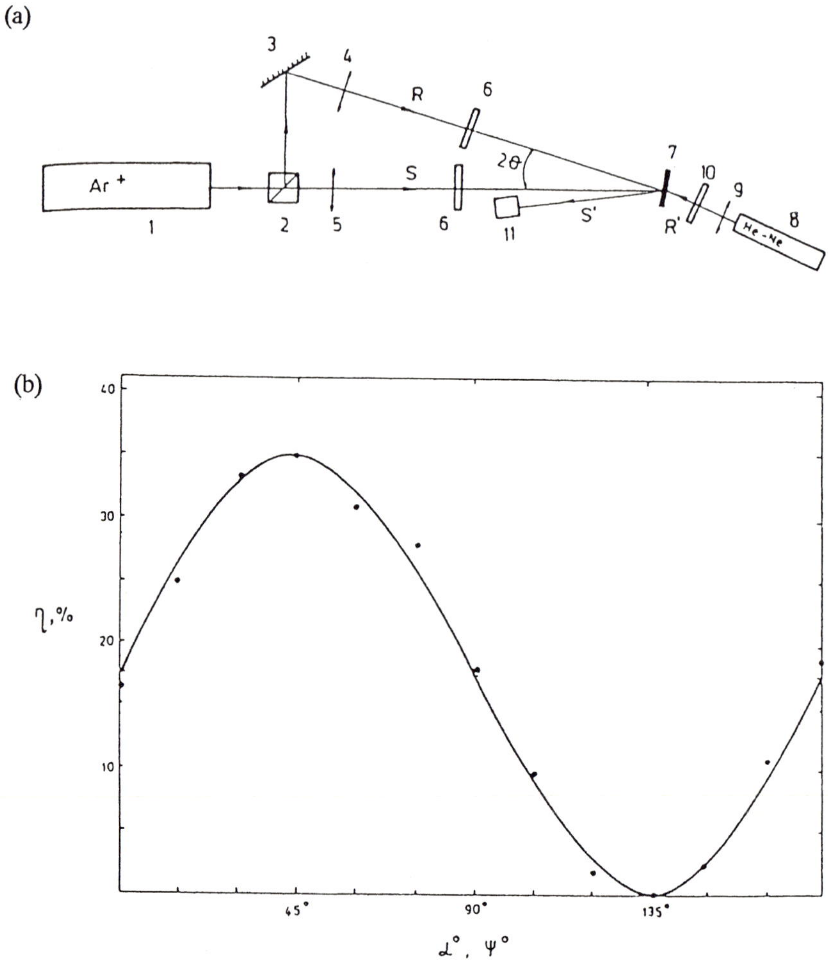

Fig. 1.22a,b. Photoinduced formation of holographic grating and birefringence for methyl orange doped in PVA [94]. (**a**) Experimental set-up; *1*, Ar^+ laser; *2*, beam splitter; *3*, mirror; *4, 5, 9*, polarization rotators; *6, 10*, $\lambda/4$ plates; *7*, recording medium; *8*, He-Ne laser; *11*, photosensor. (**b**) Dependence of the diffraction intensity, η, on the angle α between the fast axis of the $\lambda/4$ plate (*10* Fig. 1.22a) and the polarization direction of the readout beam after the polarization rotator (*9* in Fig. 1.22a)

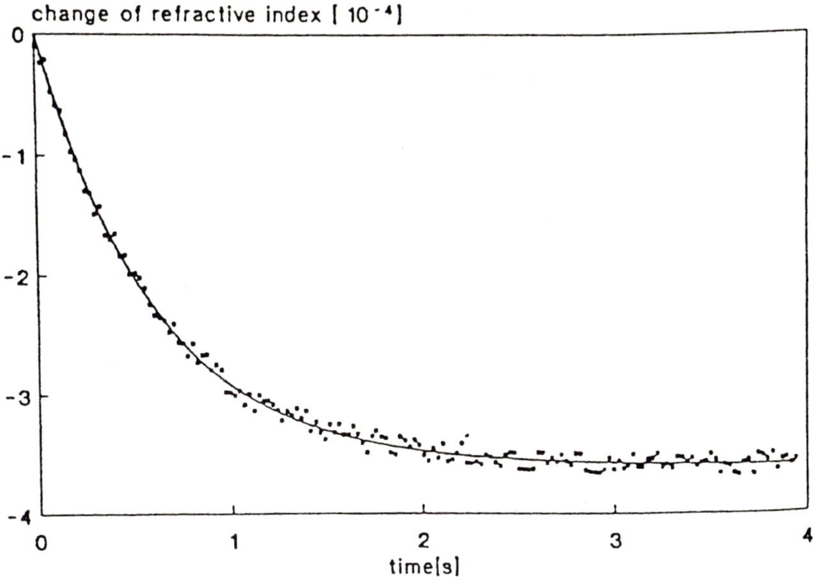

Fig. 1.23. Refractive index change induced by photoisomerization reaction of *N,N*-dihexyl-4-amino-4′-nitro-azobenzene (DHANA) doped in PMMA [98]. The sample contains 0.5 wt% of DHANA, and the irradiation intensity is 3 mW/mm^2

polarized light induces anisotropy in transmission and in the diffraction efficiency of a hologram (Fig. 1.22). The recorded anisotropy or birefringence was attributed to the orientation of the azo dyes induced by the isomerization reaction.[1]

The photooptical effects induced by the isomerization reaction of azo dyes doped in polymer systems were reported for optical waveguide systems by Lückemeyer et al. [98, 99]. Figure 1.23 shows the refractive index change induced by the isomerization reaction caused by guided light [98]. The light-induced birefringence for an azo dye doped in a polymer waveguide was observed as a difference in effective refractive indices between TE and TM modes [99].

The doped azo dyes in polymer systems can be used for optical phase conjugation [100]. The optical phase conjugation is described in the next section.

Kirkby et al. [101] reported optical nonlinearity and bistability using two photochromic fulgide compounds in poly(methyl methacrylate) (PMMA) film. The fulgide compounds undergo photoisomerization of ring-closing reaction to the colored form by UV irradiation. The transmittance of the sample at the wavelength of Ar^+ laser showed optical nonlinearity and bistability depending on the flux ratio of Ar^+ laser to Hg lamp (Fig. 1.24). This behaviour was attributed to the photoinduced change of both absorption and refractive index.

[1]The photooptical effects in azo dye-containing polymers are considered in detail in chapter 2 of this book.

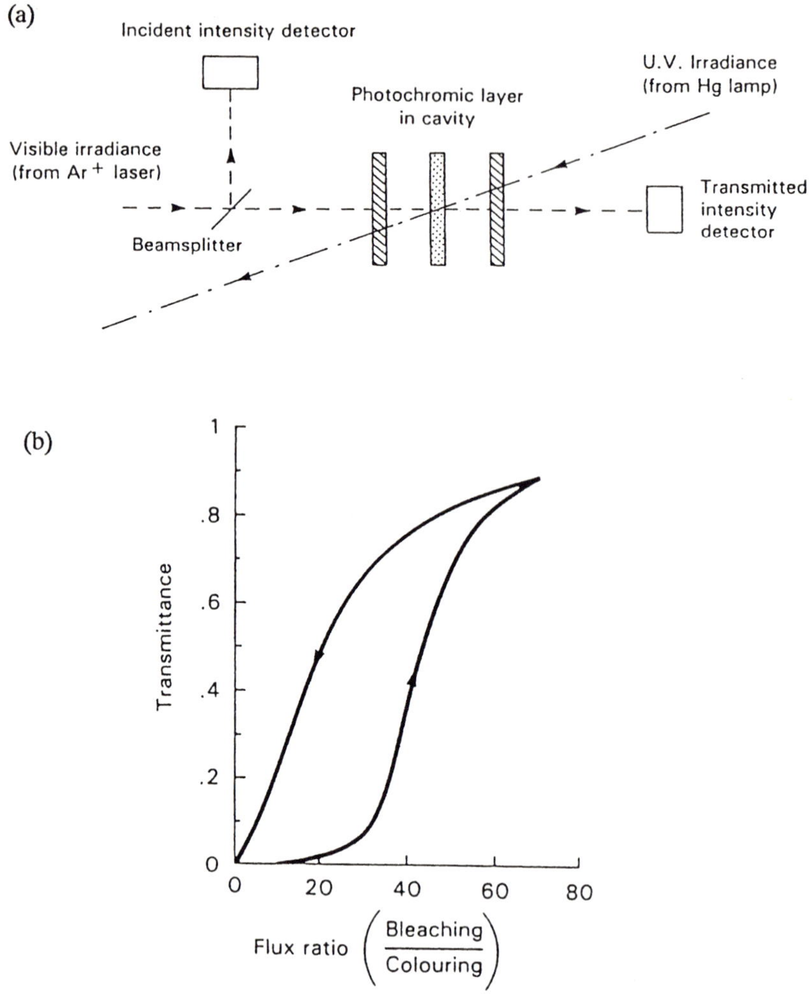

Fig. 1.24a,b. Optical nonlinearity and bistability using a fulgide compound doped in PMMA [101]. (**a**) Experimental arrangement. (**b**) Cavity transmittance of Ar^+ laser versus incident bleaching intensity with constant UV irradiance of approximately 2 mW/mm^2

1.4.2
Optical Phase Conjugation Using Saturable Absorber Dyes

Since some organic dyes have long lifetimes in the lowest excited triplet state (T_1), transmission of polymer or glass films containing organic dyes show a nonlinear dependence on excitation intensity due to the bottleneck of T_1 state. Figure 1.25 shows two examples of the nonlinear intensity dependence of transmission observed for eosin-end erythrosin-doped in gelatin film [102]. This

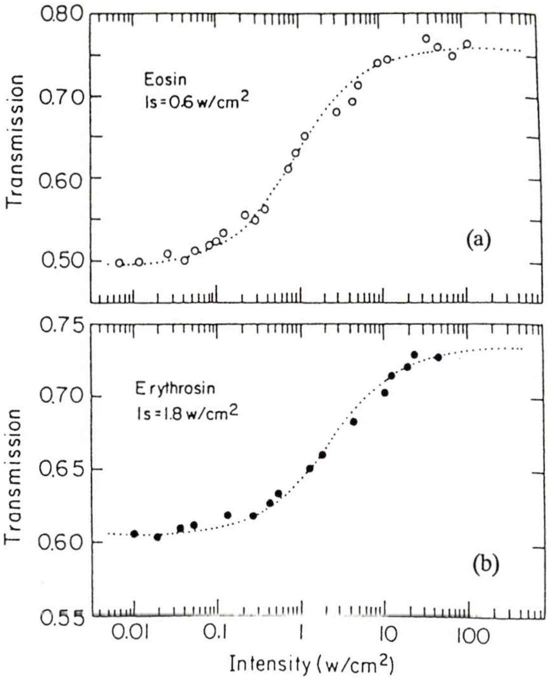

Fig. 1.25a,b. Transmission of (**a**) eosin and (**b**) erythrosin doped in gelatin films as a function of light intensity of 514.5 nm line of Ar^+ laser [102]

phenomenon can be used for optical phase conjugation with relatively low pump power [100]. The optical phase conjugation is very important for image processing such as correction of distorted light. The optical phase conjugation is schematically illustrated in Fig. 1.26. In the optical phase conjugation, a diffraction grating is formed only when three beams are used to irradiate the sample. For this reason some people call the optical phase conjugation "realtime

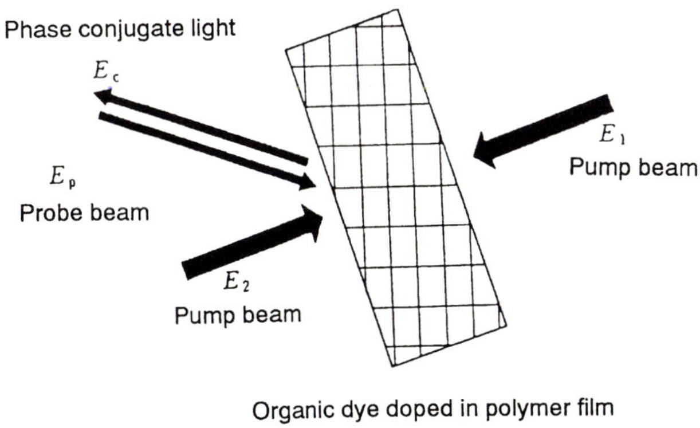

Fig. 1.26. Schematic representation of optical phase conjugation

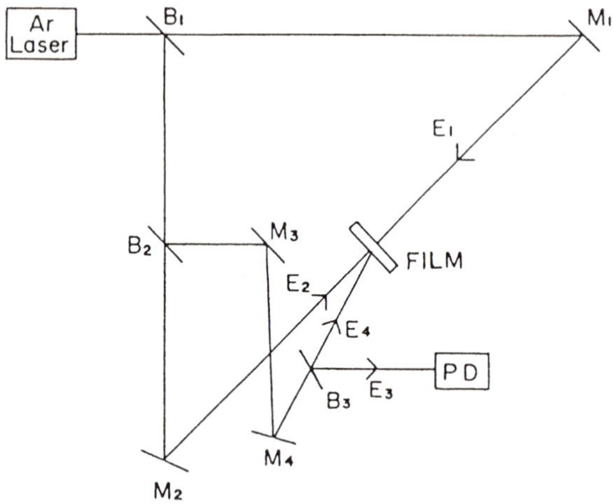

Fig. 1.27. Experimental arrangement for degenerate four wave mixing [104]. E_1, E_2, and E_4 are the incident beams, and E_3 is the phase conjugation beam of E_4

holography" [103]. A typical experimental set up for the phase conjugation, which is called "degenerate four wave mixing", is shown in Fig. 1.27 [104].

The excitation intensity which induces the absorption saturation I_s, and the response time for reaching the photostationary state, τ, can be represented by the following equations (1.9) and (1.10) [100]:

$$I_s = h\nu/\sigma QT \tag{1.9}$$

$$\tau = T/(1 + I_p/I_s) \tag{1.10}$$

where $h\nu$ is a photon energy, σ is an absorption cross section, Q is a quantum efficiency of intersystem crossing, T is a T_1 lifetime, and I_p is an intensity of the pump beam. According to the above equations, doped dyes with large σ and high Q are desirable for getting the materials with low pump intensity and fast response. The efficiency for the generation of phase-conjugated light, $R = I_s/I_p$, is affected by the T–T absorption of dyes [100].

Sometimes the phase-conjugated light does not disappear after the pump and probe beams are stopped for longer than the lifetime of T_1 [105]. This part of the phase-conjugated light is caused by the irreversible chemical change in the dyes and is called the holographic component. The holographic component can be used for the interferometry of the conjugated waves which are produced at different times [106]. If one wants to use only the phase-conjugated light by degenerate four wave mixing, the holographic component should be suppressed. The cooling of a sample suppresses the holographic component [104].

The phase-conjugated wave can be produced in the case where the polarization of the pump and probe beams are perpendicular to each other because of the generated birefringence in the medium [107].

Self diffraction, which is an energy transfer phenomenon between two light waves crossing to each other, is also possible by using saturable dye doped systems [108].

1.4.3
Photorefractive Effect of Photoconducting Electro-Optic Polymers

The photorefractive effect is a photoinduced change in the refractive index of material based on the redistribution and trapping of charge and on the linear electro-optic effect [109]. Inorganic electro-optic crystals such as $BaTiO_3$ are the most investigated as photorefractive materials.

Recently, it was suggested that polymer films which are both electro-optic and photoconductive could be used as a new class of photorefractive materials [110, 111]. The observation of the photorefractive effect in such systems was reported first by Ducharme et al. [112]. The polymer films contain a chromophore with a high second-order nonlinear susceptibility, a sensitizer for photocharge generation, and a hole transport molecule. Figure 1.28 shows the effect of an applied electric field on intensity of diffracted light for the newly found photorefractive polymer [112]. The photoirradiation under the applied electric field induces the charge generation and the charge transport, which produces a grating of charge distribution. The local electric fields associated with the trapped charge alter the refractive index of the material through the electro-optic effect.

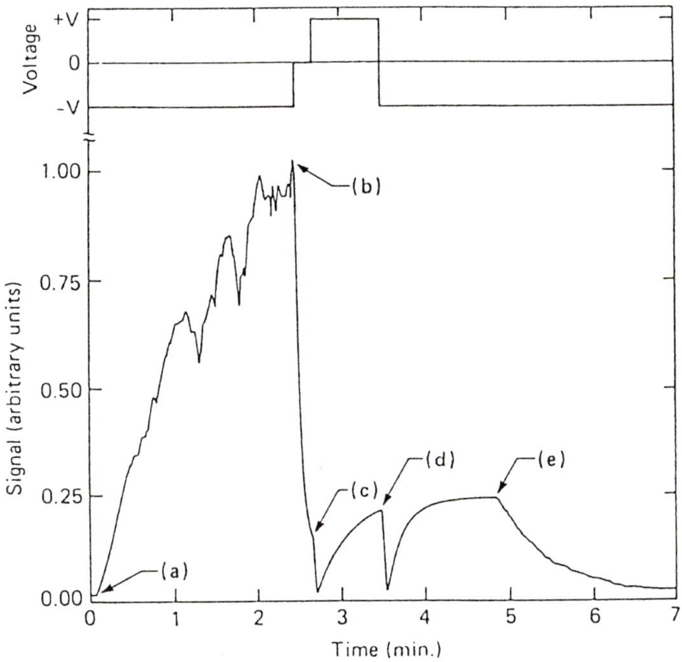

Fig. 1.28. Photorefractive effect observed for a photoconductive electro-optic polymer [112]. *a*-*b*, writing the grating in the presence of a (previously applied) negative voltage; *b*-*c*, writing beams and applied voltage truned off; *c*-*d*, positive voltage applied; *d*-*e*, negative voltage applied; and *e* on, grating erasure by one strong beam

Since the photoconductive electro-optic polymers have low dielectric constants and can be easily fabricated to a thin-film waveguide, the number of photorefractive materials of this type has been increasing rapidly [113–115].

The fullerene molecule C_{60} has recently been shown to act as a useful sensitizer of photoconduction in a photorefractive polymer [116].

1.5 References

1. Gorokhovskii AA, Kaarli RK, Rebane LA (1974) JETP Lett 20: 216
2. Kharlamov BM, Personov RI, Bykovskaya LA (1974) Opt Commun 12: 191
3. Friedrich J, Haarer D (1984) Angew Chemie 23: 113
4. Moerner WE (ed) (1988) Persistent spectral hole-burning: science and applications. Springer, Berlin Heidelberg New York
5. Horie K, Furusawa A (1992) In: Rabek JF (ed) Progress in photochemistry and photophysics. CRC Press, Boca Raton, vol 5, chap 2
6. Castro G, Haarer D, Macfarlane RM, Trommsdorf HP (1978) Frequency selective optical data storage system. US Patent No 4, 101: 976
7. Bjorklund GC, Lvenson MD (1980) Laser pulse shaping device based on Fourier synthesis using optical anisotropies produced by spectral hole burning. IBM Tech Discl Bull 23: 2517
8. Schatz P, Bogner U, Maier M (1986) Appl Phys Lett 49: 1132
9. Case SK (1979) Appl Opt 18: 1890
10. Renn A, Meixner J, Wild UP, Burkhalter FA (1985) Chem Phys 93: 157
11. Wild UP, Bucher SE, Burkhalter FA (1985) Appl Opt 24: 1526
12. Wild UP, Caro CD, Bernet S, Traber M, Renn A (1991) J Lumin 48 & 49: 335
13. Murase N, Horie K (1992) Rep Prog Polym Phys Jpn 35: 439
14. Boxer SG, Lockhart DJ, Middendorf TR (1986) Chem Phys Lett 123: 476
15. Rebane LA, Gorokhovski AA, Kikas JV (1982) Appl Phys B29: 235
16. Furusawa A, Horie K, Mita I (1989) Chem Phys Lett 161: 227
17. Köhler W, Friedrich J (1987) Phys Rev Lett 59: 2199
18. Furusawa A, Horie K (1991) J Chem Phys 94: 80
19. Horie K, Ikemoto M, Suzuki T, Machida S, Yamashita T, Murase N (1992) Chem Phys Lett 195: 563
20. Murase N, Ikemoto M, Horie K (1993) J Chem Phys 99: 6313
21. Suzuki T, Horie K, Yamashita T (1993) Chem Mater 5: 366
22. Völker S, Macfarlane RM, Genack AZ, Trommsdorff HP, van der Waals JH (1978) J Chem Phys 80: 3496
23. Bjorklund GC (1980) Opt Lett 5: 15
24. Moerner WE, Gehrtz M, Huston L (1984) J Phys Chem 88: 6459
25. Furusawa A, Horie K, Kuroki K, Mita I (1989) J Appl Phys 66: 6041
26. Völker S, Macfarlane RM (1979) IBM J Res Dev 23: 547
27. Graf F, Hong HK, Nazzal A, Haarer D (1978) Chem Phys Lett 59: 217
28. Lee HW, Gehrtz M, Marinero EE, Moerner WE (1985) Chem Phys Lett 118: 611
29. Carter TP, Bräuchle C, Lee VY, Manavi M, Moerner WE (1987) J Phys Chem 91: 3998
30. Hochstrasser RM, King DS (1975) J Am Chem Soc 97: 4760
31. Adamson G, Gradl G, Köhler BE (1989) J Chem Phys 90: 3038
32. Machida S, Horie K, Yamashita T (1992) Appl Phys Lett 60: 286
33. Jankowiak R, Small GJ (1987) Science 237: 618
34. Zollfrank J, Friedrich J, Tani T (1989) Polymer 30: 231
35. Maslov VG (1977) Opt Spectrosk 43: 388
36. Suga T, Furusawa A, Uchikawa K (1993) Chem Phys Lett 210: 411
37. Alshits EI, Kharlamov BM, Personov RI (1988) Opt Spectrosk 65: 548
38. Iannone M, Scott GW, Brinza D, Coutler DR (1986) J Chem Phys 85: 4863
39. Prass B, von Borczyskowski C, Stehlik D (1987) J Lumin 38: 48

40. Fearey BL, Stout RP, Small GJ (1983) J Phys Chem 87: 3590
41. Dubs M, Ermanni L, Günthard HH (1982) J Mol Spectr 91: 458
42. Jaaniso R, Bill H (1991) Europhys Lett 16: 569
43. Hirao K, Todoroki T, Soga N (1992) J Lumin 55: 217
44. Winnacker A, Shelby RM, Macfarlane RM (1985) J de Phys C7: 543
45. Sakoda K, Kominami K, Iwamoto M (1988) Jpn J Appl Phys 27: L1304
46. Furusawa A, Horie K, Kuroki K, Mita I (1989) J Appl Phys 66: 6041
47. Furusawa A, Horie K, Suzuki T, Machida S, Mita I (1990) Appl Phys Lett 57: 141
48. Arnold S, Liu CT, Whitten WB, Ramsey JM (1991) Opt Lett 16: 420
49. Winnacker A, Shelby RM, Macfarlane RM (1985) Opt Lett 10: 350
50. Machida S, Horie K, Yamashita T, Furusawa A (1993) J Phys Chem 97: 8234
51. Suzuki H, Shimada T (1991) Appl Phys Lett 59: 1814
52. Mahcida S, Horie K, Yamashita T (1993) J Lumin 56: 85
53. Takahashi J, Tsuchiya J, Kawasaki K (1993) Chem Phys Lett 209: 479
54. Nishimura Y, Kaneko Y, Arai T, Sakuragi H, Tokumaru K, Kiten M, Yamamura S, Matsunaga D (1990) Chem Lett 1935
55. Moerner WE, Levenson MD (1985) J Opt Soc Amer B2: 915
56. Lenth W, Moerner WE (1986) Opt Commun 58: 249
57. Murase N, Horie K, Terao M, Ojima M (1992) J Opt Soc Amer B9: 998
58. Murase N, Horie K (1993) Chem Phys Lett 209: 42
59. Murase N, Horie K (1993) Jpn J Appl Phys 33: 1053
60. Feary BL, Carter TP, Small GJ (1986) Chem Phys 101: 279
61. Kador L, Haarer D, Personov R (1987) J Chem Phys 86: 5300
62. Sesselman TH, Richter W, Haarer D (1987) J Lumin 36: 263
63. Bogner U, Beck K, Maier M (1985) Appl Phys Lett 46: 534
64. Meixner AJ, Renn A, Bucher SE, Wild UP (1986) J Phys Chem 90: 6777
65. Wild UP, Caro CD, Bernet S, Traber M, Renn A (1991) J Lumin 48 & 49: 335
66. Schwörer H, Gygax H, Rebane A, Wild UP (1992) Tec Dig OSA 22: 135
67. Moerner WE, Kador L (1989) Phys Rev Lett 62: 2535
68. Orrit M, Bernard J (1990) Phys Rev Lett 65: 2716
69. Moerner WE, Basché T (1993) Angew Chemie Int Ed Engl 32: 457
70. Ambrose WP, Basché T, Moerner WE (1991) J Chem Phys 95: 7150
71. Wild UP, Güttler F, Pirotta M, Renn A (1992) Chem Phys Lett 193: 451
72. Bernard J, Fleury L, Talon H, Orrit M (1993) J Chem Phys 98: 850
73. Basché T, Moerner WE, Orrit M, Talon H (1992) Phys Rev Lett 69: 1516
74. Tchénio P, Myers AB, Moerner WE (1993) J Phys Chem 97: 2491
75. Güttler F, Sepiol J, Plakhotnik T, Mitterdorfer A, Renn A, Wild UP (1993) J Lumin 56: 26
76. Basché T, Moerner WE (1992) Nature 355: 335
77. Orrit M, Bernard J, Zumbusch A, Personov RI (1992) Chem Phys Lett 196: 595
78. Basché T, Ambrose WP, Moerner WE (1992) J Opt Soc Amer B9: 829
79. Abella ID, Kurmit NA, Hartmann SR (1966) Phys Rev 141: 391
80. Mossberg TW (1982) Opt Lett 7: 77
81. Mitsunaga M (1991) Oyo Buturi 60: 21
82. Babbitt WR, Mossberg TW (1986) Appl Opt 25: 962
83. Bai YS, Babbitt WR, Carlson NW, Mossberg TW (1984) Appl Phys Lett 45: 714
84. Rebane A, Kaarli R, Saari P, Anijalg A, Timpmann K (1983) Opt Commun 47: 173
85. Saikan S, Nakabayashi T, Knematsu Y, Tato N (1988) Phys Rev B38: 7777
86. Saikan S, Kishida T, Kanaematsu Y, Aota H, Harada A, Kamachi N (1990) Chem Phys Lett 166: 358
87. Nakatsuka H, Sugiyama H, Matsumoto Y (1987) J Lumin 38: 31
88. Carlson NW, Rothberg LJ, Yodh AG, Babbitt WR, Mossberg TW (1983) Opt Lett 8: 483
89. Kim MK, Kachru R (1987) J Opt Soc Amer B4: 305
90. Yano R, Mitsunaga M, Uesugi N (1992) J Opt Soc Amer B9: 992
91. Xu EY, Kröll S, Huestis DL, Kachru, Kim MK (1990) Opt lett 15: 562
92. Saikan S, Kishida T, Imaoka A, Uchikawa K, Furusawa A, Oosawa H (1989) Opt Lett 14: 841

93. Uhcikawa K, Ohsawa H, Suga T, Saikan S (1991) Opt Lett 16: 13
94. Gygax H, Rebane A, Schwörer H, Wild UP (1992) 1992 Tec Dig OSA 22: 74
95. Todorov T, Tomova N, Nikolova L (1983) Opt Commin 47: 123
96. Todorov T, Nikolova L, Tomova N (1984) Appl Opt 23: 4309
97. Todorov T, Nikolova L, Tomova N (1984) Appl Opt 23: 4588
98. Lückemeyer T, Franke H (1991) Polym Eng and Sci 31: 912
99. Lückemeyer T, Franke H (1988) Appl Phys Lett 53: 2017
100. Fujiwara H (1990) Oyo Buturi 59: 756
101. Kirkby CJG, Cush R, Bennion I (1985) Opt Commun 56: 288
102. Silberberg Y, Bar-Joseph I (1981) Opt Commun 39: 265
103. Yariv A (1978) IEEE J Quantum Electron QE-14: 650
104. Nakatsuka H, Masuoka D, Yamamoto T (1991) Opt Commun 80: 215
105. Fujiwara H, Nakagawa K (1985) Opt Commun 55: 386
106. Nakagawa K, Fujiwara H (1989) Opt Commun 70: 73
107. Tompkin WR, Malcuit MS, Boyd RW, Sipe JE (1987) J Opt Soc Amer B6: 757
108. Kramer MA, Tompkin WR, Boyd RW (1986) Phys Rev A34: 2026
109. Günter P, Huignard JP (eds) (1988) Photorefractive materials and their applications. Springer, Berlin Heidelberg New York, vol 1
110. Moerner WE, Silence SM (1994) Chem Rev 94: 127
111. Schildkraut JS (1991) Appl Phys Lett 58: 340
112. Ducharme S, Scott JC, Twieg RJ, Moerner WE (1991) Phys Rev Lett 66: 1846
113. Tamura K, Padias AB, Hall HD Jr, Peyghambarian N (1992) Appl Phys Lett 60: 1803
114. Cui Y, Zhang Y, Prasad PN, Schildkraut JS, Williams J (1992) Appl Phys Lett 61: 2132
115. Kawakami T, Sonoda N (1993) Appl Phys Lett 62: 2167
116. Silence SM, Walsh CA, Scott JC, Moerner WE (1992) Appl Phys Lett 61: 2967

CHAPTER 2

Comb-Shaped Polymers with Mesogenic Side Groups as Electro- and Photooptical Active Media

V.P. Shibaev, S.G. Kostromin, and S.A. Ivanov

2.1 Introduction

The creation of the materials with desirable as well as locally-regulated optical properties, which can be easily and reversibly changed under an action of external electro-magnetic fields, is presently one of the important problems of the physics and chemistry of solids both from the theoretical and applied points of view.

The practical significance of such research is associated with the possibility of creating new materials, such as passive optical elements, optically active media for reversible optical data storage, and optically non-linear media. Such materials might be used for making some elements of high-speed electronic circuits and radio in microelectronics as well as electro-optically active waveguide devices (including modulators and switches) for integrated optics and creation of optical printed circuits.

Referring the readers to comprehensive reviews dedicated to inorganic and organic low-molar-mass optical materials [1–4][1] we would like to attract special attention to macromolecular compounds. This chapter deals with liquid-crystalline (LC) and amorphous polymers containing anisometric side chains chemically linked to the backbone via spacers. Such polymers, which are very often called comb-shaped polymers, have attracted much attention during the last few years as electro- and photo-optical regulated media [7–19].

The goal of these investigations, which are being actively carried out in different countries, is the search for new polymeric materials for write-once and multiple reversible recording and long-term (archive) storage of information, creation of new photochromic materials for holography, study of the possibilities of developing and applying planar optical waveguide devices and optical disks. Comb-shaped LC polymers may also be used to produce various kinds of so-called passive optical elements (such as polarizers, phase plates, polarization wedges etc.).

High-quality laser-addressed thin films based on comb-shaped LC polymers are a new type of material for reversible optical data storage, which will probably soon find rather interesting and unusual applications.

---

[1]There are many publications on low-molar-mass liquid crystals as optical data storage elements; some of them [5, 6] we recommend to the reader.

In this chapter we have tried to summarize our main recent results from the study of electro-, thermo- and photooptical phenomena in comb-shaped LC polymers bearing mesogenic side groups.

Because the electrooptical effects observed in these polymers have already been the subject of several reviews (with participation of one of the authors of this chapter) [14, 15, 20, 21], we will only consider very briefly some of the principal data necessary for the following discussion. Taking into account the applied approach aspect of the series "Macromolecular Systems - Materials Approach", our main attention will be paid to thermooptical and pohotooptical phenomena.

The experimental data on photooptical study presented here were obtained as a result of close collaboration between researchers of the Polymer Laboratory of Chemistry Department at the Moscow State University and Laboratory of Photochemistry Department of the Humboldt University in Berlin. However, before we start with the consideration and analysis of the main subject of this chapter, let us consider the specific features of the molecular structure and some properties of comb-shaped LC polymers.

2.2 Specific Features of the Molecular Structure of Functionalized Comb-Shaped Liquid-Crystalline Polymers

The main peculiarities of molecular structure of functionalized comb-shaped liquid-crystalline polymers is the existence of anisometric mesogenic side groups similar to the molecules of low-molar-mass liquid crystals which are chemically attached to the main chain (backbone) by means of flexible (e.g., aliphatic, oxyaliphatic or siloxane) fragments, the so-called spacers. The schematic representation of comb-shaped macromolecules of homopolymers and copolymers and some concrete examples of the main chains, mesogenic groups and spacers of LC polymers synthesized by us in the Laboratory of Chemical Transformations of Polymers at the Moscow State University are shown in Fig. 2.1a–c.

The spacer concept, suggested for the first time, in [22] was further elaborated in [7–9] and is now the standard and most convenient method for the preparation of thermotropic LC polymers with mesogenic side groups. Distancing of mesogenic groups from the main chain with the flexible chains gives the mesogenic groups relatively high autonomy with respect to the main chain and allows their cooperative interaction resulting in the formation of a mesophase.

Within the framework of the above-mentioned approach, hundreds of new LC polymers displaying virtually the same mesophase types as low-molar-mass liquid crystals have been synthesized. In the first studies published, relatively simple mesogenic groups, like the rod-like fragments ("calamitic") [7, 8, 22], were used, however, the reports that followed described structurally most varied comb-shaped polymer, containing disk-like ("discotics") [23] and board-like ("sanidics") mesogenic side groups [24]. Moreover, even more complex LC polymers containing mesogenic groups in the main chains and in the side branches have been obtained as well as polymers consisting of macromolecules

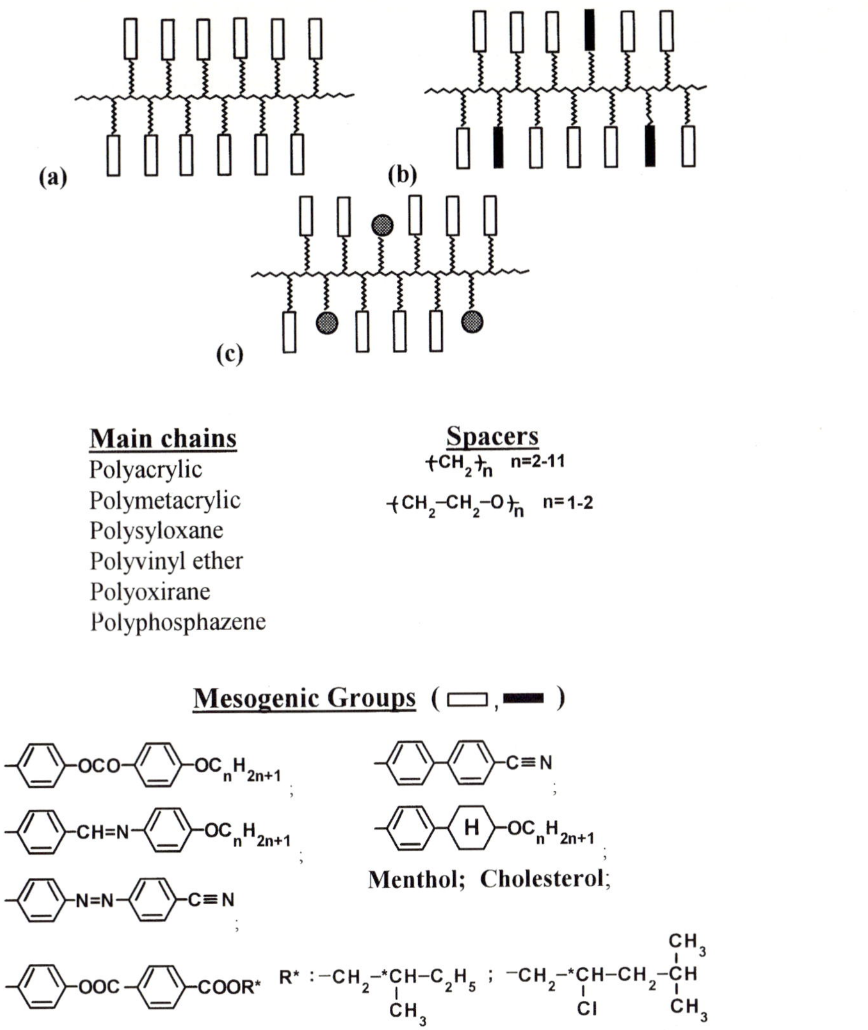

Fig. 2.1a-c. Schematic representation of comb-shaped macromolecules: **a** homopolymer; **b** copolymer containing different mesogenic groups; **c** copolymer containing mesogenic and non-

with alternating laterally bound rod-like and disk-like groups. The detailed consideration of molecular architecture and the concrete chemical formulas of such polymers are given in the literature [10, 25, 26].

The possibility, in principle, of synthesizing macromolecules consisting of any combination of mesogenic and nonmesogenic fragments opens up the richest possibility for the molecular construction of new polymeric LC compounds. The only questions are for which purpose is it necessary to synthesize one or other LC substance and how the features of the structure of the macromolecules are connected with their macroscopic properties determining the potential area of applications of such compounds. In the present chapter we shall consider LC polymers which are of interest as electro- and photooptical active media. It means that the mesogenic groups of such molecules should perform certain "functional loading" providing display of their electro- and photooptical properties under certain conditions. In other words the anisometric fragments possessing the specific series of properties responsible for the functionality of polymeric materials, are incorporated in the macromolecules in different ways (Fig. 2.2).

Thus an existence of the dipole moment μ, responsible for non-zero values of dielectric susceptibility, $\Delta\varepsilon$, is necessary for orientation of mesogenic groups under the action of electric field (Fig. 2.2a). For example, depending on the sign of $\Delta\varepsilon$, mesogenic groups can be oriented along the electric field ($\Delta\varepsilon > 0$, homeotropic orientation) or perpendicularly to the field ($\Delta\varepsilon < 0$, planar orientation).

Other examples of "functional fragments" are mesogenic side moieties, containing the chiral and/or noncentrosymmetrical groups (Fig. 2.2b,c). Incorporation of these fragments into the macromolecules leads to obtaining the polymeric cholesterics, ferroelectrics and materials with nonlinear optical properties

Finally, incorporation of photochromic fragments (Fig. 2.2d), "tuned" to a certain wavelength ensures their photooptical sensitivity and opens up prospects for obtaining photochromic LC polymer systems.

Of course, besides the functional fragments of LC low-molar-mass compounds mentioned above, other functionalized mesogenic groups such as for example physiologically and pharmaceutically active ones can be used.

It is absolutely obvious that the mesogenic fragments presented in Fig. 2.2a-e can often combine the certain properties inherent to each of the fragments

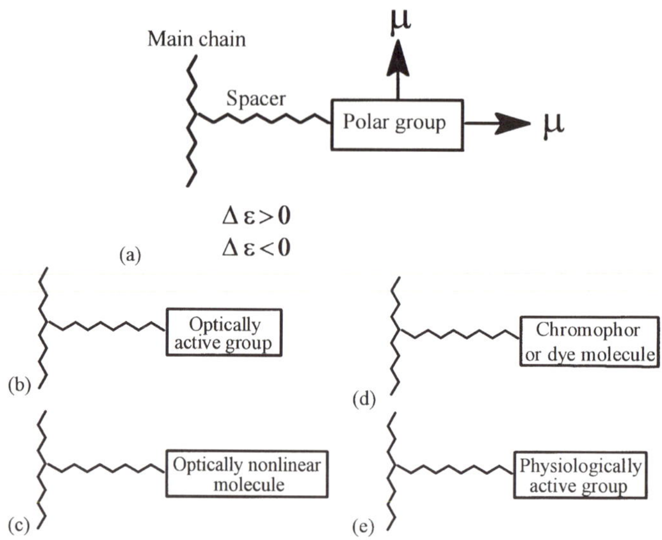

Fig. 2.2. Schemes of comb-shaped macromolecules of LC polymers containing the functionalized mesogenic groups

separately chosen in Fig. 2.2, transforming them into more complex multi-purpose systems. Moreover, the possibility of incorporation of the various "functionally-tuned" mesogenic groups and non-mesogenic groups as the side moieties of such coploymers (Fig. 2.1b,c) considerably extends the synthetic possibility of the molecular design of LC polymers.

In this field, the so-called intelligent[2] LC polymers though slowly but quite surely are beginning to move into the sphere of their practical uses. It is important to note that, apparently, because of a certain inertia of researchers, the spheres of application of these polymers are mainly being sought precisely in those areas of use where low-molar-mass liquid crystal traditionally and successfully "work". Neverthless, the last few years has seen an ever-growing number of studies in which the emphasis is one of the use of the polymeric specifics of LC polymers in terms of their possible uses.

The basic-features of LC polymers – their dual nature, enables one to combine in a single material the properties of macromolecular compounds (with their ability to form films, glasses, fibers and coatings) and the mesomorphic properties of liquid crystals.

As with any liquid crystals, LC polymers have anisotropy of physical properties. This anisotropy in due to orientational ordering of the mesogenic fragments of the macromolecules, in particular, the refractive index, extinction, viscosity, dielectric and magnetic constants are anisotropic. The characteristic values of these parameters of comb-shaped LC polymers essentially coincide in order of magnitude with those of the corresponding magnitudes of low-molar-mass liquid crystals with the exception of viscosity. As in the case of low-molar-mass liquid crystals, LC polymers can be also aligned in the LC phase under the action of electro-magnetic fields changing their structure.

A combination of two factors – dielectric (magnetic) susceptibility of the medium and high mobility of the mesogenic fragments, realized above the glass transition temperature (T_g) allows us to control their supramolecular structure in the mesophase by means of external electric (and magnetic) fields.

Figure 2.3 schematically illustrates the dual nature of comb-shaped LC polymers. Since the majority of comb-shaped LC polymers are not crystallized, their LC state is characterized by the high mobility of the mesogenic groups and only realized above the glass transition temperature. In other words the LC state embraces the temperature interval between the glass transition temperature and the clearing temperature T_{cl}, i.e. the temperature of the transition from the mesophase to the isotropic state. In the LC state the macromolecules (that is their mesogenic groups) or organized spontaneously forming a certain structural type of the mesophase in line with the chemical structure of the polymer and its molar mass.

A peculiarity of this state is that the polymer when exposed to a certain external influence in the mesophase (for example, of the nematic, smectic or

---

[2]It should be noted that the studies on the evolution of the new functionalized intelligent material stimulated the appearance of a new International Journal – "Journal of Intelligent Material Systems and Structures".

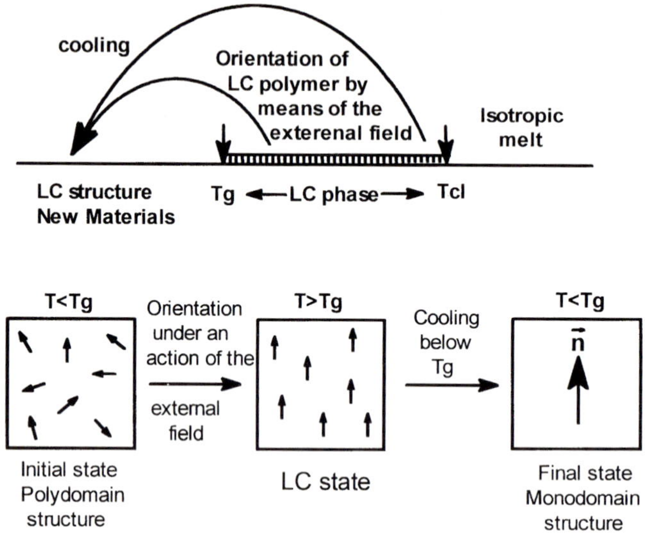

Fig. 2.3. Schematic illustration of the phase state feature of comb-shaped LC polymers and their use for obtaining LC materials with a highly oriented structure

cholesteric type) "memorizes" the structure imparted to it on its subsequent cooling (Fig. 2.3). Herein, lies one of the most interesting and important features of LC polymers making it possible by using the anisotrophy of the "flowing" LC phase and cooling the polymer below T_g to fix the structure type (inherent to the mesophase) in a solid with the anisotrophy of the physical properties peculiar to the LC polymer.

The scheme in Fig. 2.3 shows it more clearly. In the initial non-oriented LC polymer below T_g the structure of the sample can be regarded as the polydomain one, where the directors $\vec{n}$ in microdomains have a statistical distribution. Above T_g the sample is subjected to the external field and the directors are aligned, transforming the polydomain structure into a monodomain one, characterized by the direction of the directors $\vec{n}$. This high oriented monodomain structure is "frozen" in the glassy state after the cooling of LC polymer below T_g.

This peculiarity opens up new possibilities for regulating the polymer structure, using non-traditional methods of polymers processing such as electric and magnetic fields or the action of light enabling us to create new types of functionalized polymeric materials.

Since the structural transformation type of side mesogenic groups is determined by their function (see Fig. 2.2), as specific "set" of physico-chemical processes can be realised for each case as shown in Fig. 2.2, and this leads to different supramolecular structures being fixed in the solid.

Description, interpretation and understanding of these complex structural and conformational transformation of the functional mesogenic and the neighboring side groups as well as the main chains, also taking part in these processes – is one of the current, complicated and at the same time most attractive problems in polymer science.

Only the understanding of the regularities of all the strucutral transformations on the molecular and supermolecular levels, the description of dynamics and kinetics of these transformations will give an opportunity to purposefully control polymer structure having in mind both a creation of the new polymeric materials and the new methods of regulation of their physical properties.

2.3 Comb-Shaped Dye-Containing Polymers with Mesogenic Side Groups

As shown in the previous section,the functionalization of comb-shaped polymers is achieved by the "chemical incorporation" of the appropriate molecular fragments or groups responsible for the necessary "functional loading" in the side branching of the macromolecules.

Thus, for example, the existence of polar mesogenic fragments is necessary for creation of "electro-regulated" LC polymers, the existence of dye-containing monomeric units is needed for obtaining "photoregulated polymers". Incorporation of the polar mesogenic groups in the macromolecules is usually performed by both copolymerization of polar and non-polar monomers and by polymer-analogous reactions between the reactive macromolecules (main chains) and mesogenic molecules with polar substituents (mesogenic side groups). The examples of such macromolecular compounds capable of being oriented in the electric fields can be found in numerous publications [7–10, 15, 19, 27, 32].

In this part of the chapter, our main attention will be concentrated on the comb-shaped LC polymers with dye-containing monomer units and some aspects concerning the mixture of polymers and organic dye will also be touched.

Depending on the functional action of dyes on the polymer matrix one can roughly distinguish two groups of organic dyes.

The first group includes the dyes, playing the passive part of the "guest" in the host polymeric matrix and not changing the molecule shape under the action of an external field (mechanical forces, electrical or light). Some examples of such dichroic dyes are the naphto- and anthraquinone dyes, the Schiff- base dyes etc. The molecules of such dyes can be dissolved as guest components in the LC polymer or covalently attached via flexible spacers to the backbone. In both the cases the dye molecules are the "guests" of the host polymeric matrix aligning in the polymeric films according to the alignment of the side mesogenic groups. This "guest-host" effect which is well-known in low-molar-mass liquid crystals has also been shown in polymeric compounds, in particular during the study of electrooptical behaviour of LC polymers [14–15, 32–35].

Another group of organic dyes consists of compounds whose molecules undergo configurational and conformational transformations when exposed to light. Mono- bis-, tris-, azo dyes, stilbene and spiropyran- dyes belong to this group [36]. For example, azobenzene dye molecules display a *trans-cis* isomerization under light illumination; the spiropyran groups are transformed into the merocyanine form. Independently of the phase state of such dye-containing

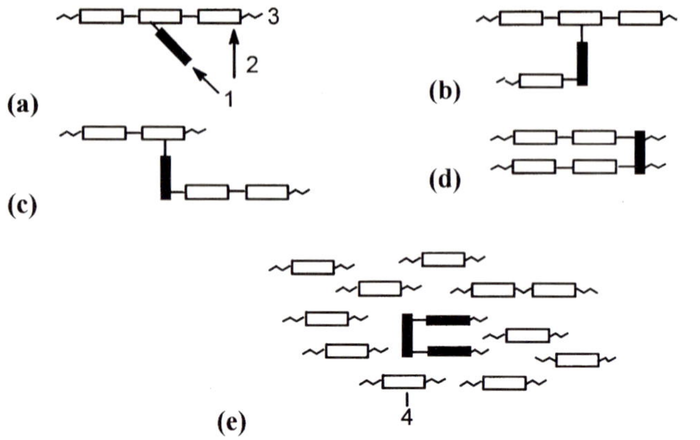

Fig. 2.4a-e. Model structures of dichroic dye-containing molecules (**a-d**) and one example of their packing in the liquid crystal layer (**e**). *1*- dye-containing block; *2* - mesogenic rigid block; *3*- alkyl group responsible for the solubility of dye molecules in liquid crystal; *4* -molecule of liquid crystal

copolymers, irradiation of the copolymers leads to photochemical transformations of the photosensitive units. This results in the refractive index of the medium being changed and very often structural changes of the host-matrix take place. In other words the "host-molecules" of the dyes exposed to light in their turn are capable of affecting the structure of the copolymer as a whole.

The mechanism of the dye action on the polymeric matrix will be considered in detail a little later; here we consider, in broad terms, the possible variants of incorporation of the dye molecules in low-molar-mass and polymeric liquid crystals.

Tthe mixture of polymeric or low-molar-mass liquid crystals and organic dyes[3] is the most general case.

As a rule, for this purpose, dichroic dyes having rod-shaped molecules with chemically linked LC fragments are used. The molecular structure of such complex compounds can be represented in the form of one or several blocks with chromophore groups alternating with a different number of LC blocks (Fig. 2.4). Each block of such a dye performs a certain function, providing "photochemical" and "liquid-crystalline" properties. The main feature of this sort of dyes lies in the fact that the structure gives them good solubility in liquid crystals without the perturbation of the LC phase. Using this approach to the synthetic design of dichroic dyes in combination with quantum - chemical calculations, a lot of azo- and anthraquinone dyes having practical application have been obtained. Some of them are shown in Table 2.1 [38].

[3]The pure organic dyes are practically not used for creation of photooptical materials and in general they are only of academic interest while the organic photochromic materials are created as a rule by the use of different polymeric compounds. However in recent years some new photochromic materials have been produced only the basis of organic dyes; they are obtained either by the vacuum deposition, or have been produced by Langmuir-Blodjett technique (multilayer films); see the review [37].

Table 2.1. Examples of the liquid crystalline dichroic dyes, composed of liquid crystal and dye-containing blocks [38]

Model structure	Chemical structure
	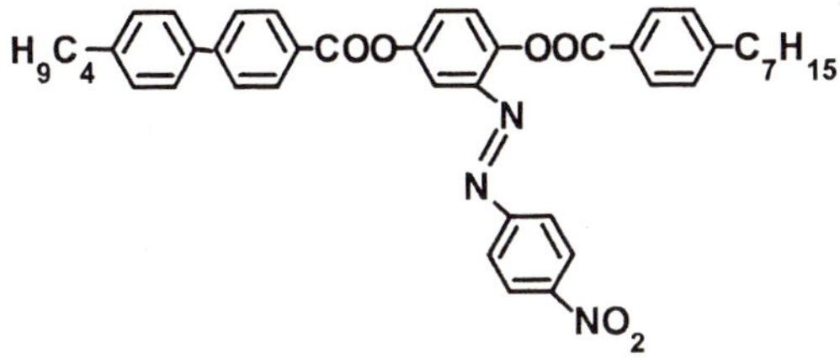 $T_m = 160\,°C; T_{cl} = 198.5\,°C$
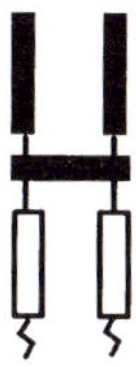	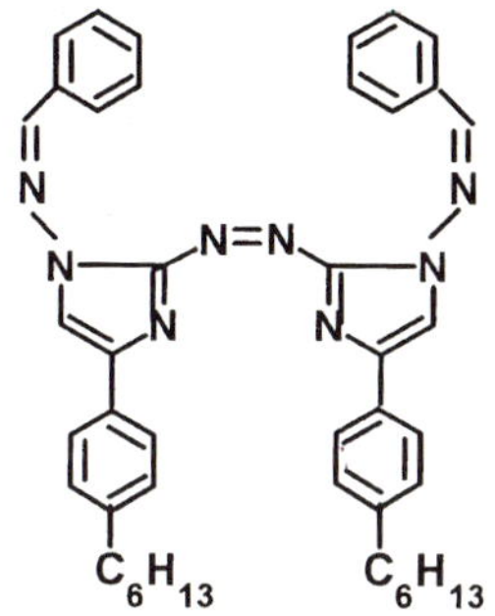

Going on to the consideration of LC polymer systems containing dye fragments in the macromolecules, first of all the analogous nature of the molecular structure of comb-shaped polymers and the complex dichroic dyes mentioned above should be emphasized.

The synthesis of such complex polymers is more often carried out via the copolymerization of mesogenic and dye-containing monomers (Fig. 2.5a) or the attachment of dye molecules to a polymeric backbone via polymer-analogous reactions is used (Fig. 2.5b). The polycondensation of the photochromic fragments with other blocks bearing the functional groups is chosen much more rarely (Fig. 2.5c). Using all these methods one can "chemically incorporate" the dye-containing fragments in the polymeric chains. As can be seen from Fig. 2.5 one can vary the connection type of dye-containing fragments widely including them in the main and the side chains as well as changing their distribution type along the macromolecule. Some types and properties of copolymers obtained by us via copolymerization, namely mesogenic and azo dye-containing monomers, are shown in Tables 2.2 and 2.3.

Polymers with covalent-connected dye molecules have essential advantages over polymer-dye-mixtures[4]. First of all, this concerns the solubility of dyes in a

[4]The properties of the mixtures of dyes and some amorphous polymers (polycarbonate, polymethylmetacrylate, polyvinyl alchohol and others) were considered a long time ago; these data can be found, for example, in [39–42]. Some of these data are discussed in chapter 1 of this book.

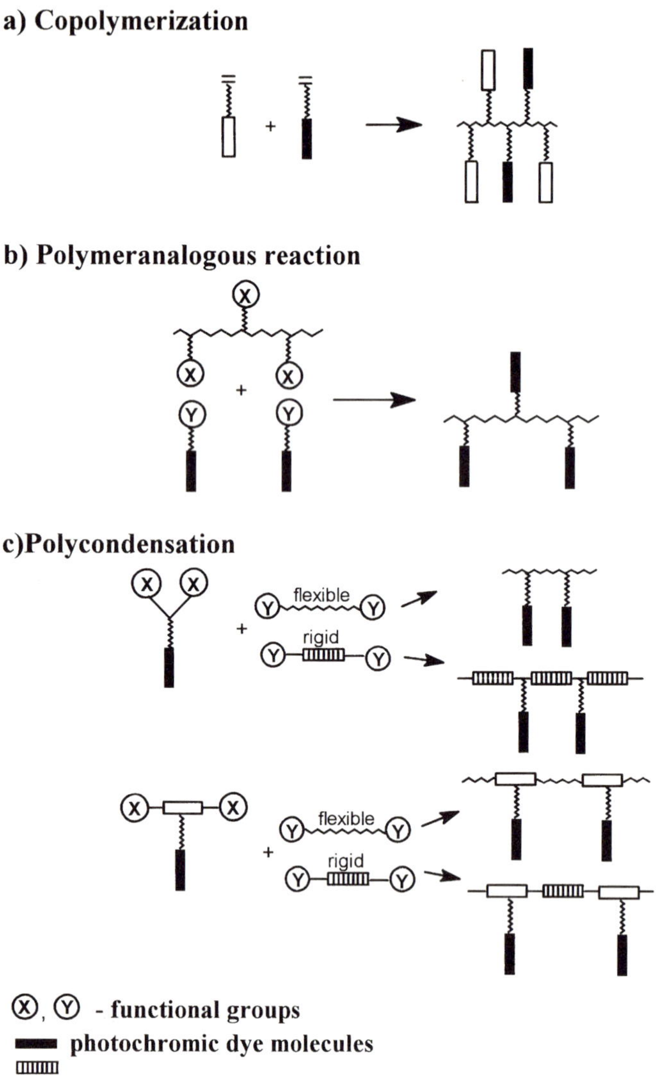

Fig. 2.3. Approaches to creating dye-containing copolymers

polymer matrix. As a rule the maximum content of "host-dyes" in the polymer-dye-mixtures in general does not exceed 1–5 mol %; a higher concentration of dyes leads to the phase separation of the mixture or to the disturbance in the LC phase. In addition, these mixtures are very sensitive to temperature; a decrease in the temperature increases the tendency to phase separation. In this connection, the dye-containing copolymers are considerably stable due to the covalent bonds of dyes with a LC "mother-matrix". As seen from Tables 2.2 and 2.3, the content of dye-containing monomeric units in the copolymers may reach 60–80 mol %

Table 2.2. Chemical structures and phase transition temperatures of the dye-containing acrylic and methacrylic copolymers with the different spacer length [43]

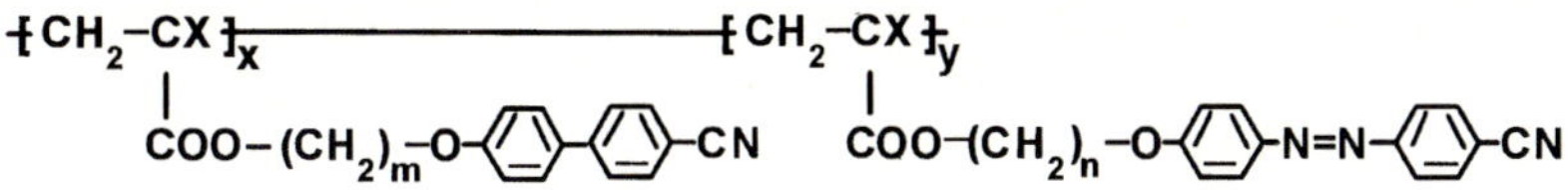

N	X	m	n	x mol %	y mol %	T_g, °C	Phase* type	T_{cl}, °C
1	H	2	2	0.6	0.4	85	N	106
2	H	2	6	0.6	0.4	67	LC	108
3	H	3	6	0.8	0.2	61	S_A	91
4	H	4	6	0.8	0.2	55	N	113
5	H	5	6	0.8	0.2	38	S_A104N	110
6	H	5	6	0.6	0.4	38	S_A105N	110
7	H	6	6	0.8	0.2	40	RN71S_A106N	121
8	CH_3	2	2	0.6	0.4	107	LC	128
9	CH_3	2	6	0.6	0.4	93	LC	119

*LC - LC phase, type of LC phase is not defined; N - nematic phase, RN - reentrant nematic phase; S_A - smectic A phase.

without destruction of the LC phase. Of course, the phase state of a dye-containing copolymers is defined by numerous factros, including such factors as the chemical similarity of mesogenic groups and dye molecules, the length of the spacer of each of the comonomers, the type of main chain, the molecular weight etc.

If the side mesogenic groups and the dye molecules have a similar chemical structure (and a closely similar shape) and the dye-containing homopolymer

Table 2.3. Chemical structure and phase transition temperatures of the dye containing cyanobiphenyl acrylic and methacrylic copolymers with the different spacer length [43]

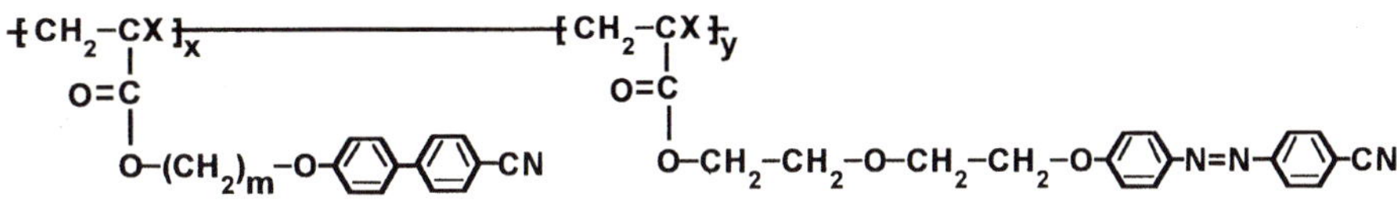

N	X	m	x mol %	y mol %	T_g, °C	Phase* type	T_{cl}, °C	$\overline{P_n}$	$\overline{P_w}$
10	CH_3	2	0.8	0.2	101	am	–	42	167
11	CH_3	2	0.6	0.4	88	am	–	45	145
12	CH_3	2	0.4	0.6	77	am	–	43	105
13	CH_3	2	0.2	0.8	67	LC	98.7	45	96
14	CH_3	6	0.4	0.6	57	LC	102.4	–	–
15	H	2	0.8	0.2	77	N	91.3	50	272
16	H	2	0.6	0.4	70	N	87.2	–	–
17	H	2	0.4	0.6	60	LC	78.1	–	–
18	H	2	0.2	0.8	54	LC	83.1	–	–
19	H	6	0.4	0.6	43	S87°N	97.0	–	–

*am - amorphous; LC - the type of liquid crystalline phase is not defined; N - nematic phase; S - smectic phase.

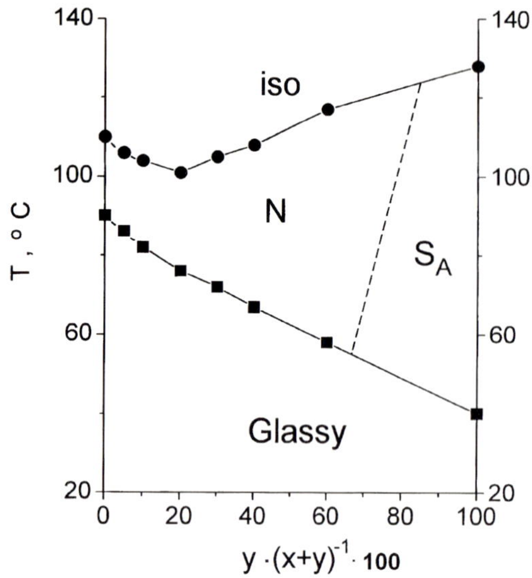

Fig. 2.6. Phase diagram of copolymer **I** series [43]

itself forms the LC phase, the mesophase can be realized for all compositions of the copolymer.

It can be easily seen in Fig. 2.6 which shows the phase diagram of the Copolymer **I** series [43].

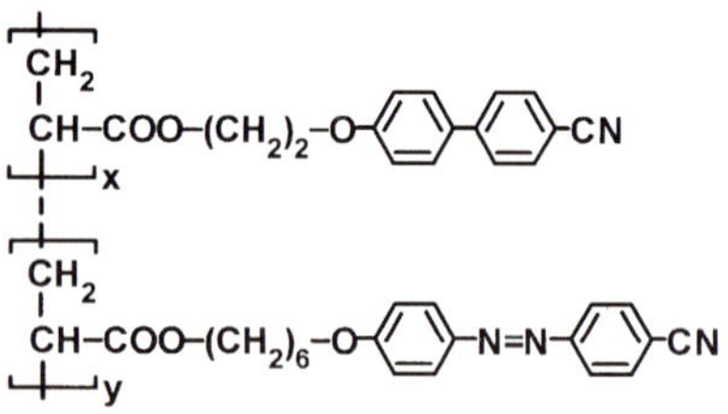

Cop. (I)

In this case even in spite of the different spacer lengths of the two comonomeric units, the LC phase is observed for all compositions – the S_A-phase is observed at high concentrations of dye-containing monomer, the concentration decrease of the latter leads to nematic phase formation. So the use of the dye containing copolymers opens up wide possibilities for the synthesis and selection of photochromic copolymers with a desirable mesophase type and temperature range (T_g and T_{cl}) of the LC state.

The second example demonstrates the opposite case, when one can change the phase state of the polymer by varying the dye concentration in the copolymers and main chain flexibility. As seen from Fig. 2.7a and Table 2.3, the acrylic copolymers display a LC phase over the whole range of dye concentration. However, the methacrylic copolymers (Fig. 2.7b) (polymers 10–12), with the same cyanobiphenyl groups and a dye content of 20–60 mol % do not form any mesophase [43]. A further increase in dye-containing comonomer induces the LC

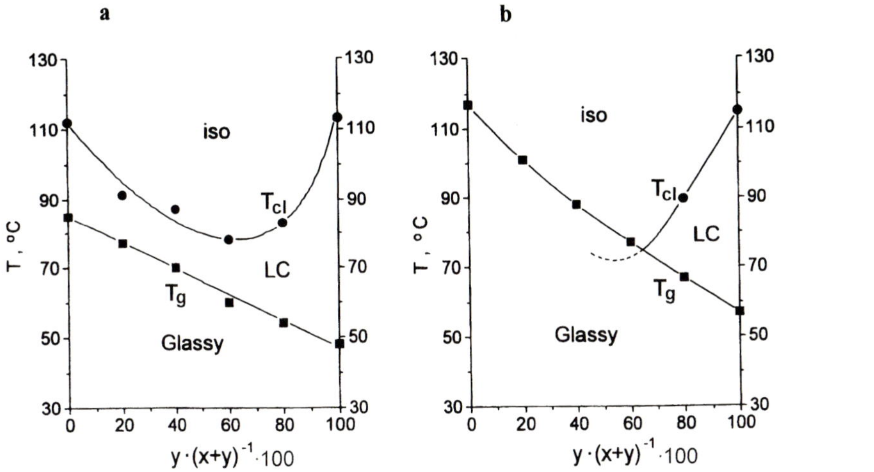

Fig. 2.7a-b. Phase diagram of the dye-containing acrylic (a) and methacrylic (b) copolymers [43]:

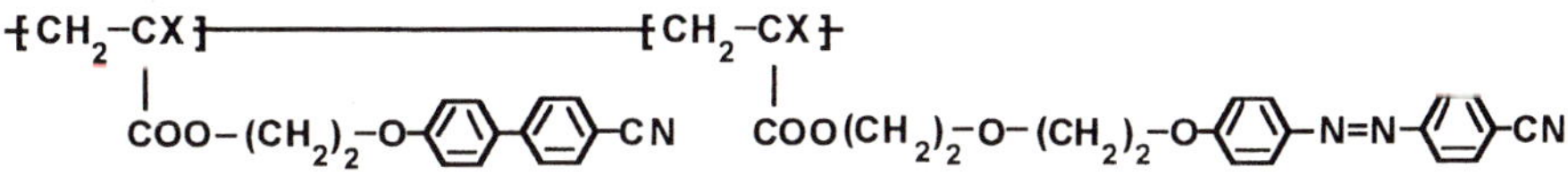

phase formation due to the higher anisotropy of azobenzene groups as compared with cyanobiphenyl fragments.

On the other hand, the difference between these two phase diagrams (Fig. 2.7a and b) may also be explained from the formal point of view taking into account the relation between glass transition temperatures and their clearing temperatures. In the case of the acrylic copolymers, the LC phase is formed over a wide range of dye concentration owing to the fact that $T_g < T_{cl}$ for all copolymers. In the methacrylic copolymers the values of glass transition temperatures are higher

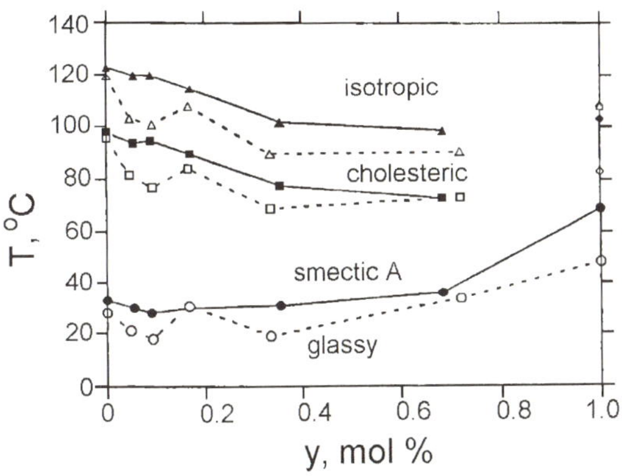

Fig. 2.8. Phase diagram of copolymer II series: *full lines* and *full symbols* represent the high-molar-mass samples; *broken lines* and *empty symbols* represent the low-molar-mass samples [34]

than their hypothetic clearing temperatures and the mesophase is not realized. A further increase of azobenzene monomeric units in methacrylic copolymers results in changing the relation between T_g and T_{cl} (T_{cl} becomes higher than T_g) and the mesophase is formed.

In general, it is necessary to note that the azobenzene fragments are well "built" into the LC phase of copolymers and owing to their mesogenic rod-like shape do not disturb the LC structure of the initial matrix [33–35]. The high values of the order parameters S of such copolymers show strong evidence for the above-mentioned data. For example, the order parameter of the nematic copolymers 5 and 6 (Table 2.2) are in the range of 0.5–0.6, that is close to the values of the order parameter of low-molar-mass liquid crystals in the nematic phase [35].

The attractive chiral azo dye monomers incorporated in the macromolecules of the acrylic copolymers **II** are described in [34].

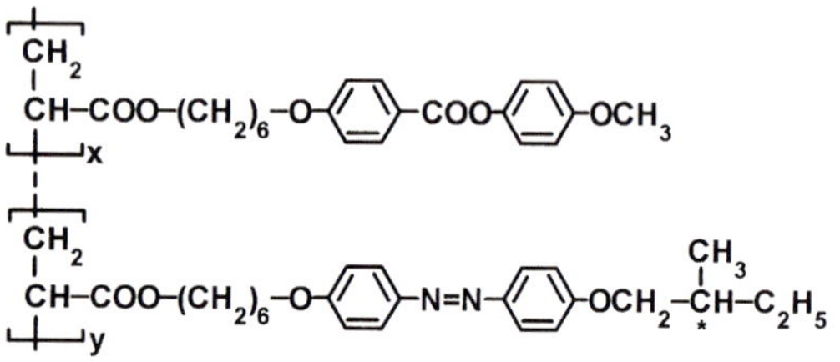

Cop. (II)

Such copolymers, side by side with the smectic A phase, form the cholesteric mesophase with a high concentration of azo dye. In this case, as can be seen from Fig. 2.8, the mesophase is realized over a very broad range of copolymer compositions regardless of the molar mass of copolymers.

On the other hand, the incorporation of bulky substituents such as anthraquinone and spiropyran dyes into the side chain of macromolecules generally destroys the mesophase. Figure 2.9 clearly shows that the concentration increase of anthraquinone dye in acrylic copolymers **III** causes a disappearance of the smectic mesophase accompanied by a decrease in the clearing temperature [33].

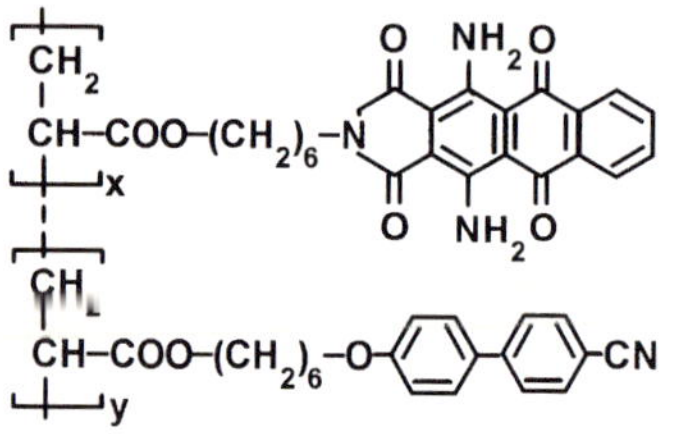

Cop. (III)

Nevertheless, the copolymer containing 36 wt % anthraquinone dye is still nematic. At the same time it should be mentioned that similar anthraquinone dyes in low-molar-mass nematic hosts in general have poor solubilites (usually no more than 0.5 wt %) [33].

Highly unusual LC polymers containing a couple of azo dyes in each monomeric units were synthesized by South Korean researchers (polymer **IV**) [44].

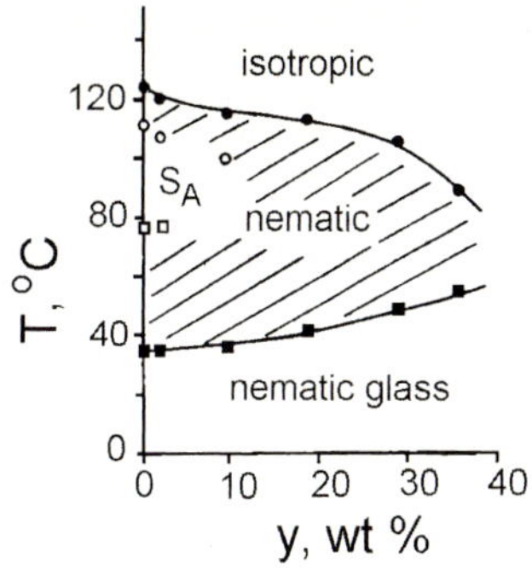

Fig. 2.9. Phase diagram of copolymer **III** series [33]: □ -transition reentrant nematic mesophase to smectic A(S_A); ○ -transition S_A to nematic mesophase; ● -transition nematic mesophase to isotropic melt; ■ -glass transition

Table 2.4. Transition temperatures of polymalonic esters with paired mesogens (Polymers **IV**) [44]

Spacer length n	Phase Transitions, °C (on cooling)	
	T_{cl}	$T_{N-cryst}$
1	130	117
2	89	56
3	59	32

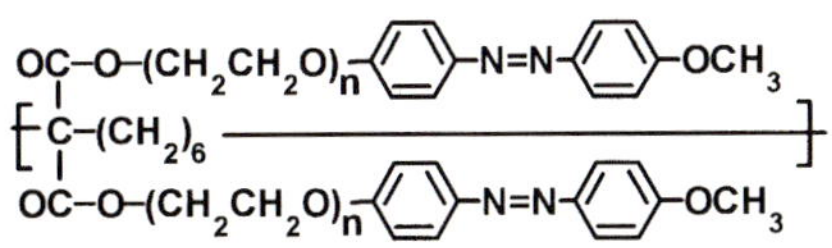

(**IV**)

where n = 1 − 3

Such polymalonic esters with n=1–3 form a crystalline and montropic LC phase (Table 2.4); they are of considerable interest from a practical point of view for optical storage applications.

From this point of view, the hydroxystyrene derivatives **V** and **VI** containing azobenzene fragments with polar terminal substituents are also of interest; the homopolymers of these compounds have been synthesized recently for studying their orientation in an electrical field [45].

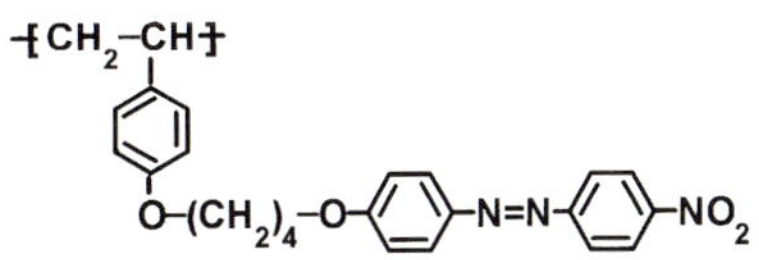

(**V**)

$T_g = 70\ ^\circ C; T_{S \to I} = 155\ ^\circ C$

Table 2.5. Chemical structures and phase transition temperatures (°C) of dye containing monomers used as a host of liquid-crystalline comb-shaped polymers

N	Chemical formula of the monomer	Ref.
1	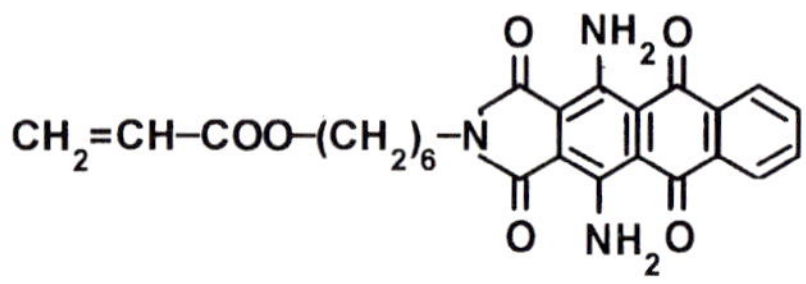 Cryst 162 Isotropic	[46]
2	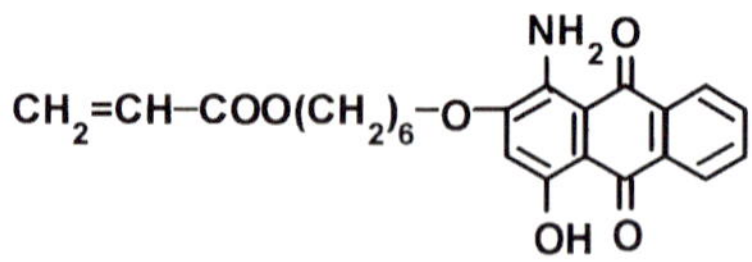 Cryst 119 Isotropic	[46]
3	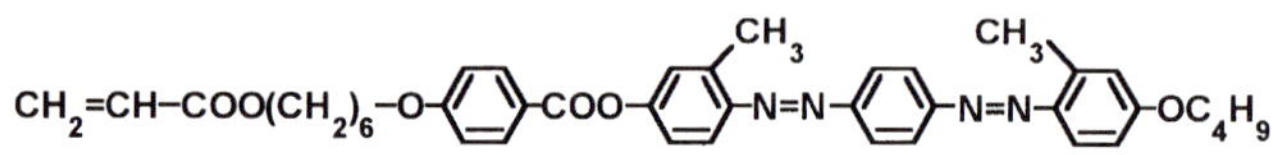 Cryst 114 N > 230 Isotropic	[46]
4	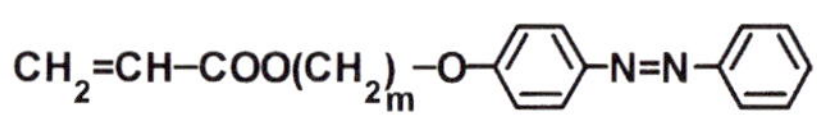 m=2 Cryst 80-85 Isotropic m=6 Cryst 94-96 Isotropic	[47]

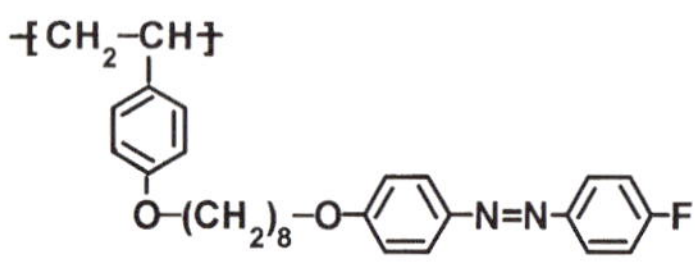

(VI)

$T_g = 55\ ^\circ C$; $T_{S \to I} = 110\ ^\circ C$

Some other types of dye-containing monomers used for the copolymerization with mesogenic monomers are listed in Table 2.5.

All these facts demonstrate that it is possible to synthesize LC copolymers containing dichroic dyes with a very high dye concentration. These high dye concentrations allow us to prepare thin LC polymer films (down to 1 μm thickness) with adjustable optical densities and absorption properties.

In addition to comb-shaped LC polymers containing mesogenic photochromic groups in the side chains, there also exist (as cited above) the so-called combined LC polymers, which contain the mesogenic groups both within the main and the side chains (Fig. 2.5c). In this case, in principle, photosensitive chromophore groups can be included both in the main chain and the side

Table 2.6. Combined LC polymers, containing azobenzene photosensitive mesogenic groups in the side chains

N	R	m	Phase transitions, °C	Ref.
	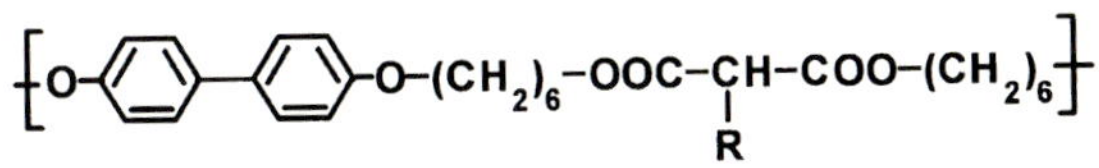			
1	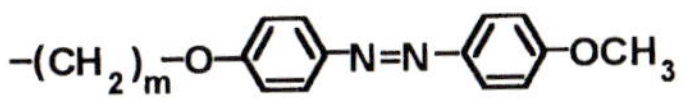	2	$K_1$109 $K_2$120 N153 I	[48]
		6	$K_1$108 $K_2$112 S_C131 S_A136 N155 I	
		10	G72 $S_1$120 S_C139 S_A154 I	
		6	G61 S_A112 N184 I	
2	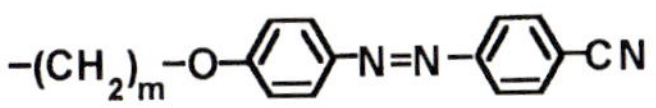			
	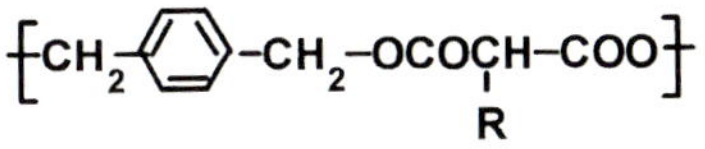			[49]
3	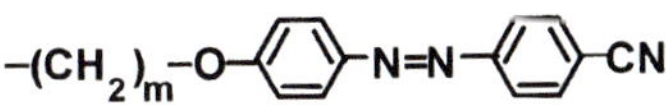	6	G43 S_A94 N104 I	
	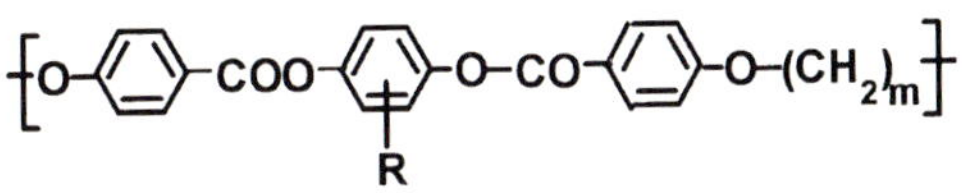			[50]
4	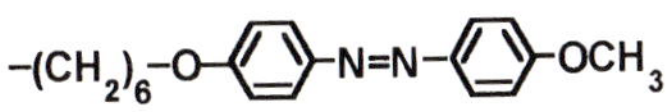	2	G90 N310 I decomp	
		6	G56 N268 I	
		9	G45 N221 I	
	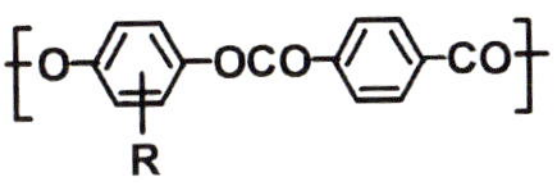			
5	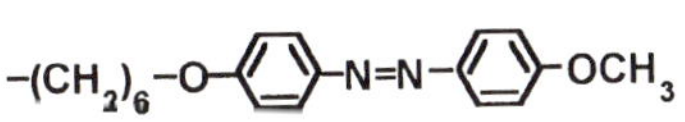		G 67 K221 N > 350 decomp	[50]

branching. However, up till now only the combined copolymers containing dye molecules in the side chains alone have been synthesized; these dye-containing fragments can be attached both to the spacers of the main chain and also to the hard mesogenic blocks of the backbone (Fig. 2.5c). Some such LC copolymers and their thermal properties are presented in Table 2.6.

As can seen from Table 2.6, among these polymers, there exist smectic and nematic types of mesophases characterized by a broad temperature range of mesophase formation; in particular, it applies equally to the polymers 4 and 5 having sufficiently high clearing temperatures, which permits us to study their electro- and photooptical properties over a broad temperature range.

The polyurethane derivatives containing "optically nonlinear groups" like 4–nitrostilbene are also worthy of attention [51].

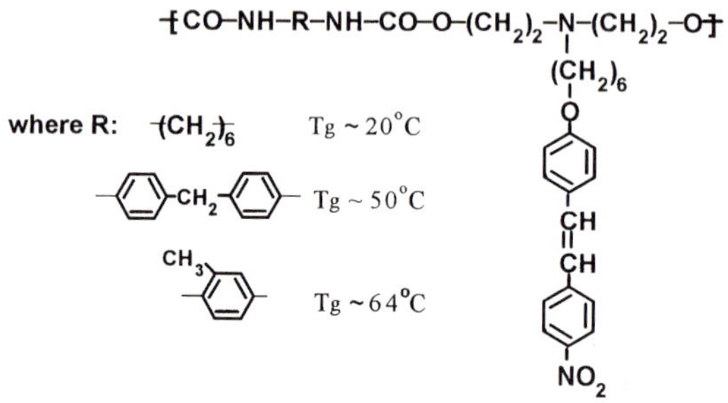

Cop. VII

The authors of [51] did not present any evidence for the existence of the LC phase with regard to their optically nonlinear properties, however, in our opinion, electro- and photooptical behavior of such systems is also of great interest.

In general, it should be noted the aspiration of the researchers is the synthesis of polymers containing azonitrobenzene fragments as the side groups, independently of whether the polymer will form a LC phase or not.

So the possibility of the use the amorphous (polymers **VIII** and **IX**) and LC homopolymer (polymer **X**), containing the above-mentioned mesogenic side moieties for photooptical recording of information was very convincingly demonstrated by Natanson et al. [52–54].

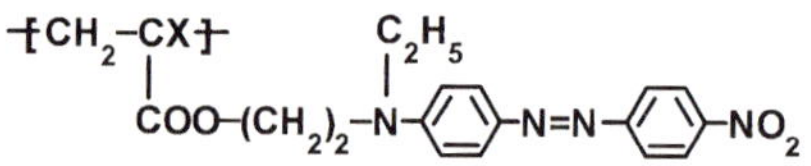

(VIII)

a) X = H; $T_g = 91$ °C b) X = CH_3; $T_g = 129$ °C

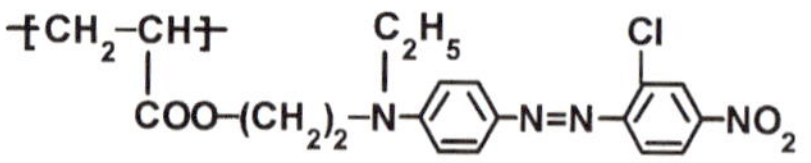

(IX)

$T_g = 87$ °C

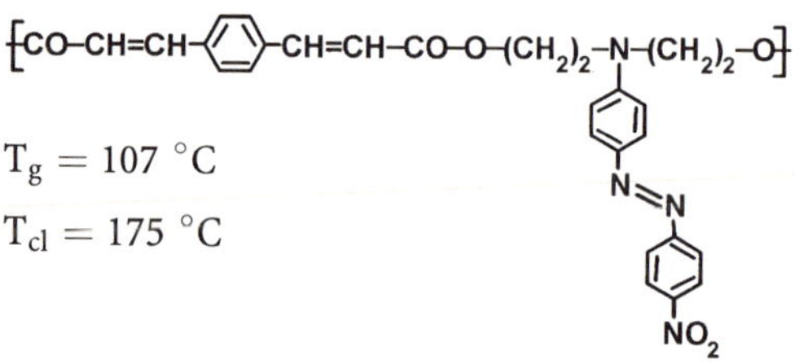

$T_g = 107$ °C

$T_{cl} = 175$ °C

(X)

As will be shown in Sect. 2.4.2, the rate of the orientational process occuring in LC polymers under the influence of an electric field essentially depends on the molar mass of the polymers, responsible for the abnormally high viscosity of LC polymers as compared with low-molar-mass liquid crystals. In this respect, the LC polymers "lose" in comparison with low-molar-mass liquid crystals – their values of threshold voltage and rise (and decay) times are substantially greater.

One can "get round" these disadvantages of LC polymers using the formulas, presented in Table 2.7. The switching times can be shortened by raising the temperature, applying higher voltages, and decreasing the thickness of the LC electrooptical cell.

Another approach which allows us to decrease the viscosity of polymeric systems and therby to accelerate the orientational process associated with electro- and photooptical effects is based on the use of the mixtures of the dye-containing copolymers with low-molar-mass liquid crystals or mixtures of LC polymers with low-molar-mass dye-containing liquid crystals (Fig. 2.10). The former of these two types of mixtures is the typical examples of the "guest-host" whose electrooptical behavior is similar to the mixtures of low-molar-mass liquid crystals with the dyes.

As for the mixtures of **b** type (Fig. 2.10) as well as the mixtures of the **a** type (Fig. 2.10), the main problem of their preparation is the problem of compatibility of the copolymer and the low-molar-mass component. The creation of an adequately compatible polymer – "monomer" system in the broad composition range requires a careful selection of high and low-molar-mass components and the detailed study of their phase diagrams for finding the temperature-composition range of compatibility. Some examples of such complex phase diagrams of the spiropyran- containing copolymers and spiropyran dyes have been presented in detail by Krongauz et al. [55].

Here we would like to emphasize that highly compatible polymeric mixtures can be obtained if the low-molar-mass component molecules (liquid crystals or dyes) have a molecular structure similar to the side chains of the polymer. A fine example of the above-mentioned is the mixture of copolymer **III** (containing 19 wt % of anthraquinone) and multicomponent commercial mixtures on the basis of alkylcyanobiphenyles.

In this case the side mesogenic groups of copolymer **III** are similar to the molecules of the two-components in the low-molar-mass mixture of liquid crystals, this allows us to obtain stable mixtures with higher dye concentrations than in the case of low-molar-mass liquid crystals [33].

Such mixtures are of considerable interest for a variation (modification) of the rheological properties and the elasticity characteristics of low-molar-mass liquid crystals. It is very important for the practical application of LC polymers in electrooptical devices and some data concerning these aspects will be considered in the next section.

Concluding this "chemical" part of this section, we would like to draw attention to one essential circumstance associated with the synthesis and creation of the polymer systems containing chemically linked dye molecules.

There has recently been considerable interest in the study of monomolecular polymeric Langmuir-Blodgett films containing dye molecules [56–57]. These investigations are usually associated with the problems of the optical data storage and the study of optically nonlinear properties. According to the theoretical predictions (see chapter 5 of this book) the polyacrylates and polysiloxanes containing different azobenzene derivatives with the polar terminal groups are widely used for these purposes. Some of them are shown below [56]:

Table 2.7. Orientational effects in nematic liquid crystals

Orientation molecules in electrooptical cell Initial state Ultimate state	Threshold voltage, E_0	Response times, τ_{on} and τ_{off}
Splay effect ($\Delta\varepsilon>0$) 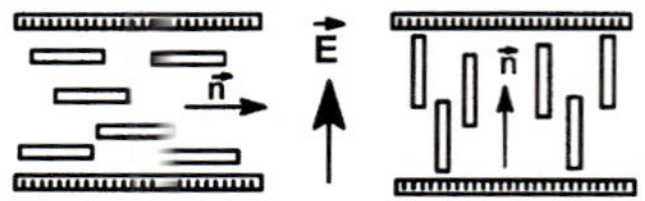	$E_0 = \pi\left(\frac{4\pi K_{11}}{\Delta\varepsilon}\right)^{1/2}$	$\tau_{on} = \frac{4\pi\gamma_1 d^2}{\Delta\varepsilon E^2 - 4\pi^3 K_{11}}$ $\tau_{off} = \frac{\gamma_1 d^2}{\pi^2 K_{11}}$
Bend effect ($\Delta\varepsilon<0$) 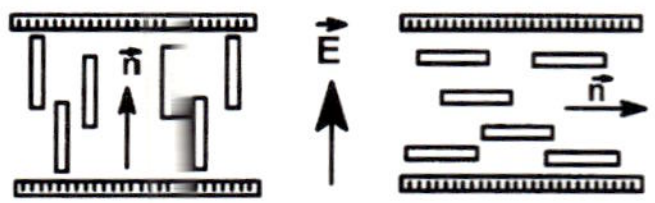	$E_0 = \pi\left(\frac{4\pi K_{33}}{\Delta\varepsilon}\right)^{1/2}$	$\tau_{on} = \frac{4\pi\gamma_1 d^2}{\Delta\varepsilon E^2 - 4\pi^3 K_{33}}$ $\tau_{off} = \frac{4\pi\gamma_1 d^2}{\pi^2 K_{33}}$
Twist effect ($\Delta\varepsilon>0$) 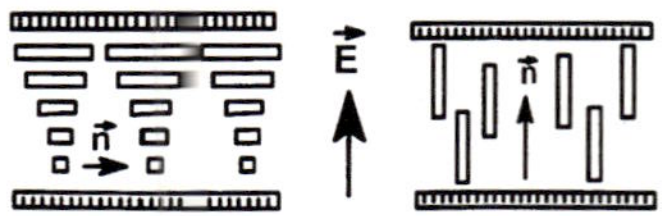	$E_0 = \pi\left(\frac{4\pi}{\Delta\varepsilon}\left[K_{11} + \frac{1}{4}(K_{33} - 2K_{22})\right]\right)^{1/2}$	$\tau_{on} = \frac{4\pi\gamma_1 d^2}{\Delta\varepsilon E^2 - 4\pi^3 K_{11} + 1/4(K_{33} - K_{22})}$ $\tau_{off} = \frac{\gamma_1 d^2}{\pi^2 K_{11} + 1/4(K_{33} - 2K_{22})}$

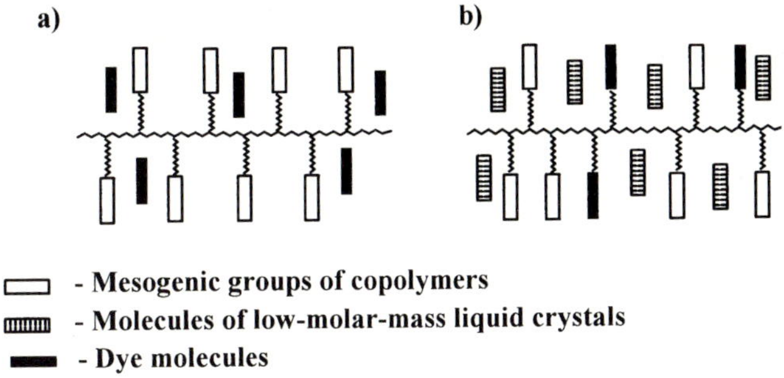

Fig. 2.10a,b. Schematic representation of "guest-host" systems based on LC-comb-shaped polymers. **a** LC comb-shaped polymer (or copolymer) in mixture with dichroic dye; **b** Dye-containing copolymer in mixture with low-molar-mass liquid crystal

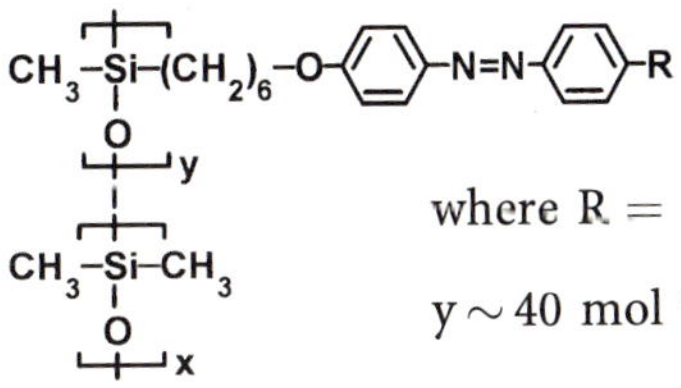

where R = $-CH_2-CH_2-OH$, $-CN$, $-COOCH_3$, $-COOH$, $-NO_2$

y ~ 40 mol % (XI)

More complex polyimide polymers with azobenzene side groups used in the form of Langmuir - Blodgett films could be especially interesting in optical data storage [57].

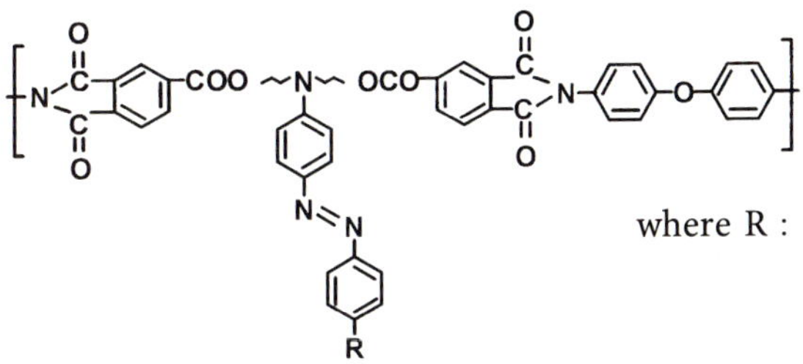

where R : $-OCH_3$, $-NO_2$ (XII)

Thus, as can be seen from the foregoing data, the majority of dye-containing copolymers used for the study of electro- and photooptical phenomena are obtained on the basis of azobenzene and spiropyran dyes [58] chemically attached to the polymer backbone. At the same time a lot of the mixtures exist containing dyes with low-molar-mass liquid crystals and with polymers. The recent review by Matsuoka [59] was dedicated to the dyes used for the optical data storage in "guest-host" systems. However, probably some of them could be used for chemical linking with the polymeric matrix. Without doubt, further success in synthetic chemistry directed towards the synthesis of new monomer dyes and dye-containing copolymers will open up the wide possibilities for obtaining new electro- and photosensitive materials.

2.4
Electrooptical Behavior of Liquid-Crystalline Compounds

2.4.1
Low-Molar-Mass Liquid Crystals in Electric Fields (Field Effects)

One of the unique and attractive properties of low-molar-mass liquid crystals is their capacity to change their optical properties under the action of an external electric field of low voltage (1–5 Volts) and low power (10^{-4}–10^{-6} W/cm^2). Namely the existence of ensembles of molecules possessing higher anisotropy of polarizability than the individual molecules predetermines their higher energy of interaction with the electric field exceeding the energy of the disorienting influence of temperature ($\sim$kT). The relatively low viscosity of liquid crystals and the considerable high energy of their interaction with the electric field exceeds kT, which ensures the macroscopic turning of the director, sharply changing the optical properties of the anisotropic layer of the liquid crystal [38, 60].

In essence all modern indicator techniques are based on the possibility of a regulation of the optical properties of the thin layer of liquid crystal sandwiched between two electroconducting glasses. Such a hermetically sealed cell, called an electrooptical cell, is the basis of the up-to-date alphanumerical indicators - these are the displays of computers, flat TV screens, calculator and watches etc.

In spite of the seeming simplicity of such a cell, tens of years were needed for commercial applicability of LC cells. It demanded not only the solution to a series of technical problems but also the fundamental problems such as the elaboration of the elastic theory of viscous anisotropic media, the study of electrooptical and magnetooptical phenomena in liquid crystals, the elaboration of the problems relating to an anisotropy of conductivity, the detailed investigation of the surface effects in the thin anisotropic films etc.

As for the electroopticals of the polymeric liquid crystals their history dates back slightly more than ten years and the first publications in this direction appeared at the beginning of the 1980s [7, 9, 14, 15, 19, 61–67]. In this section we do not intend to review the abundant literature related to the electroopitcal properties of LC polymers because the certain analysis of these aspects was done some time ago by different authors [14, 15, 163, 67, 68]. In this section, we will limit ourselves to the consideration of some electrooptical phenomena (in comb-shaped LC polymers) which can be used for the creation of new polymeric materials with unusual optical properties and in particular for the preparation of highly-oriented polymeric films for photooptical applications.

It would be sound practice to start with the consideration of electrooptiacal phenomena in low-molar-mass liquid crystals because, basically, electrooptics of LC polymers and low-molar-mass liquid crystals differs slightly.

Usually, all electrooptical effects may be rougly divided into two groups. First of all there are so called orientational effects which are due solely to the effect of the electric field (field effects) on LC polymers, but are not the result of a current flowing. The second group of electrooptical effects is attributed to thephenomena ascribed to the anisotropy of electrical conductivity of liquid crystals. In this case the flow of any ionic current is accompanied by mass transfer, i.e. a hydrodynamic flow. These processes are called electrohydrodynamic effects. However in this review we will consider only the field effects.

From the point of view of the electrooptical behavior of all LC compounds (low-molar-mass and polymeric liquid crystals), the existence of a longitudinal or transversal dipole moment (Fig. 2.2a) in the molecule (or in the side branching of the macromolecule) defining the anisotropy of dielectric susceptibility $\Delta\varepsilon = \varepsilon_{\parallel} - \varepsilon_{\perp}$ (where $\varepsilon_{\parallel}$ and $\varepsilon_{\perp}$ – are the dielectric constants of a liquid crystal measured along an external field and perpendicular to the field, or director $\vec{n}$, respectively) is necessary.

Due to the anisotropy of the dielectric susceptibility, $\Delta\varepsilon$, the free energy of an ensemble of liquid crystal [69]:

$$F = \int F_0 \cdot dV \tag{2.1}$$

(where F_0 is the free energy density; V – is a volume of liquid crystal) is a minimum one in the electric field at the definite director orientation in respect to the vector of the electric $\vec{E}$. Dircctor $\vec{n}$ tends to align itself along the field of $\Delta\varepsilon > 0 (\vec{n} \parallel \vec{E})$ and perpendicular to it $(\vec{n} \perp \vec{E})$, provided that $\Delta\varepsilon < 0$ [69].

Table 2.7 shows the schemes of the typical orientational electroopitacl effects corresponding to the three distortion types – splay, bend and twist, described by a certain combination of the elastic moduli K_{ii} [60, 69].

In the initial state (see the left part of Table 2.7), the vector of the electric field $\vec{E}$ and the director $\vec{n}$ do not correspond to the minimum of the free energy F; however, after applying the electric field, the reorientation of the director takes place and, as a result, a compromised director profile is maintained, which satisfies the condition of the new minimum free energy value (see the right part of Table 2.7)

Here, it is important to note that the principal feature of rheologcial and mechanical properties of liquid crystals is their elasticity connected with the local changing of the director orientation.

In general, in the presence of an electric field, the free energy density F_0 of the nematic liquid crystal may be described by the sum of two terms placed in brackets of Eq. (2.2) [69]:

$$F_0 = \frac{1}{2}\left[K_{11}(\mathrm{div}\vec{n})^2 + K_{22}(\vec{n}\cdot \mathrm{rot}\vec{n})^2 + K_{33}(\vec{n}\times \mathrm{rot}\vec{n})^2\right] - \frac{1}{2}\left[\varepsilon_0 \Delta\varepsilon\left(\vec{E}\cdot\vec{n}\right)^2\right] \tag{2.2}$$

where K_{11}, K_{22} and K_{33} – are the Frank constants characterizing the orientational elasticity of liquid crystal. As can be seen from Eq. (2.1) and Table 2.7 the distorted state of the nematic liquid crystal is completely described by the director orientation for the three general deformation types – splay, twist and bend, where K_{11}, K_{22} and K_{33} are the splay, twist and bend elastic constants respectively.[5]

If the distortion corresponds to the "pure" splay, "pure" twist and "pure" bend deformation, such deformations are called S (when $\operatorname{div}\vec{n} \neq 0$), T (when $\vec{n} \cdot \operatorname{rot}\vec{n} \neq 0$) and B (when $\vec{n} \times \operatorname{rot}\vec{n} \neq 0$) deformations respectively (Table 2.7).

The left part of Eq. (2.2) in brackets corresponds to the elastic deformation of liquid crystal without the external field, the second term in brackets corresponds to an interaction of the electric field with dielectrically anisotropic media.

The evaluation of the minimum of the free energy from Eq. (2.2) at the fixed boundary conditions defined by the geometry of the experiment permits us to obtain the threshold voltage E_0. These threshold values for different types of deformations are presented in Table 2.7. Here the appropriate expressions for the kinetic parameters corresponding to the orientation time (so called rise time, t_{on}) and to the relaxation time (so called decay time, t_{off}) of the electrooptical effects are given.

As can be seen from the formulas of Table 2.7 the threshold voltage E_0 is connected to the above mentioned elastic constants; the kinetic constants τ_{on} and τ_{off} are defined also by the rotational viscosity coefficient of the liquid crystal, γ_1. Thus, the typical switching times of low-molar-mass liquid crystals are 10^{-3} s, relaxation after switching off takes place much more slowly (decay time, $\tau_{off} \sim 10^{-2} - 10^{-1}$ s).

The values of the elastic constants and the rotational viscosity coefficients for some well studied LC compounds are shown in Table 2.8.

The study of electrooptical effects permits us to determine the most important elastic characteristics of liquid crystals, quantitative evaluation of which is necessary for the creation and optimization of a variety of LC materials widely used in different electroopitcal devices.

2.4.2 Electrooptical Effects in Comb-Shaped Liquid-Crystalline Polymers

During the consideration of electrooptical effects in LC polymers, it should be pointed out the essential difficulties of studying them. One can probably pick out two main reasons why an orientation of LC polymers under action of electric field is hampered.

Firstly, the existence of the main chain should lead to steric restrictions applied by the backbone to the orientation of the mesogenic side chains. The backbone flexibility and the spacer length – these are the basic parameters defining the ability of the mesogenic groups to be oriented (at the fixed $\Delta\varepsilon$ values).

The second reason is a consequence of the first one. This is the enhanced viscosity of LC polymers; the viscosity of LC melts depends on the degree of polymerization of the LC polymer. Namely this "polymeric nature" of the LC

[5] The elastic constants are of the order of 10–11N for polymeric as well as for low-molar-mass liquid crystals.

Table 2.8. Elasticity Constants and Viscosity Coefficients of Comb-Shaped LC Polymers and Low-Molar-Mass Liquid Crystals

LC Compound	$K_{11}.10^7$, dyne	$K_{22}.10^7$, dyne	$K_{33}.10^7$, dyne	K_{33}/K_{11}	γ_1, Pa.S	Ref.
1. 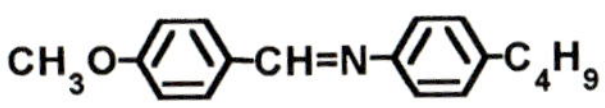	5.8	3.4	7	–	0.1	[70]
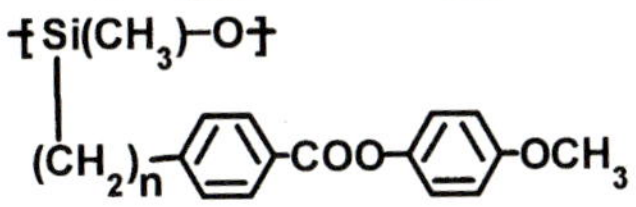						
2. n=4 ($\overline{DP}$ = 95)	7	–	13	1.8		[71]
3. n=6	8	–	7	0.9		
4. 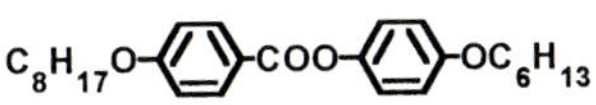	15.8	–	10	0.6		[71]
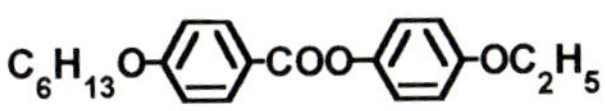	9	–	10	1.1		
6. Cholesteric Copolymers XIII ($\overline{DP}$ = 50, 80, 160)	–	4÷7	–	–		[68, 72]
7. 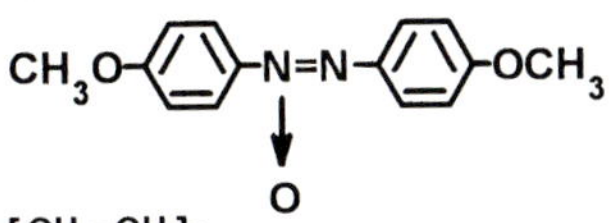	4.5	2.9	9.5 (125°C)	–	$6.7.10^{-2}$	[70]
					8.	
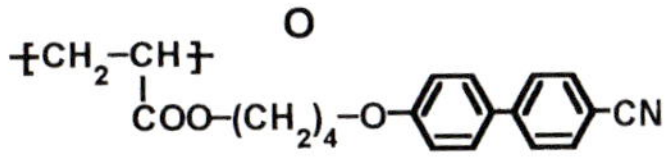 ($\overline{DP}$ = 2530; T = 0.95T_{cl})		–	–	–	10^4	[73]

phase is responsible for the slowing down of all dynamic and relaxation processes carried out in LC polymers under the influence of the external fields which makes it difficult to prepare highly homogeneously oriented samples of LC polymers.

Figure 2.11 illustrates the orientation process of the unoriented sample of an acrylic polymer with side cyanobiphenyl mesogenic groups under the influence of an electric field. Measuring the intensity of the transmitted light as a function of time at different voltages one can observe how the initial unoriented and turbid polymer film is transformed into the oriented and completely transparent film (Fig. 2.11a-d). Between the crossed polarizers of an optical microscope the sample becomes completely dark and the X-ray diffraction pattern and conoiscopic picture of the film confirm the high degree of mesogenic group orientation. Such a transparent film, in essence, is similar to an anisotropic uniaxial monocrystal with the axis direction coinciding with the electric field direction.

The reciprocal rise time is universally proportional to the square of the voltage value (Fig. 2.11b) which is in a good agreement with the general equation Eq. (2.2) (see the second term in brackets). It proves that this orientational process is a result of the field effect, when the electric field interacts with dielectrically anisotropic media (in this case $\Delta\varepsilon > 0$).

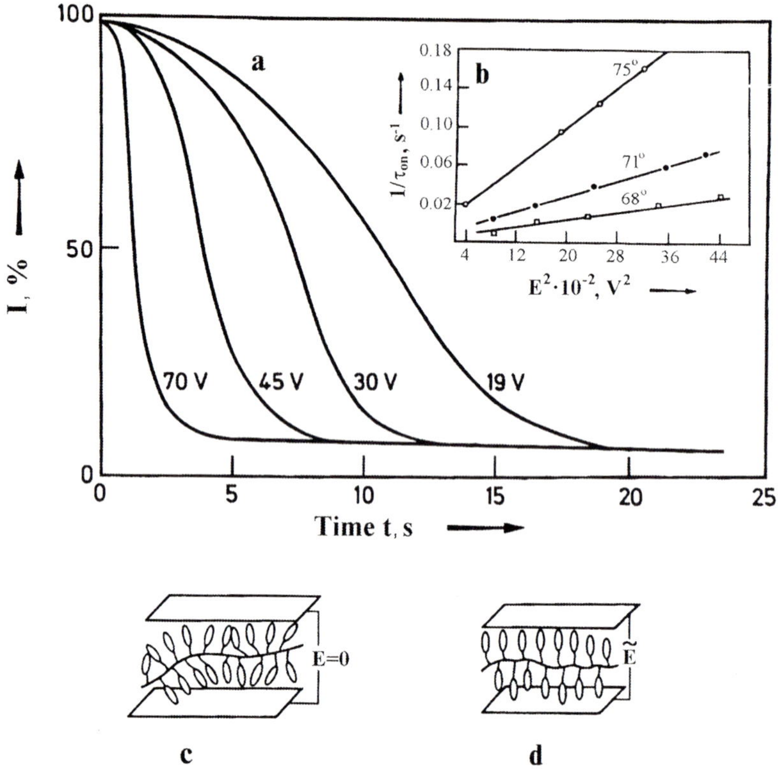

Fig. 2.11a-d. Electrooptical behavior of the nematic polymer

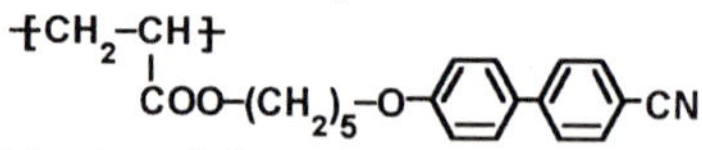

a -kinetics of the orientation of polymer upon application of an electric field (the optical transmission in crossed polarizers is given as a function of time); **b** -dependence of the reciprocal rise time on the voltage squared at different temperatures, °C; **c** and **d** -schemes of mesogenic groups orientation before and after the application of an electric field, respectively

The concrete type of field effect depends on the initial orientation of the liquid crystal and the sign of $\Delta\varepsilon_0$ in accordance with the data in Table 2.7. In spite of certain difficulties in the preparation of the perfect and the homogeneous initial texture of LC polymers (planary , homeotropic ets.), nevertheless, such a texture can be obtained with the use of the same methods of treatment for the cell surface as in the case of low-molar-mass liquid crystals. Usually these types of treatment of the glass walls of the electrooptical cell can be supplemented with the prolonged annealing of the samples (sometimes it takes tens of hours or several days; the annealing time is defined by the degree of polymerization).

Analysis of our own and the literature data clearly shows that, at the present time, any field effect (see Table 2.7) may be realized in LC polymers if the initial orientation of the mesogenic groups is strongly fixed.

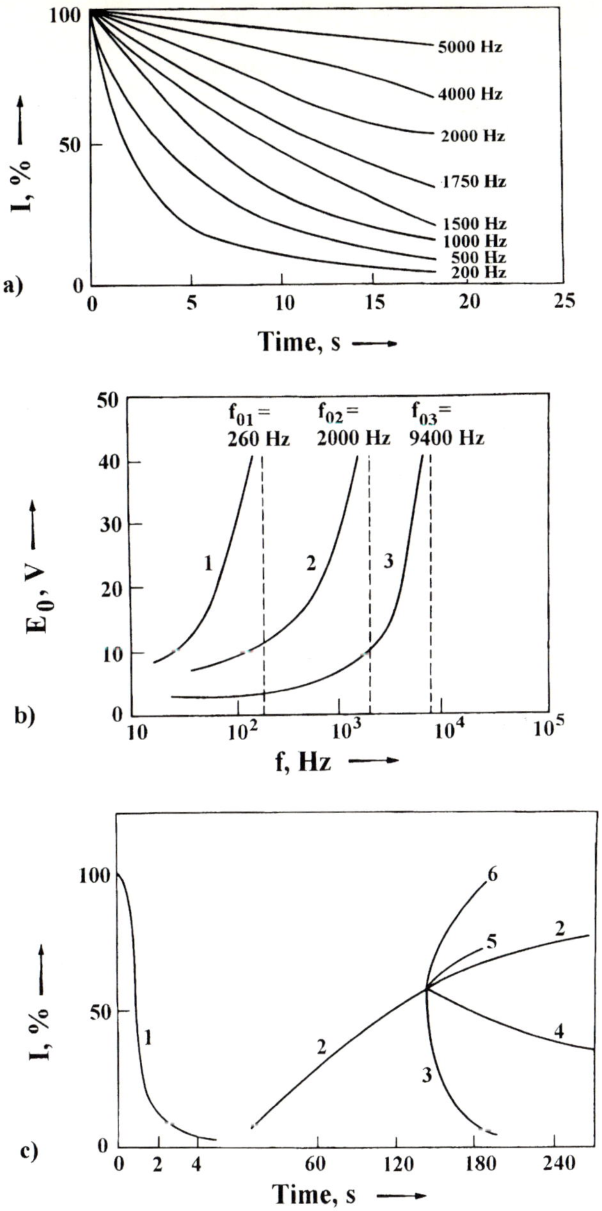

Fig. 2.12a-c. Influence of the electric field frequency on the electrooptical behaviour of the LC nematic polymer

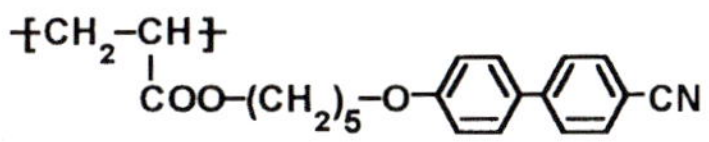

($\overline{DP_w}$ ~ 20) [61]

a -Optical transmission I, as a function of time at different frequencies (E=30 V). **b** -Threshold voltage E_0 as a function of frequency f at different temperatures ($T_1 < T_2 < T_3$). **c** -Optical transmission as a function of time upon application of an electric field at E = 85 V (f = 50 Hz) (*curve* 1). Relaxation upon switching the electric field off (*2*) upon application of an electric field (E = 80 V) of different frequency f = 1(*3*); 5(*4*); 7(*5*) and 20 kHz (*6*) during the relaxation process

The study and analysis of the regularities of the orientational behavior of LC polymers has elucidated a series of other analogies, but at the same time showed an essential difference in orientational phenomena of LC polymers as compared to low-molar-mass liquid crystals.

The certain analogy is exhibited, for example, in the frequency dependence of the mesogenic group orientation. Figure 2.12a shows the optical transmittance I of a film of a nematic polymer placed in an electrooptical cell during the application of an electric field of varying frequency. It is clearly seen that the optical transmittance decreases sharply at low frequency (200–500 Hz) which means the appearance of homeotropic mesogenic group orientation; at the same time, such an orientation is not displayed at a high frequency of the electric field (4000–5000 Hz). The asymptotic shape of the frequency dependence of the threshold voltage E_0 (Fig. 2.12b) demonstrates the existence of the so-called isotropy frequency f_0 at which $\Delta\varepsilon$ returns to zero and above which it can assume negative values (see value E_0 for the splay-effect in Table 2.7). (The effect of a change in the sign of the dielectric anisotropy is explained by a change in the portion of orientational polarization). Such an effect is accounted for by the change of $\Delta\varepsilon$ sign at f_0, and is thus an example of the frequency addressing of the splay effect.

Figure 2.12c demonstrates the influence of the electric field frequency on the process of mesogenic group orientation. Curve 1 in Fig. 2.12c corresponds to the process of homeotropic structure formation effected by a low-frequency field (f=50 Hz). Switching off the field leads to relaxation – the system is disoriented (curve 2). By repeatedly applying the field at various frequencies, one may observe either the disorientation of homeotropic structure (curves 5 and 6) or, in contrast, the additional orientation of mesogenic groups in the same direction (curves 3 and 4). Consequently by varying the electric field frequency one may vary the mode of mesogenic group alignment and then, by cooling the polymeric film below T_g one can fix their macroscopic structure formed in the LC state under an action of electric field.

Applying an electric field is a powerful method allowing us to change not only the nematic and smectic polymers but also to change strongly the supramolecular structure of cholestric LC polymers. Using cholesteric copolymers **XIII** and **XIV** containing the polar cyanobiphenyl mesogenic groups responsible for the positive $\Delta\varepsilon$ sign, the authors [68, 72] described in detail the process of the helix untwisting under the influence of an electric field.

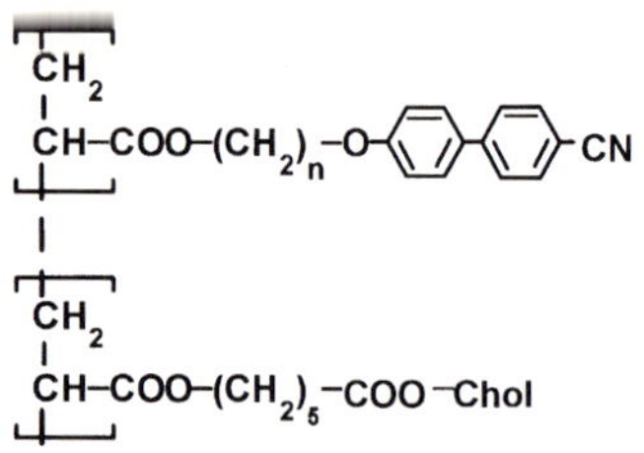

n = 4 Cop. XIII
n = 5 Cop. XIV

where **Chol** - cholesterol.

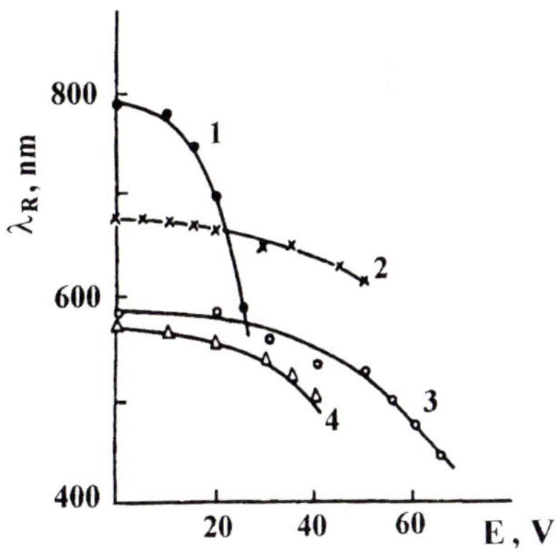

Fig. 2.13. The wavelength of selective light reflection λ_R as a function of voltage for copolymer **XIII** containing 1–21 (102 °C), 2–27 (89 °C), 3–30 (92 °C) and 4–33 (95 °C) mol % of cholesterol units [68]

Figures 2.13 and 2.14 show how the optically active planar-oriented cholesteric structure is transformed into the optically inactive homeotropically oriented texture after applying an electric field. This process takes place in two stages.

In the first stage, the axis of the cholesteric helix changes its direction; these structural rearrangements are accompanied by a shift in the wavelength of the selective light reflection maximum (λ_R) to the region of shorter wavelengths (Fig. 2.13). Simultaneously the initial planar texture is transformed into the confocal texture and the overall transparency of the sample decreases.

The second stage corresponds to untwisting of the helix and the mesogenic groups are arranged along the electric field direction in accordance with the positive sign of $\Delta\varepsilon$ stipulated by CN-groups existing in the copolymer. An additional increase in the field voltage destroys the confocal texture and a new optically inactive transparent texture is formed. In dependence on the spacer length of the cholesterolcontaining copolymers **XIII** and **XIV** (n = 4 or n = 5), the structural transition induced by the electric field may be described in the framework of cholesteric→nematic (n = 4) or cholesteric→smectic (n = 5) phase transition.

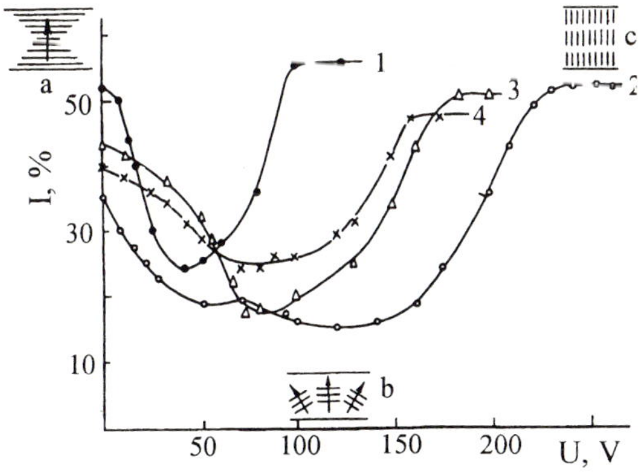

Fig. 2.14. Untwisting of the cholesteric helix. The transmittance as a function of voltage for copolymers **XIII**, containing 21 (*1*), 27 (*2*), 30 (*3*) and 33 mol % of cholesterol units. *a*, *b*, and *c*: schemes of the arrangement of the mesogenic groups and the helix (it is shown by the arrow) in the helix untwisting process; *a* - initial planar cholesterol texture; *b* -confocal cholesteric texture; *c*- homeotropic smectic texture [68]

Table 2.9. Influence of the degree of polymerization, $\overline{DP_W}$ on the rise time, τ_{on}, for the polymer: [14]

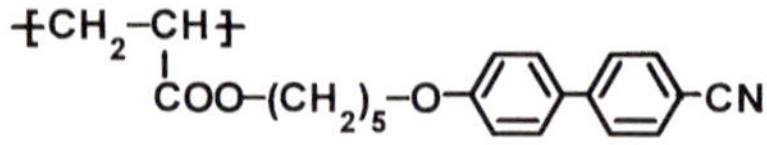

$\overline{M_w}$	$\overline{DP_w}$	T_{cl}, °C	τ_{on} at 100V (f=1 kHz), min
6.9×10^3	20	77	0.05
2.4×10^4	70	90	0.5
7.9×10^4	240	120	1.5
6.6×10^5	2000	120	10.0

Thus, under the influence of an electric field, the supramolecular texture of LC polymers may be essentially changed independent of the chemical binding of the mesogenic groups to the polymer backbone; after cooling below T_g this modified texture is fixed in a solid polymer film retaining all the optical properties inherent to the LC phase.

It should be emphasized that the description of all the orientational processes in LC polymers is a more complicated task than in low-molar-mass liquid crystals due to the dual nature of the comb-shaped polymers with due regard to the conformational transformations in both the main and the side chains.

In this chapter, we will not consider the very interesting problems associated with the regularities of the kinetics and mechanism of orientational processes (these questions have been already considered in a series of publications [6, 14–16, 19, 67–68, 74–76]). We would like to discuss here only some aspects of the electrooptical behavior of LC polymers related to the quantitative evaluation of the main electrooptical parameters such as orientational times, elasticity moduli and the rotational viscosity coefficient. On the one hand, these characteristics are the fundamental constants of LC compounds [77] (including LC polymers), on the other hand, these values determine the principal parameters of the electro-optical devices used for the practical applications.

First of all, let us consider the time characteristics of the orientational process such as the rise time and the decay time. Analysis of the orientation kinetics of LC polymers has convincingly shown that the rise time increases with increasing DP (Table 2.9). At the same time there are some data [74] indicating lower values of the response time (~200 ms) at the low threshold voltage ($E_0 \sim 8$–10 V) that it is close to the appropriate response times of low-molar-mass liquid crystals.

Unfortunately, authors do not always compare the time characteristics with the molar-mass parameters of LC polymers; the low values of the rise and decay times mentioned above probably correspond to the low values of DP and the broad molar-mass distribution. Besides, when considering the kinetics of orientation of LC polymers, special attention should be given to the temperature at which the electrooptical measurements are carried out. It is sufficient to assume in a first approximation an exponential temperature dependence of rotational viscosity coefficient γ_1 according to an Arrhenius function:

$$\gamma_1 \sim \exp(B/T) \text{ and } \log \gamma_1 \sim B/T \tag{2.3}$$

where T – is the absolute temperature and B is a constant. Taking into account an exponential temperature dependence of the polymer viscosity and the fact that the temperature dependence of γ_1 predominates over $\Delta\varepsilon$ and E_0 (see the equations in Table 2.7) one can affect the rise time considerably, increasing the temperature of measurement, i.e. bringing the measurement temperature nearer to the clearing temperature.

Indeed, authors of [78] succeeded in significantly decreasing the rise time (about three orders of magnitude!) for the LC acrylic polymer containing cyanobiphenyl and anthraquinone monomeric units only by increasing the temperature. So, for example, they obtained a rise time of around 2 ms (E_0 = 16 V) at the temperature of 1 °C below the N→I transition. The doubling of the applied voltage may shift the time parameters of orientational process into microsecond range.

Now we turn to the consideration of the viscoelastic characteristics. Usually LC polymers are characterized by elevated values of the rotational viscosity coefficient γ_1. The calculation of γ_1 for the acrylic polymer

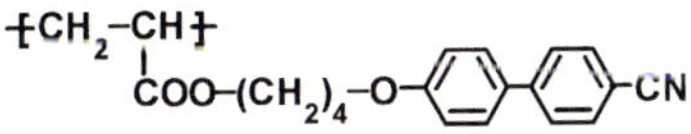

(XV)

from NMR data [73, 76] has shown that the values of γ_1 varied from 9.6×10^2 Pa·s ($\overline{DP}$ = 560) to 1.6×10^4 Pa·s (DP = 2530) at $T^* = T/T_{cl}$ = 0.95–0.96. These γ_1 values exceed the appropriate values of γ_1 of low-molar-mass liquid crystals by 2–4 orders of magnitude.

At the same time, it appears from the scanty literature data that the values of the elastic moduli for different distortion types of LC polymers (K_{11}, K_{22} and K_{33}) are very close to the appropriate elastic moduli of low-molar-mass liquid crystals (Table 2.8).

Unfortunately, at the present time, there is no certain theoretical correlation between the molecular structure of liquid crystals and the quantitative values of their elastic moduli. In accordance with the continuum theory of elasticity one can consider only some approximate relationships showing that temperature dependence of the elastic moduli correlated with that of the square of the order parameter [6].

$$K_{ii} \sim S^2(T), \quad i = 1, 2, 3 \tag{2.4}$$

It is now well-known that the values of K_{22} are usually less than K_{11} and K_{33} for low-molar-mass liquid crystals. According to [6, 77] the typical values of the elastic moduli ratios K_{33}/K_{11} and K_{22}/K_{11} lie in the range:

$$\begin{aligned} 0.5 &< K_{33}/K_{11} < 3.0 \\ 0.5 &< K_{22}/K_{11} < 0.8 \end{aligned} \tag{2.5}$$

It is most frequently considered that the elastic moduli depend on the geometrical dimensions of the mesogens, i.e. on the ratio of the effective length of the molecule to its width. However, the flexibility of the alkyl chains, the existence of the lateral substituents, the dimer formation and some other factors are also very important; together they cause considerable difficulties for finding the specific correlations between the molecular dimensions of mesogens and the elastic moduli.

With respect to LC polymers, only isolated, unsystematic data are available in the literature on the experimental evaluation of some viscoelastic characteristics (Table 2.8). As can be seen from the Table, the values of the elastic moduli of low-molar-mass liquid crystals and polymer liquid crystals are very close in respect to the order of the magnitude.

One can assume that the absence of a contribution of the backbone to the elastic moduli is due to the high autonomy of the mesogenic groups caused by the high flexibility of the main chain (polyacrylates, polysiloxanes) which together with the spatial remoteness of the mesogens create the favourable conditions for the appearance of elasticity inherent to the LC phase alone, formed by the packing of the mesogenic side groups. However, it has been reported [71] that a decrease in the spacer length, which does not significantly affect the K_{11} value, results in a significant increase in the ratio K_{33}/K_{11} (see polymer 2 in Table 2.8). This fact serves as an indication of the contribution of the polymer chain to the elastic properties of comb-shaped LC polymers.

Unfortunately, the absence of the quantitative data relating to the elastic moduli for a series of LC polymers already studied, hampers even the phenomenological description of the regularities obtained. The study of the influence of the degree of polymerization on the elastic moduli is also worthy of our attention. The electrooptical investigations of the cholesteric copolymers with different molar masses performed in [68, 72] have shown, that the change of the $\overline{DP}$ with the limits of 50 to 160 has almost no effect on the value of $k_{22} \cong (4-7)10^7$ din, which coincides in order of magnitude with the twist elasticity constants of low-molar-mass nematic-cholesteric mixtures.

At the same time, the detailed measurements carried out by Coles et al. [78] have convincingly shown that the mixing of LC polysiloxane copolymer **XVI**

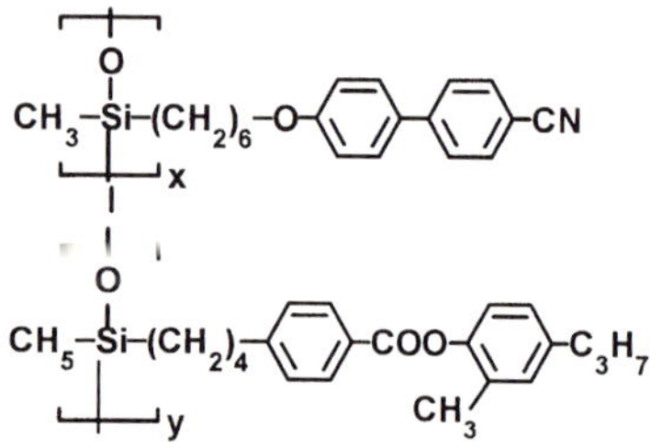

Cop.XVI

$T_g = 4\ °C,\ T_{S-1} = 86\ °C,\ m = n,\ \overline{DP} \sim 50,\ x:y \sim 1:1$

with the mixture of low-molar-mass nematic liquid crystals leads to a decrease in the values of both elastic constants K_{11} and K_{22} with an increase in polymer concentration [78] (Fig. 2.15).

The contrary influence of the LC polymer, i.e. an increase of the bend elastic constant K_{33} was observed for the mixture of an acrylic LC polymer and pen-

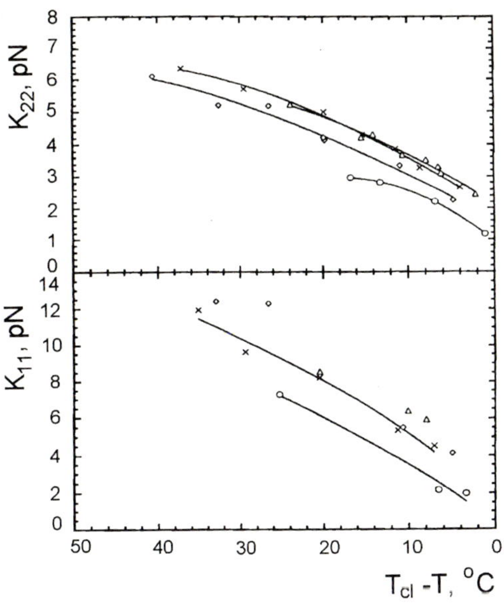

Fig. 2.15. Splay (K_{11}) and twist (K_{22}) elastic constants for low-molar-mass nematic mixture E-7 (BDH Ltd and DRA-Malvern) and for siloxane copolymer **XVI** containing E-7 as a function of reduced temperature [78]: x-E-7; △ - 5% polymer solution ◇ - 13% polymer solution and ○ - 48% polymer solution

tylcyanobiphenyle (5 CB) with increasing polymer concentration for the samples with various degrees of polymerization (DP=65, 26 and 130) [79] (Fig. 2.16).

Everything taken together shows that the currently available experimental data do not permit us to draw any definite conclusions yet concerning the roles of the chain structure of the macromolecules and to understand the effect of the "polymeric nature" of the liquid crystal on the viscoelastic properties of the LC phase. It is necessary to develop the broad studies aimed at evaluating the conformational state of macromolecules (which changes in the process of polymer orientation), the study of the molar-mass charateristics of LC polymers, as well as the flexibility of the main chain and spacers, which are responsible for "the autonomy degree" of the mesogenic groups and backbone.

However, in spite of the existence of some unsolved questions, the use of the electric field for orientation of LC polymers suggests a new and wide range of

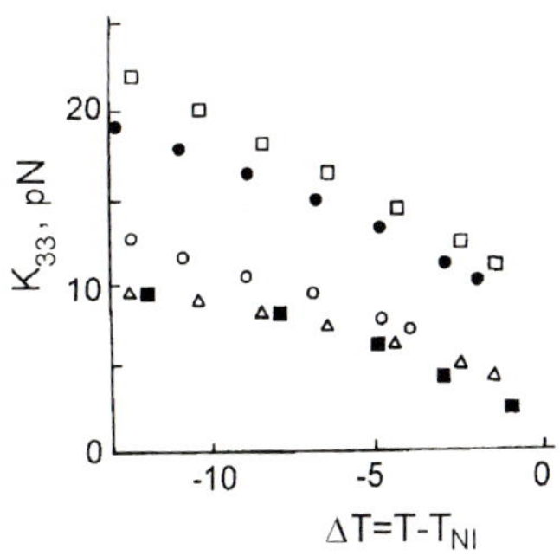

Fig. 2.16. Temperature dependences of the elastic constants K_{33} for pentylcyanobiphenyl (5 CB) (△) and solution of the fractions of LC polymer

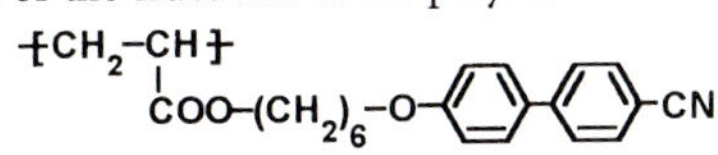

with DP=25 (○), 65 (□) and 130 (○) in 5CB [79]. The data obtained for 5CB in [80] are shown by (■)

possibilities for the creation of new types of polymeric films with interesting and desirable optical properties. Such highly-oriented films may be obtained by the action of an electric field in the LC state in the appropriate temperature range with the successive fixation of such structures in the solid by cooling the sample below the glass transition temperature.

In Table 2.7 are presented the main types of nematic structures, well-known for low-molar-mass liquid crystals – homeotropic, planar and twist textures.

Such structures may also be obtained in LC polymers by certain specific methods. So, for example, polymeric films with planar texture constitute the phase plates; their optical properties are characterized by the difference in the course $\Delta\Phi = \Delta nd$, where $\Delta n = n_{\parallel} - n_{\perp}$ is the optical anisotropy, i.e. the difference in the refractive indices of light for polarization parallel ($n_{\parallel}$) and the perpendicular ($n_{\perp}$) to the optical axis; d is the thickness of the plate.

In such polymeric films, the values of $\Delta\Phi$ may be modulated on the coordinate both in the thickness of the sample and along the plane of the film, this gives us the possibility of obtaining polarization wedges, phase cylindrical or spherical lens. Some examples of different mesogenic group arrangements in such films, corresponding essentially to the supramolecular structure of the "frozen mesophase" are presented in Fig. 2.17. All these structures shown in this Figure may be created by applying to the layer of the LC polymer an electric field the voltage of which changes along the plane of the layer, for example, by the use of electrodes with distributed resistance, and subsequent cooling of the film below the glass transition point.

An important advantage of LC polymers over low-molar-mass liquid crystals is the possibility of superimposing one formed layer of the LC polymer with certain optical properties to the surface of another which allows one to construct complex optical systems of low thickness. The possibility of obtaining thin polymeric films, exotic and various variants of the mesogenic groups arrangements allow us to consider the LC polymers as the prospective materials for the creation of new passive optical elements.

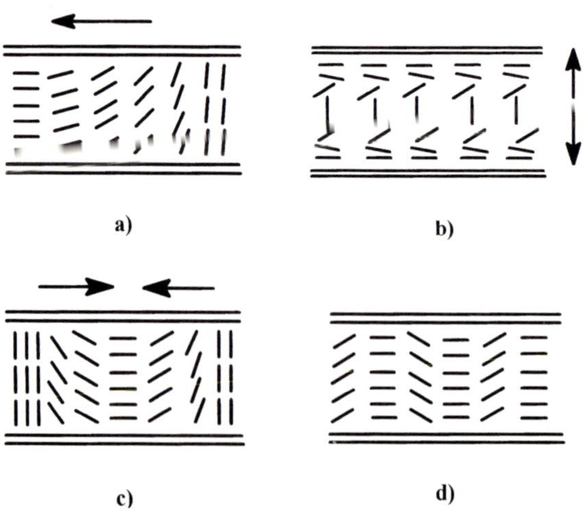

Fig. 2.17. Various type of mesogenic groups arrangements, which may be used to produce passive optical elements. *Arrows* indicate direction of the Δn gradient. (The main chains are omitted for simplicity)

However, in addition to the possible use of LC polymers as materials of passive optical elements, LC polymers may be used in active controlled elements. In this case external factors such as temperature, electric field and light act as controlling signals. In this way, such LC films may be used as materials for optical data storage.

2.5 Thermooptical Recording of Information on Films of Comb-Shaped Liquid-Crystalline Polymers

As already stated, one of the specific features of polymer liquid crystals is the ability to keep ("to freeze") the LC oriented structure (formed in the LC state) in the glassy state below the glass transition temperature (Fig. 2.3). Such highly oriented polymeric films are of considerable interest, from the practical point of view for optical data storage, working according to the principle: orientation of the polymer in LC phase (by the influence of an electric field or by the walls of the electrooptical cell) – disorientation and transition in the isotropic melt on heating – formation and fixation of a strongly light-scattering polydomain structure on cooling below the Tg. In so far as the process of image formation is induced by the heat field, this recording method is usually called "thermooptical recording" or "thermal addressing" [6, 11].

The use of LC polymers for thermooptical recording of information was demonstrated for the first time in the joint work of scientists of the Chemical and Physical Departments of Moscow State University as illustrated by polyacrylate polymer readily orienting in an electric field (see Fig. 2.11) [81–84].

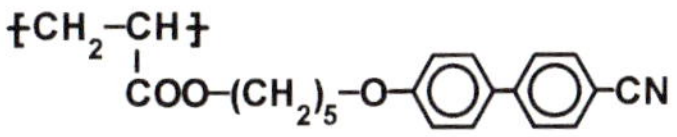

(XVII)

The principle scheme for information recording is shown in Fig. 2.18. On a transparent film of a homeotropically oriented LC polymer (20–40 μm thick), using a focused laser, sites of local overheating of the polymer above its clearing temperature (T_{cl}) are created. At these sites, the LC polymer melts to form an isotropic melt with disturbance of the monodomain orientation of the mesogenic

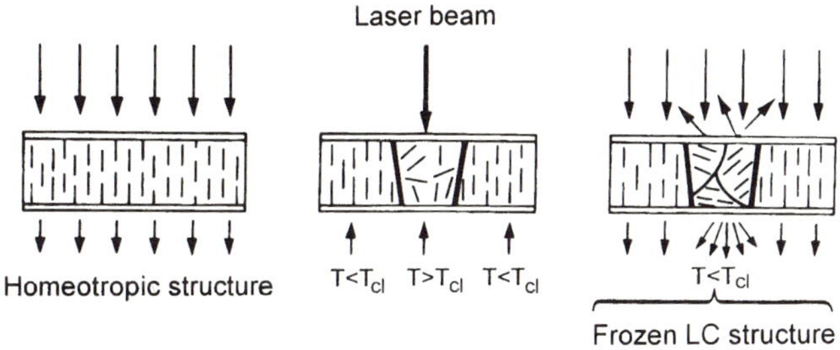

Fig. 2.18. Scheme of thermorecording using the film of the hometropically oriented LC polymer [18, 82]

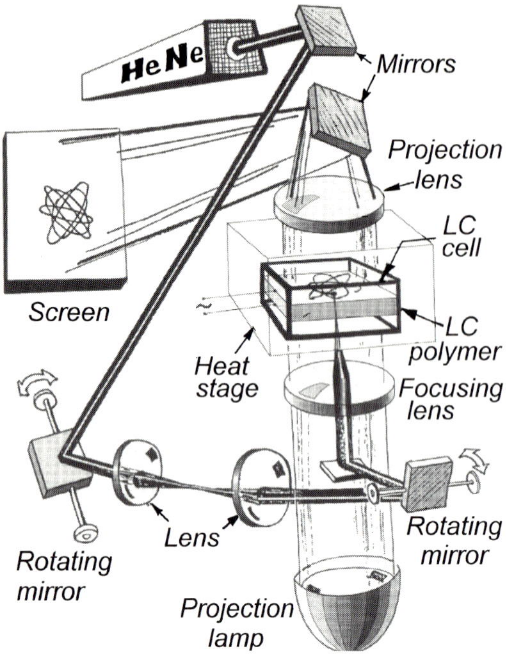

Fig. 2.19. Schematic representation of experimental optical set-up for thermorecording

groups. Cooling the sample, for example, through simple removing the laser beam leads to the formation on the transparent film of a strongly light-scattering polydomain texture which corresponds to recording certain bits of information. Illumination of the polymeric film with a defocused laser beam allows the recorded image to be projected onto a screen.

The scheme of the experimental optical set-up for studying the laser addressing of LC polymers is shown in Fig. 2.19. The He-Ne laser with a power of 10 mW and a diameter of the laser beam focus $\sim$ 70 μm was used as the light

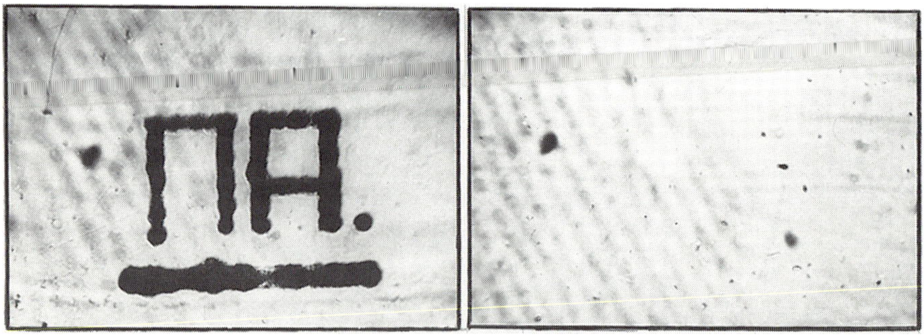

a b

Fig. 2.20a-b. Photomicrographs showing (**a**) letters thermorecorded on the film of polymer **XVII** (film thickness 40 μm) (**b**) - texture of polymer after erasure process by applying electric field [18, 82]

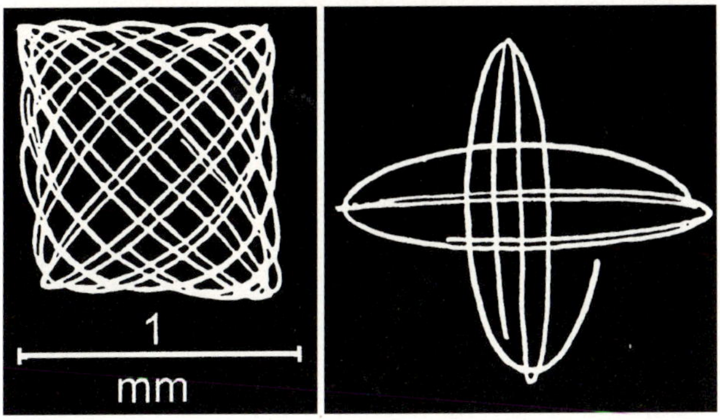

Fig. 2.21. Typical examples of thermooptical recording on a homeotropically oriented film of polymer **XVII**. Film thickness 30 μm, laser power 2.6 mW, scanning speed 10 mm/s [20, 84]

source. The laser beam passed through an attenuator, the mirror systems and lens was directed on the homeotropically aligned sample which was placed in an electrooptical glass cell. The cell was kept in a special thermostat.

After several seconds under the action of the laser beam on the polymer sample (close to the clearing temperature) and on subsequent illumination of the polymeric film by an unfocused laser beam the contrast dark spot is shown on a screen.

The diameter of this spot exceeds the diameter of the unfocused laser beam by two-three times. By subsequently moving the laser beam along the polymer film one can record letter symbols which can be projected onto a screen corresponding to zones produced by the action of the focused beam. Figure 2.20a shows Russian letters recorded on a polymeric film. In the places on the film where illumination was absent or was of low power (less than 80 W/cm^2) the sample kept homeotropic orientation and looked transparent. The symbols obtained in the above mentioned way are erased by switching on an alternating electric field applied to the electrooptical cell (U = 100 V; f = 50 Hz) (Fig. 2.20b) or by heating the sample above the clearing point.

The information (or image) thus recorded may be kept for a long time if the sample of the LC polymer is preserved below the glass transition temperature where the relaxation rate is negligibly small.

Figure 2.21 gives examples of recording different geometrical figures on polymeric film containing a dye for increasing the laser illumination absorption [20, 84].

The introduction of a dye of the anthraquinone series into polymer **XVII** and the use of a laser beam with $\lambda = 1.06$ μm allowed information to be recorded with high contrast (1:120) and quite good resolution of the order 1000 and more elements per line at the scanning rate of the laser beam ~ 0.45 cm/s [3][6].

[6]This work was undertaken by A.N. Drozhin and S.G. Kostromin.

Our investigations [81–83] have clearly demonstrated the possibility of using LC polymers for thermoaddressing and long-term information storage over a broad temperature range and also the possibility of its erasure by means of an alternating electric field or by heating. As can be seen from the above-mentioned data, in these cases, the electrooptical effects play only an auxiliary role insofar as they are mainly used for the creation of the initial orientation of the mesogenic groups and for the erasing of information.

In a series of investigations and publications by Coles [78, 85, 86] and McArdle [11, 87], siloxane polymers and copolymers containing terminal nitrile groups in the side mesogenic fragments including a small amount of azo- and anthraquinone dyes "tuned" to a certain wave length of laser irradiation (see polymers **XVI** and also copolymers **XVIII** and **XIX**) were used for thermo-optical recording.

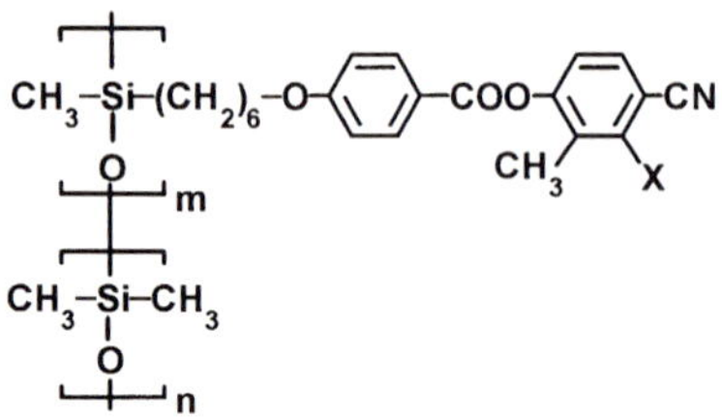

$X = H;\ T_g = -12\ ^\circ C,\ T_{S-1} = 100\ ^\circ C,$ **Cop.XVIII**

$X = F;\ T_g = -19\ ^\circ C,\ T_{S-1} = 98\ ^\circ C,$ **Cop.XIX**

$m = n,\ \overline{DP} = 35, \overline{M}_w/\overline{M}_n = 1.5 - 1.9$

A most detailed investigation close to a technical solution was reported by McArdle et al. [11, 87] the results of which are briefly considered below.

Using a laser, information recording was carried out on a homeotropically oriented film of polymeric LC sample (9 μm thick) placed in a electrooptical cell to which may be fed both an alternating and a constant electric field.

The use of a He-Ne laser (λ=633 nm) with a power of about 7 mW working in the pulse regime (about 200 μs) enables one to register and reversibly erase (by applying an electric field with a voltage of 60 V and a frequency of 3 kHz) the recorded information as is well illustrated by Fig. 2.22. Undoubtedly an important and attractive feature of LC polymers is their quite high stability in repeated cycles of re-recording as may be seen comparison of the curves in Fig. 2.23 for low-molar-mass and polymeric liquid crystals. Even without the use of any stabilizers the "fatigue optical strength" of an LC polymer far surpasses the corresponding parameter of a low-molar-mass LC compound.

As for the "sensitivity" of LC polymers, according to the results of various authors, it depends significantly on the type of polymer used, temperature and other specific conditions of the experiment. Thus, in Ref. [11] for copolymer **XVIII** containing about 2% by weight of a dye at 24 °C sensitivity was 1.2 J/cm^2 with the use of 100-microseconds pulses sensitivity was 0.3 mJ per point. For the polysiloxane copolymer **XIX** for a power of the laser beam 0.5 mW, sensitivity and resolution were 0.1 J/cm^2 and 0.5 μm respectively.

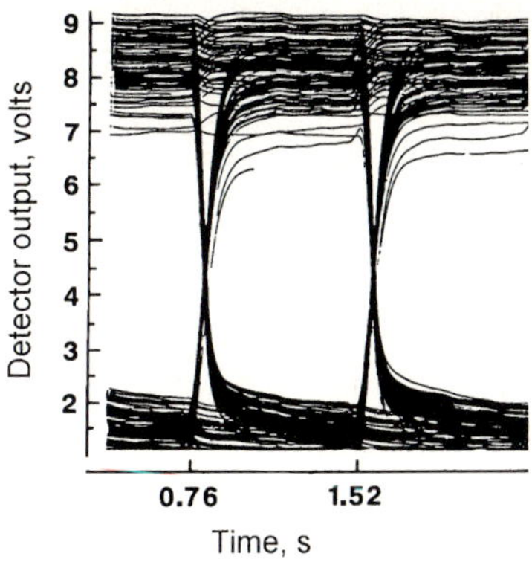

Fig. 2.22. Digital eye diagram for copolymer **XIX** dyed to 3 wt % recorded after 500 write-erase cycles. Conditions employed were: write pulse and erase pulse equal to 200 μs (T=55 °C); erase field (60 V, f=3 kHz) sine wave applied just before laser pulse and continued for ~ 300 ms thereafter [11]

Although the fast action for most polymeric LC materials is not great they are of interest of recording and long-term (archive) storage of information, microfiches, for cartography and use in systems of micrography.

In Ref. [87] the photomicrograph of a military-topographical map recorded on a homeotropically oriented sample of an LC polymer in a cell measuring 8×8 mm with a scanning speed of the laser beam of 60 mm/s at 11 mW onto a focused point demonstrated very high resolution.

The use of LC polymers for information recording is not confined to polymeric smectics; in some studies, Japanese scientists described the use of oriented cholesterics based on cholesterol-containing comb-shaped polysiloxanes for the same purposes [88]. In this case, with planar orientation of the mesogenic groups, when the axis of the cholesteric helix is directed perpendicularly to the walls of the cell, the film of the LC polymer is colored if the wavelength of the reflected light lies in the region in relation to the size of the pitch of the cholesteric helix.

Figure 2.24 demonstrates the principle of thermooptical recording on a cholesteric layer of an LC polymer deposited on carbon dispersed black polyethylene terephthalate film. In this case, the authors of Ref. [88] instead of a laser beam,

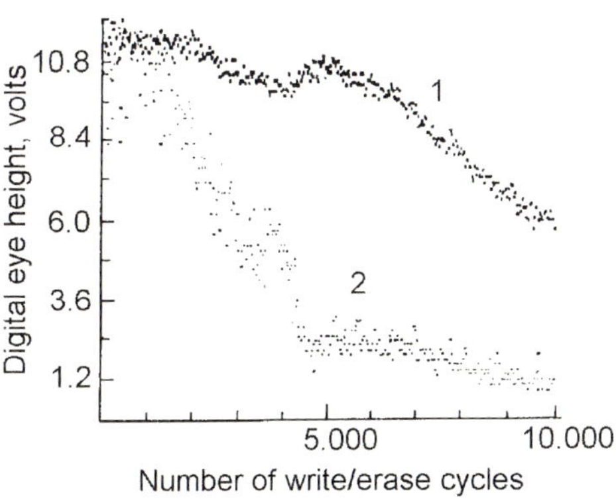

Fig. 2.23. Plot of digital eye height against number of write-erase cycles for LC copolymer **XIX** (*1*) and sample of low-molar-mass liquid crystal S2 (BDH) (*2*) [11]

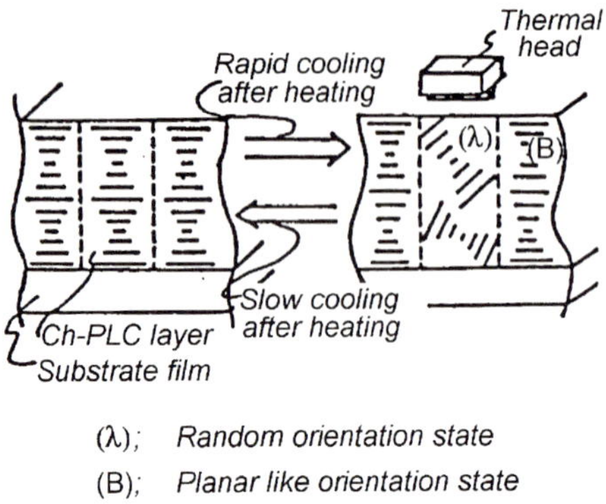

Fig. 2.24. Scheme of thermooptical recording on cholesteric film [88]

used a thermal head, which locally heated the cholesteric layer. The written state is induced by a rapid heating and cooling cycle due to heat pulse application and rapid thermal head scanning. In the activated, heated region, the initial planar alignment becomes random and polydomain, so that light scattering occurred, i.e. the written spot is in a light scattering area. As a result, a white image appears on the uniformly colored (green or red) background. The sensitivity of such cholesteric film is around 0.15 mJ/dot input energy density. The resolution is more than 8 dots/mm. The constant ratio gradually increased with increasing input energy density and saturated at about 0.3 mJ/dot. The saturated contrast ratio was about 5.

The use of such cholesteric films combined with circular "right" and "left" polarizers makes it possible to change both the color of the symbols recorded and that of the background against which the recording is made. Unfortunately, the authors pointed out some difficulties of erasing the written information that can be achieved by heating the samples to 90–100 °C. Though most of the regions in the written spots are erased, complete erasure could not be obtained. Especially the erase state in the written spot edge region becomes imperfect. This incomplete erasure is caused by surface damage to the cholesteric film, due to the cholesteric polymer film sticking to the thermal head. According to the authors' opinion, the surface damage can be overcome by coating a suitable material on the cholesteric layer.

In spite of the difficulties mentioned above, these investigations are undoubtedly of considerable interest in terms of creating miniature coloured optical elements of a varied nature, "strip" and other kinds of filters and diffraction patterns with readily varied geometry.

Comparing the potential merits of polymeric LC films used as recording media with other optical media, McArdle [11] has given the prognosis of the technology of producing LC microfilms and other microfilms (he calls them micrographic media) as follows: "The most likely scenario is that microfilm, optical disk and other media will coexist for some time. The cost per Megabyte for optical disk technology will only approach that for micrographic media by 1995. Currently microform masters at ×48 reduction are about 14 times less

Table 2.10. Comparison of some characteristics of low-molar-mass cholesteric (Chol) and smectic LC materials [89] as well as LC polymer, used for thermooptical recording [11, 18]

Parameters	Chol	Smectic	LCP
Working temperature of information recording, °C	48	38	40–100
Temperature range of optical data storage, °C	5–48	5–38	from low temperature to T_{cl}
Diameter of bit recording, μm	20	15–20	10–20
Power of laser ($\lambda = 633$ nm), mW	3	6–10	5–10
Electric voltage at local erasure, V	–	17–25	
Electric voltage at total erasure, V	60	60	60
Constrast	10:1	35–50:1	120:1
Resolution in line, number of dots	1000	1000	1000
Speed recording, $cm \cdot s^{-1}$	10	5	0.5–1.0
Sensitivity, $J \cdot cm^{-2}$			10^{-2}
Bit stability, months	1–2	4	years
Dwell time of one point, ms	–	0.5–0.6	

expensive, and microform duplicates some 110 times cheaper, than optical disk technology".

The creation, on the base of polymeric LC films, of reversive recording media and special devices readily compatible with computers and fitted with laser sources give grounds for looking forward to the wide use of LC films for storage of diverse archive information. The possibility of selective erasure and recording is attracting special attention to these materials.

Some comparative data for low-molar-mass cholesteric and smectic liquid crystals, LC polymers and other reversive optical materials used for information recording are given in Tables 2.10 and 2.11.

Table 2.10 shows some average characteristics of a hypothetical LC polymer adopted from McArdle's data [11] for polymers **XVIII** and **XIX** . The first

Table 2.11. Comparison of some important characteristics of a number of selected erasable optical storage media and LC materials [11]

N	Medium	Fatigue after: (Write/erase cycles)	Bit stability	Write pulse / Erase pulse	Contrast*, $(A_1 - A_0)/A_1$
1	$Te_{96.8}\ As_3\ Ge_{0.2}$	20	3 weeks	2×10^{-3}	50
2	$Te_{88}\ As_5\ Ge_7$	20	long term	0.1	–
3	$Te_{87}\ Ge_8\ Sn_5$	1×10^6	20 weeks	1.6×10^{-3}	33
4	Sb_2Se	1×10^4	long term	0.25	10
5	Magneto-Optic	1×10^6	long term	1	low
6	Dyed LC Polymer (Cop. XIX)	1×10^4	long term	1	> 80
7	Dyed Low-Molar-Mass Liquid Crystal S2 (BDH, UK)	5×10^3	long term	1	> 80

*A_1 - Signal amplitude of erased state; A_0 - Signal amplitude of written state

remark to this Table is concerned with the temperature range of recording and retaining the information. In the case of LC polymers, the most suitable range of recording is the temperature close to the clearing temperature which provides a high recording rate due to the lower viscosity of the LC polymer. So far as the values of clearing temperature may be easily varied by the DP change, this temperature range for polyacrylic and polysiloxane LC polymers can be set within the range suitable for the experimental investigation and practical application (for example, in the temperature range 40–100 °C).

As for the temperature range of the recorded information storage, there are no limits to the low temperature range because the majority of LC polymers are vitrified on cooling. In contrast, low-molar-mass liquid crystals usually crystallize on cooling down which hampers their use for long-term optical data storage. The same is probably true for the stability of the recorded information. As can be seen from Table 2.10, the low-molar-mass liquid crystals have a limited bit stability. In contrast, LC polymers are characterized by long-term bit stability, so, for example, the authors of Refs. [11, 87] did not observe any changes in polymeric LC films over a period of two years. According to our data, the information recorded on LC acrylic polymer films has already been kept for eight years without any changes.

With regard to other characteristics presented in Table 2.10, they are practically not inferior to the same parameters of low-molar-mass liquid crystals and are better than the latter with their high contrast.

It is interesting to compare some important characteristics of thermotropic LC materials with other optical materials used for reversible information recording. Ideal media must fulfill the following requirements: high stability (number of recording-erasure cycles not less than 10^6), long-term storage of the recorded information, comparable short times for recording and erasure cycles, and high image contrast. As may be seen from Table 2.11, LC polymers having slightly lower "optical fatigue" than inorganic materials (the recording principle of these materials is based on the phase transition: amorphous-crystalline) are distinguished by the high stability of information storage, the comparable times of recording and erasure as well as by the high image contrast. As can be seen from the same Table, tellurium-based alloys have either good optical fatigue but poor stability or poor optical fatigue but high stability. Sb_2Se optical media working on the phase-transition base are quite impressive, offering good kinetic characteristics but low cycloability.

Taking into account that the results mentioned above have, so far, only been realized for a limited number of polymeric LC materials and probably not always with an optimum chemical composition and structure, the comb-shaped LC polymers should apparently be regarded as the highly promising materials for reversible optical recording and long-term information storage. One can suppose that an application of LC polymers as thermooptical media has considerable advantages over low-molar-mass liquid crystals in respect to the control of their phase state and transition temperatures as well as physico-mechanical properties.

2.6
Photooptical Effects in Comb-Shaped Polymers

2.6.1
General Consideration

The thermooptical effects mentioned above have been studied on LC homopolymers and copolymers not containing chemically linked dye molecules. Such dye molecules are usually dissolved in LC polymers with low dye concentration (about 1–2 mol %) only to increase absorption of the writing beam at a certain lengthwave (see chapter 1 concerning mixtures of polymers and dyes dissolved as guest molecules).

In the case of photooptical recording on LC polymers, comb-shaped copolymers containing chemically-bound dye molecules are usually used. Under the influence of polarized light or a laser beam, the dye molecules undergo configurational (and conformational) changes leading to the appearance of a photoinduced birefringence and to changes of optical properties of the polymeric film. In this case, the phase image of the object is a consequence of spatial modulation of the refractive indices. The values of the photoinduced birefringence stored in the polymer matrix can be easily changed in respect to the electric vector of the incident light. Sometimes this method is called polarization information recording [90–93].

The first experiments concerning the applications of comb-shaped LC polymers to phase and holographic recording were described by Wendorff and Eich [49, 94–96]. Using the films of nematic and smectic LC polymers with azobenzene mesogenic groups (Copolymer **XX** and **XXI**), the authors observed the induced birefringence Δn of the order of 7×10^{-3}–1×10^{-2} caused by polarized laser beam irradiation and briefly described the registered diffraction and holographic patterns.

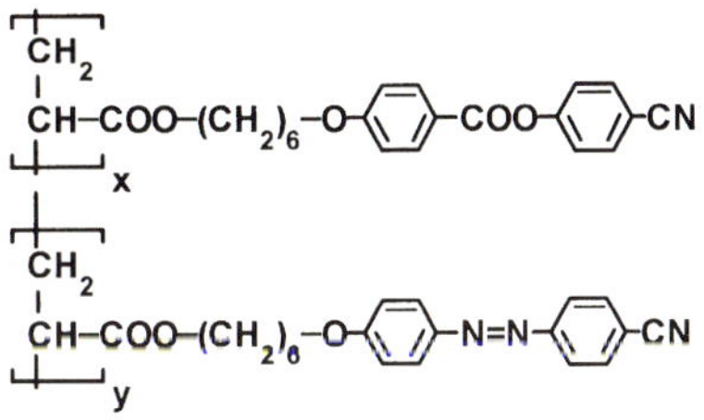

Cop. XX

$x = 0.7,\ y = 0.3,\ T_g = 30\ ^\circ C,\ T_{N\to I} = 103\ ^\circ C$

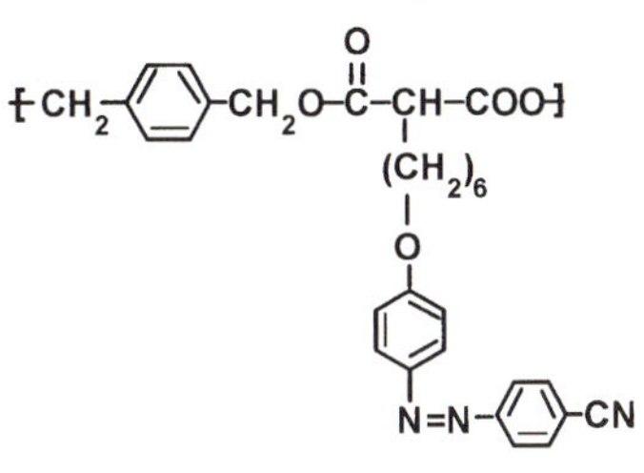

(XXI)

$T_g = 43\ ^\circ C,\ T_{S_A \to N} = 94\ ^\circ C,\ T_{N \to I} = 104\ ^\circ C$

The data presented in these publications [94–96] indicated a relatively large spatial resolution of more than 3000 lines/mm which corresponded to a resolution of 0.3 μm; the maximum diffraction efficiency amounted to 50% in thick phase gratings.

A little later, on the same principle, a holographic recording in cyclic siloxanes, also containing azo dye groups and forming a mesophase of cholesteric type was described in Ref [97]. In this case, the maximum diffraction efficiency was 38% with photosensitivity of 0.001 to 0.1 cm^2/J (for λ=476 nm and 413 nm).[7]

The interesting results related to the photooptical recording of information using azobenzene-containing LC comb-shaped copolymers were presented in Japanese [98–101] and German [102] publications. Krongauz et al. [58] provided an elegant study of photooptical phenomena on a series of comb-shaped spiropyran-containing copolymers on the basis of acrylates and siloxanes.[8] And finally it is necessary to call attention to a series of the novel publications by Canadian authors who investigated not only liquid crystalline but also amorphous polymers which might be used as photo-optical media [52–54].

All these data demonstrate the potential of comb-shaped LC polymers as highly sensitive reversible materials and prove the necessity of a detailed investigation of mechanism of processes associated with the interaction of light with LC polymers, the study of the kinetics and dynamics and the configurational and conformational transformations caused by these processes. Some of these aspects, especially in respect to optically nonlinear phenomena are considered in the review by Wendorff and Eich [96]; some photooptical phenomena in spiropyran-containing photochromic materials are reported by Krongauz in "Applied Photochromic Polymer Systems" (McArdle (ed)) [58]. The existence of this book, in which the various photochromic (non liquid-crystalline) compounds, containing spiropyranes, fulgides and viologan-based systems are discussed, releases us from our duties to consider these questions in this chapter.

The main attention of this chapter will be dedicated to our own results concerning the study of photooptical phenomena in the LC and amorphous comb-shaped polymers and the investigation of the possibility of using these materials as optical information recording media. These data are the results of our collaboration with German colleagues from the Humboldt University in Berlin.

2.6.2
Origin of Photoinduced Optical Anisotropy

The majority of the works concerning photooptical phenomena in comb-shaped polymers were carried out on the copolymers containing different azo dye molecules as the photosensitive fragments. These kinds of copolymers were also the

[7] These investigations are described in details in chapter 3.
[8] These investigation are summarized in the review of the book [58].

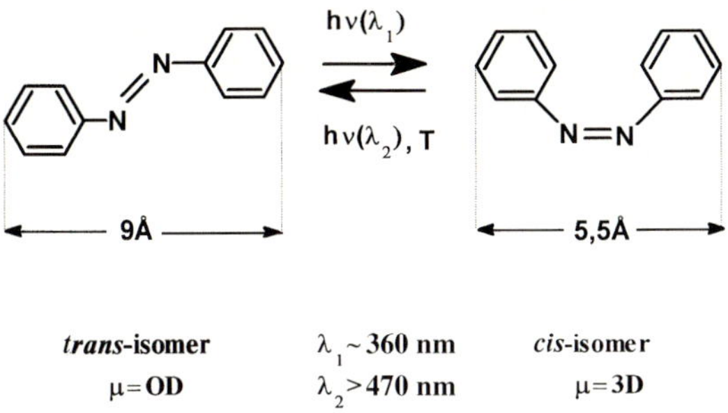

Fig. 2.25. Photochemical isomerization of azobenzene (μ-dipole moment)

subjects of our investigations (Tables 2.2 and 2.3) and they will be considered here in detail.

The photochemistry of low-molar-mass azo compounds is well-elaborated [39–40, 103–104]; there is also voluminous literature associated with the study of the mixtures consisting of amorphous or LC polymers and different derivatives of azo compounds (so called "guest-host" systems), which have been studied in solutions as well as in solids [41, 105, 106].[9]

Less attention has been given to the study of the comb-shaped polymers containing dichroic dyes covalently attached to the polymer backbone via flexible spacers. However, before we begin with the consideration and analysis of the photooptical phenomena in comb-shaped copolymers containing the dye, let us consider what happens with azo dyes under an action of the light.

The reversible photochromism of aromatic azobenzene compounds is a well-known phenomenon which can be performed in solutions and in polymer films [103, 104]. The photochromism of azo dyes is due to the *trans-cis* isomerization of molecules with respect to the N=N double bond (Fig. 2.25).

Usually the absorption of one photon raises azo dye molecules to an excited state from which a non-radiative decay brings them back to the ground state either in the *trans*-form or in the cis-form. The reverse transition from the cis- to the more stable trans- form is usually accelerated by light with $\lambda_2 > 470$ nm. As can be seen from Fig. 2.25, the photochromic *trans*→*cis* isomerization is accompanied by strong geometric changes, the azobenzene rod-shaped molecules are transformed into bent-shaped molecules, which can undergo a reverse transition according to the data given above. The establishment of photostationary equilibrium between *trans*- and *cis*-forms which are formed during and after irradiation plays an essential role in the mechanism of information recording on such photochromic polymers.

Among the photosensitive systems subject to photoisomerization, there are a number of organic compounds containing the C=C double bonds (for example, stilbene, olefins) and other photochromes such as azomethynes, spiropiranes, fulgides, salicylidenanilines etc. However, the number of polymers containing chemically bonded fragments of these compounds are limited to some vinylic,

[9]Some of these systems are considered by Horie and Mashida in the chapter 1 of this book.

acrylic and siloxane types containing photosensitive groups such as azobenzene derivatives, azomethyne and spiropyran. The detailed information of photo-optical behavior of the latter was given in Ref. [58]. Here the main attention will be focused on the comb-shaped polymers containing azobenzene mesogenic fragments (Tables 2.2 and 2.3).

We have used LC comb-shaped polyacrylic and polymethacrylic copolymers with different chemical natures and lengths of the spacer containing varied concentrations of chemically-attached azo dye molecules [16, 81–84, 107–112]. Depending on the chemical structure of the spacer and the ratio between the mesogenic and the dye-containing units these copolymers could exist in the amorphous or the LC state [108–112].

Depending on the phase state of the copolymers, the samples for photooptical measurements were prepared in different ways [108–112]. Samples for optical measurements were prepared by placing an isotropic melt of the copolymer into the slit of a sandwich-type electrooptical cell consisting of two planar, transparent In_2O_3 electrodes on a glass substrate, separated by Teflon spacers of various thicknesses (from 6 to 100 μkm).

Homeotropic orientation was achieved by applying an a.c. voltage (50 Hz; 30 to 100 V) to the electrodes. The temperature was then gradually lowered to bring the copolymers to the LC state; after that the sample was annealed below the clearing temperature for an appropriate time. Then the temperature was gradually lowered to an ambient one and the voltage was switched off. In such a way a homeotropically oriented monodomain polymeric film with the optical axis perpendicular to the glass surface was prepared.

To achieve the planar orientation, the surface of the glass plates of the electrooptical cell was first polished unidirectionally. The special electrooptical cells coated with a polyimide film and rubbed along in a given direction were also used. After prolonged (several hours) annealing of the copolymer in the LC phase at a temperature of several degree below the clearing temperature, the film attained a planar orientation where the axis was parallel with the glass surface and coincided with the direction of the preliminary polishing.

The amorphous copolymers were sandwiched between two glasses, heated above glass temperature and slowly cooled down.

In all cases, optically transparent films were obtained.

The experimental optical set-up for photooptical measurements is shown in Fig. 2.26 [16]. Vertically polarized laser beam I (Argon laser, λ=488 nm) of 1 mm diameter (power density P = 40–300 mW/cm^2) was directed to a selected area of the polymer sample placed in the electrooptical cell which was incorporated in a thermoregulated block ("Mettler"-FP-82). To register the photoinduced structural changes, the same part of the cell was simultaneously irradiated with a He-Ne laser beam (λ = 632.8 nm); it was established by preliminary experiments that the light of this laser does not affect the copolymer properties. The angle between the polarization planes of "writing" (I) and "reading"(II) beams was 45°. To facilitate the measurements for the homeotropic orientation of polymers, a quarter-wave plate with optical axis oriented vertically was inserted.

The intensity of the reading beam (II) that passed through the cell and the analyzer was registered by the special photo-detector and registering device. For

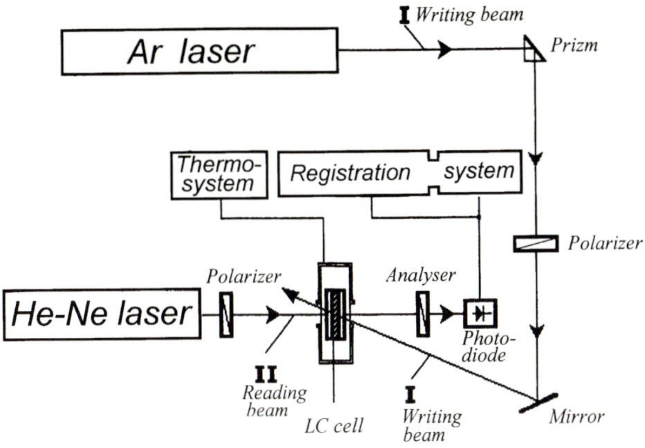

Fig. 2.26. Experimental optical set-up for photooptical recording on polymer films [16]

measuring the intensity and absorption coefficient of the "writing" beam which had passed through the sample, the monitoring channel was used. This channel was also used for checking the sample thickness.

The intensities measured with the analyzer, axis oriented perpendicularly to and in parallel with the polarization plane of beam II, were used to evaluate the photoinduced birefringence of the copolymer, Δn_{ind}. The sign of birefringence was determined by the compensation method using a quartz plate and wedge in the polarizing microscope.

The irradiation of azobenzene-containing copolymers induced the photoisomerization of azobenzene side fragments according to the scheme presented in Fig. 2.25. The UV spectrum of the investigated azobenzene- containing copolymer 5 (Table 2.2) clearly demonstrates the increase in the concentration of the *trans*-isomers during irradiation (Fig. 2.27). An increase in the irradiation time leads to the establishment of the photostationary equilibrium between *trans*-and *cis*-isomers; meanwhile the equilibrium concentration of each isomer depends on the wave length of the irradiation. For example, the concentration of *trans*-isomer sharply decreases under irradiation with a wave length of λ=365 nm coinciding with λ_{max} corresponding to maximum absorption of the *trans*-isomer (Fig. 2.27). However, after an irradiation of copolymer 5 (Table 2.2) with an argon laser with λ=488 nm the equilibrium concentration of *trans*-isomer noticeably exceeds the *cis*-isomer concentration.

The establishment of the photostationary equilibrium initiates a slower process of the rotational diffusion of side dye fragments. Simultaneously the neighboring side nonphotochromic rigid mesogenic groups are also involved in this process of reorientation (it will be considered in more detail below) which leads to the appearance and increase of photoinduced birefringence Δn_{ind} in the sample.

It is well-known that the phenomenon of photoinduced anisotropy was discovered by Weigert [113] many years ago in viscous solutions of azo dye. Later the photoinduced birefringence phenomena in films obtained by evaporation of

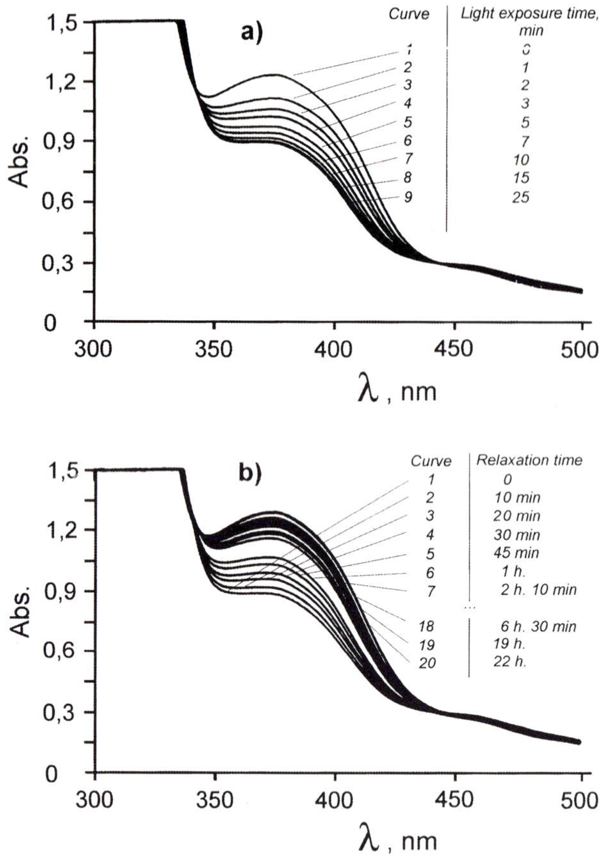

Fig. 2.27a-b. UV-Vis absorption spectra of copolymer 5 (Table 2.2). **a** -during irradiation by monochromatic light of $\lambda = 365$ nm (Mercurylamp, P = 1.7 mW/cm^2) [108] and **b** -after relaxation (*cir-trans* isomerization). Light exposure time (**a**) and relaxation time (**b**) are indicated in the figure

dye solutions and in the polymer dye mixtures were studied in a number of works [4, 90–92, 105–106].

According to Ref [93], the appearance of photoinduced anistropy in isotropic media containing azo dyes under an illumination by a polarized light is explained by the *trans-cis* photoisomerization of azo dye with respect to the N=N double bond. The mechanism of this process includes the transition from the *trans-* form into the *cis*-form and from the *cis-* form into other *cis-* form (with other direction of the axis of the dye molecule in respect to the electric vector of actinic light). A rotation of the terminal groups of the azo dye molecule around the "weakened" N=N double bonds at transition from the *cis-* to the *cis*-form causes an appearance of the rotational moment turning the central part of the molecule in such a way that the direction of the absorption oscillator of the molecule is changed to the orthogonal one. This leads to the orientational redistribution of the absorption oscillators under illumination by linear-polarized light and it explains the appearance of photoinduced anisotropy. However it is doubtful whether such explanation based on the relatively free rotation of dye molecules in the solid may be completely applied to the polymer systems with chemically linked dye molecules.

Let us consider in detail what happens under the laser irradiation of azo dye-containing polymer in the oriented "frozen" LC phase or the amorphous glassy state.

Figure 2.28 shows the electrooptical cell containing an homeotropically oriented sample of an LC copolymer. This cell is irradiated by a vertically polarized laser beam. The electric vector of the incident laser beam coincides with the Y-direction; the Z-direction corresponds to the long axis of the mesogen. The left part of the cell is covered by a metal plate; in other words this part of the cell is not illuminated. On the right, the microphotos of polymeric films, recorded in the crossed polarizers of an optical microscope before and after irradiation are shown.

The optical scheme of such experiments is presented in Fig. 2.29. The refractive indices of the initial polymer film in the framework of the coordinate scale are designated as n_x^o, n_y^o and n_z^o. As can be seen from Fig. 2.29, the "writing" beam is placed in the horizontal plane (X, Z) at a small angle to the optical axis and this beam is vertically polarized, i.e. the electric vector of the electromagnetic wave of the laser beam is parallel to the co-ordinate $Y(\vec{E}_y^I \parallel Y)$. The "reading" beam II spreads along the optical axis of the sample, i.e. along axis Z.

At the initial moment i.e. before the laser beam is switched on, the optical properties of the homeotropically oriented film are similar to the characteristics

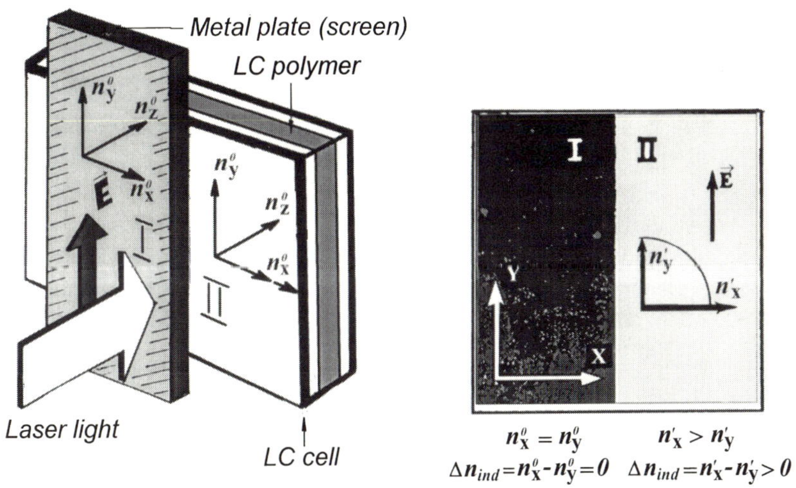

Fig. 2.28. Schematical illustration of the optical recording on the homeotropically oriented LC films. On the *left* the electrooptical cell is shown. On the *right* the microphoto of the homeotropically oriented film after laser-beam irradiation in crossed-polarizers is shown. The *left part* of the cell was not irradiated

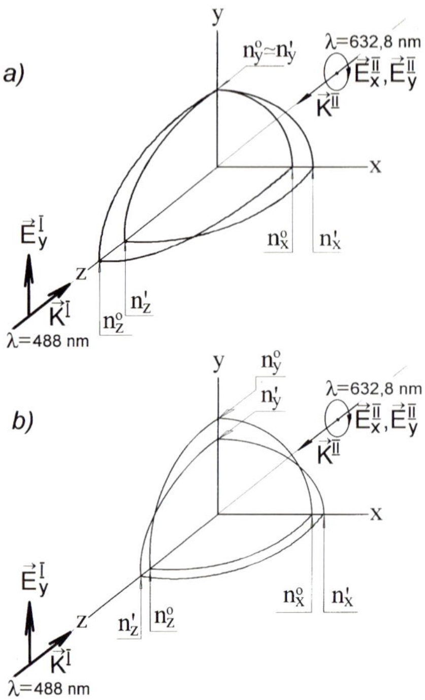

Fig. 2.29.a-b. Optical scheme of the photo-optical experiment for observation the photo-induced birefringence (see also Fig. 2.28) $\vec{K}^I, \vec{K}^{II}$-are the wave vectors; $\vec{E}_y^I$, $(\vec{E}_x^{II}, \vec{E}_y^{II})$ – are the electric vectors of electromagnetic wave for the writing (I) and for the reading (II) beams. (n_z^o, n_x^o, n_y^o) –are the principal refractive indices before and (n_z', n_x', n_y') -after irradiation. The ratio between refractive indices: **a** -the smectic LC polymers – $n_z^o > n_z' > n_x' > n_x^o = n_y^o \cong n_y'$; **b** -the amorphous polymers- $n_x' = n_z' > n_y^o > n_x^o = n_y^o = n_z^o > n_y'$

of the positive uniaxial monocrystal (see Sect. 2.4). In this case the refractive indices are described by the following relations:

$$n_z^o = n_{max}$$
$$n_x^o = n_y^o \qquad (2.6)$$
$$n_z^o > n_x^o = n_y^o$$

The value of anisotropy of the film is defined by the n_z^o value and birefringence Δn in the XY plane is equal to 0 ($\Delta n = n_x^o - n_y^o = 0$). According to this reasoning, the left (non-irradiated) part of the electrooptical cell is dark (in crossed polarizers) as shown in Fig. 2.28.

Let us consider how the optical properties are changed after the irradiation of the film by the laser beam. The special optical measurements [16] show that the switching on the "writing" beam I leads to the changing of the main refractive indices, after a certain period of illumination, which are related in the following way:

$$n_z^o > n_z'$$
$$n_y^o \cong n_y' \qquad (2.7)$$
$$n_x^o < n_x'$$

In other words, irradiation of a homeotropically oriented polymer film causes an increase in the refractive index in the X direction, i.e. in the direction per-

Fig. 2.30. Photomicrograph of a film of LC copolymer **5** (Table 2.2) obtained with the test-pattern. Film thickness 20 μm, distances between the stripes -5 μm [107]

pendicular to the electric vector of the "writing" beam I. Thus the polymer film becomes a biaxial one with the photoinduced birefringence equal to $\Delta n_{ind} = n'_x - n'_y$. An appearance of the biaxiality is clearly manifested in the microphoto on the right side of the cell. As can be seen from a comparison of the left and right sides of the microphoto (Fig. 2.28), the right side of the film becomes practically light after irradiation.

If we place a diffraction grating (test pattern) on the LC cell surface and illuminate this sandwich construction with a laser, one can obtain the image of this test pattern, which is shown in Fig. 2.30 [20, 107]. It should be emphasized that the stored image is invisible to the naked eye and can be "read" only in polarized light. The optimum visibility of the irradiated regions was attained when the cell was situated so that the angle between the direction of the polarization plane of the recording ("writing") beam and the polarizer axis was $\pm 45°$; when the angle was 0 or $\pm 90°$, the image was no longer visible.

A simpler situation is observed in the initially amorphous samples of comb-shaped polymers (Table 2.3, polymers 10–12). In this case there is the following relation between the refractive indices before an irradiation (Fig. 2.29b):

$$n_x^o = n_y^o = n_z^o \tag{2.8}$$

The polarized irradiation of isotropic film generates the appearance of the photoinduced birefringence Δn_{ind}. Optical measurements show that the polymer film acquires the properties of the negative uniaxial crystal with the optical axis oriented along the Y-axis, i.e. parallel to the electric vector of the "writing beam" $\vec{E}_y^I$. In this case, the following relation between the refractive indices occurs:

$$n'_z = n'_x > n'_y \tag{2.9}$$

Thus, one can affirm that the polarized irradiation of the amorphous comb-shaped polymers containing mesogenic photochromic (azobenzene) side groups leads to considerable changes of their optical properties in the solid.

It is manifested in the increase of the refractive index of the amorphous and LC films in the direction perpendicular to the electric vector of the laser beam; in other words the irradiation process is accompanied by a change in the direction

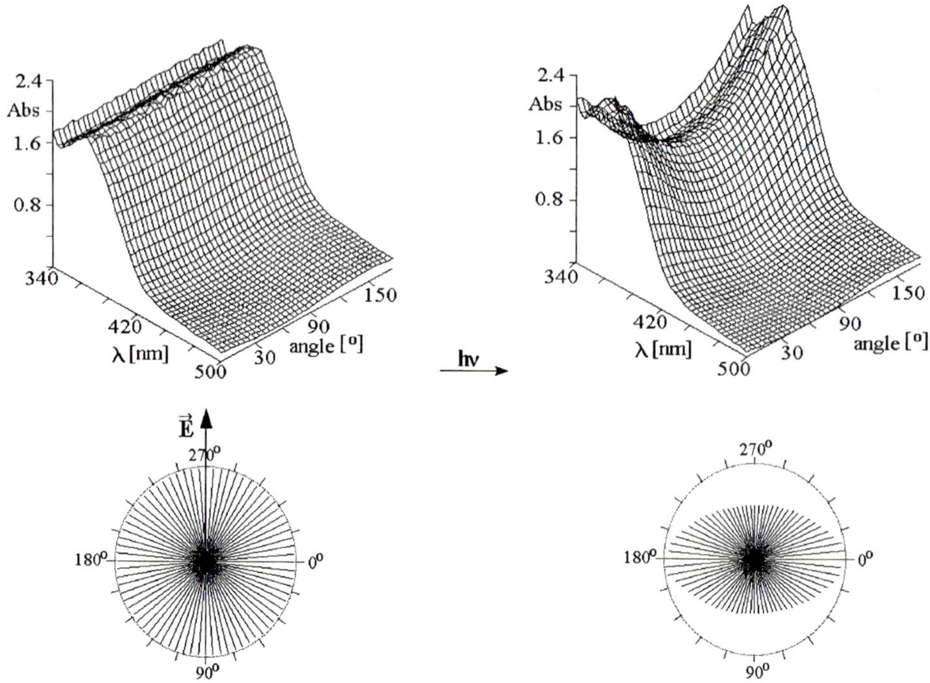

Fig. 2.31. Polarized UV/Vis spectra and angular dependence of the absorbance at 380 nm of copolymer **XXII** in the initial state (optically isotropic; *left side* and after irradiation and relaxation (*right side)* [112]

of the optical axis of homeotropically and planary oriented LC films or the appearance of the orientational order in the initially amorphous samples.

The study of the polarized IR and UV/VIS spectra as well as angular dependence of the appropriate absorbance bands of copolymers synthesized (Table 2.2 and 2.3) confirms the above mentioned statement.

Figures 2.31 and 2.32 show the polarized IR and UV/VIS spectra and angular dependence of the intergrated areas of the CN-stretching vibration at 2229 cm^{-1} and the vibration of N=N double bonds of azobenzene groups at 380 nm (the $\pi - \pi^*$ transition of the *trans*-azobenzene groups) before and after the irradiation of the copolymer **XXII** (see copolymer 3 in Table 2.2)[111–112].

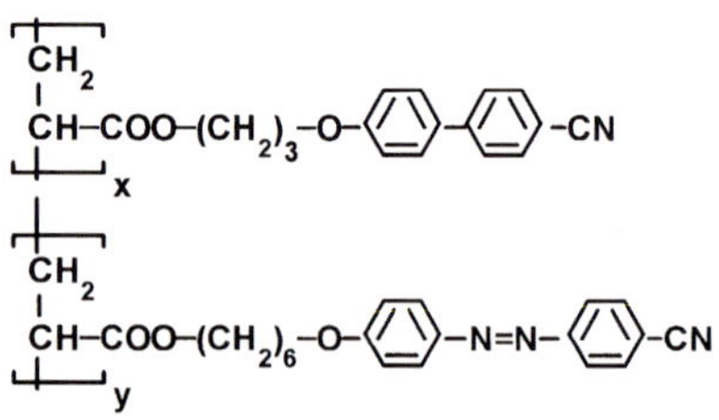

Cop XXII

y = 0.2; x = 0.8

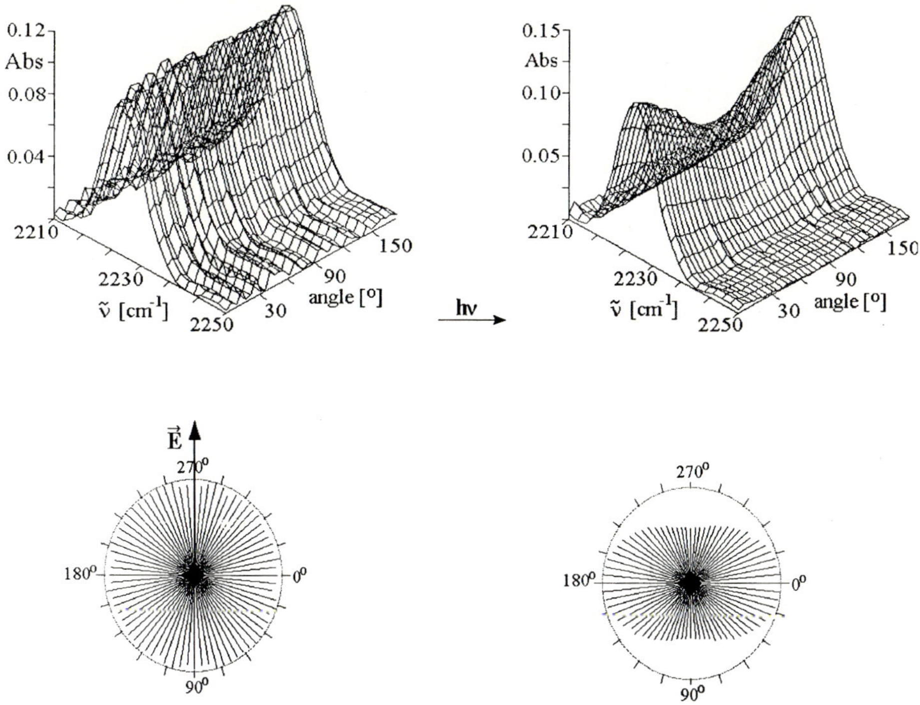

Fig. 2.32. Polarized IR spectra and angular dependence of the integrated area of the CN stretching band of copolymer **XXII** in the initial state (optically isotropic; *left side*) and after polarized irradiation and relaxation (*right side*) [112]

As can be seen from Fig. 2.31 and 2.32 the isotropic distribution of UV and IR absorbance in the initial state are characteristic for both vibration types. It means that there is no preferential orientation of the side groups of LC copolymers.

However, after the polarized irradiation of the polymeric films (right side of Fig. 2.31 and 2.32) the isotropic character of the polar diagrams showing angular dependence of the appropriate spectra is changed sharply and dichroism of the absorbance bands appears. Analysis of these data in combination with the data obtained for amorphous copolymers containing azobenzene dye fragments without terminal CN groups [111–112] gives evidence that the long axes of the mesogenic groups containing and not containing photochromic fragments are oriented perpendicularly to the electric vector of the laser beam.

The change of the optical axis direction of the oriented samples is also observed after irradiation of homeotropically and planary-oriented films of LC polymers. Figure 2.33 shows, as an example, the polarized UV/VIS spectra and angular dependence of the absorbance (at 380 nm) of an uniformly planary oriented film of copolymer **XXII** in the initial state and after polarized irradiation by a laser. As can be seen from Fig. 2.33, in the initial state, the electric vector of

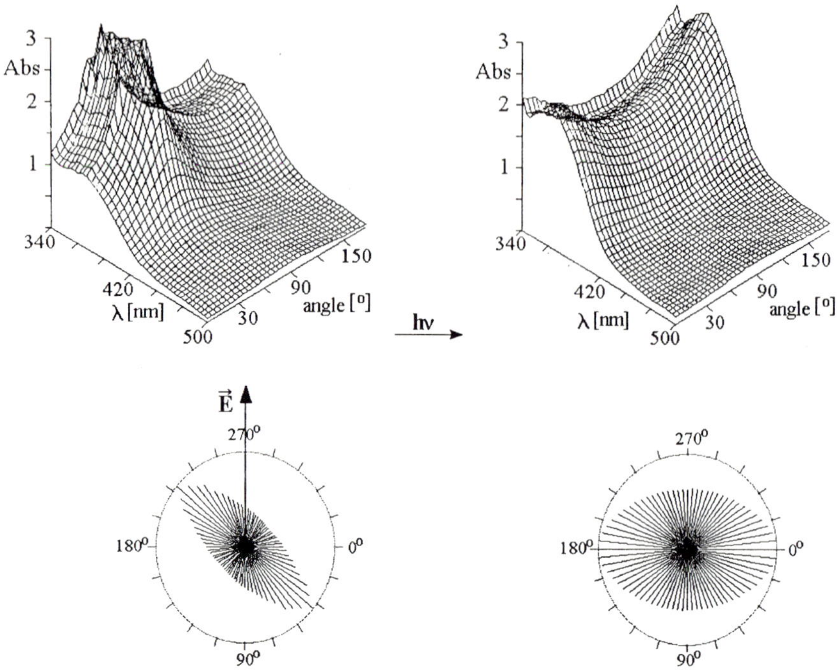

Fig. 2.33. Polarized UV/V is spectra and angular dependence of the absorbance at 380 nm of an uniformly planar oriented film of copolymer **XXII** in the initial state (*left side*) and after polarized irradiation and relaxation (*right side*) [112]

the actinic light forms an angle of 45° to the director orientation of the polymer film. After irradiation (and relaxation), the optimal axis rotated by 45°, i.e. a reorientation of the side mesogenic groups occurs.

2.6.3 Kinetic Regularities of Photoinduced Processes

Detailed studies of the orientation kinetics of the comb-shaped polymers under light irradiation has been considered in Refs. [16, 21, 107–112] for a series of acrylic and methacryclic copolymers containing various concentration of azo dyes as the mesogenic side groups (see Tables 2.2 and 2.3).

The typical curves of the photoinduced birefringence as a function of exposure (I·t, where I – is the power density of the laser beam and t is the time) for amorphous copolymers- **XXIII** and copolymer 12 (Table 2.3) at room temperature are shown in Fig. 2.34.

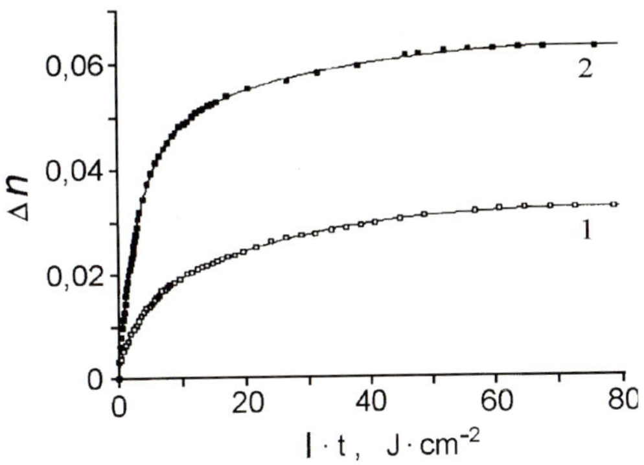

Fig. 2.34. Increase of birefringence of films of amorphous copolymer **XXIII** (*1*) and LC copolymer **12** (Table 2.3) (*2*) due to light induced reorientation ($\lambda = 488$ nm; Intensity of the laser beam, I = 40 m W/cm^2; d = 10 μm) [112]

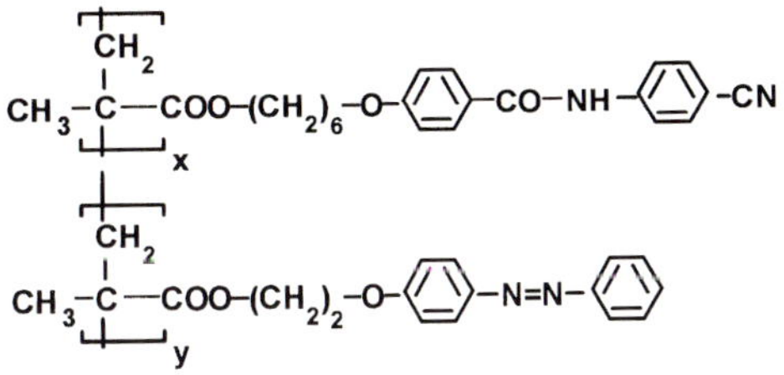

Cop. XXIII

y = 0.24; x = 0.76; $T_g = 97$ °C.

As can be seen from Fig. 2.34, in both the cases, the appearance of the photoinduced birefringence Δn_{ind} is observed. At the initial period of irradiation the values of Δn_{ind} increase rapidly and the rate of the Δn increase becomes smaller on con-tinued irradiation achieving a stationary value of Δn_{ind} at the values of appropriate exposure. The detailed study of kinetic features of this process at different temperatures corresponding to different physical state of the polymers revealed the following regularities.

Figure 2.35 shows, the time and temperature dependence of the photoinduced birefringence of some acrylic copolymers containing different concentrations of dye fragments (Table 2.2; Copolymer 4–6) as well as the methacryclic copolymer **XXIV**.

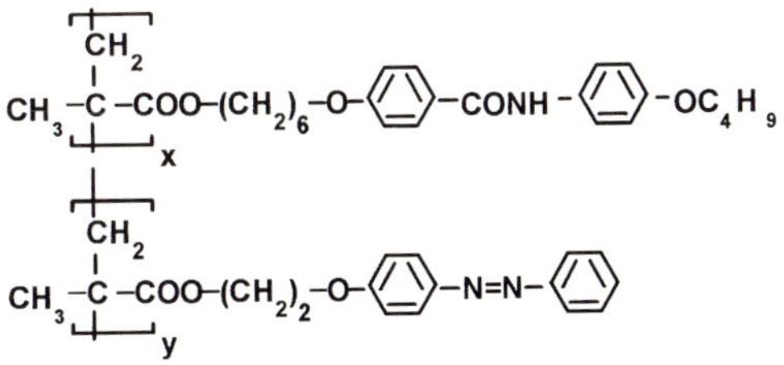

Cop. XXIV

x = 0.8, y = 0.2, $T_g = 70$ °C, $T_{cl} = 106$ °C

We would like to draw attention to the following points.

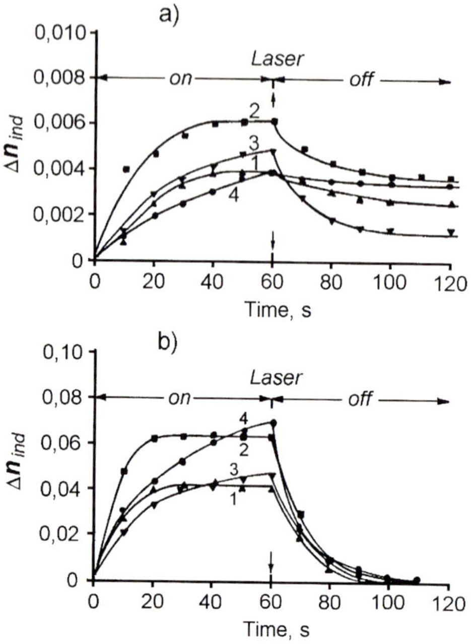

Fig. 2.35. Time dependence of laser-induced birefringence Δn_{ind} in homeotropic films of copolymers 5(*1*), 6(*2*), 4(*3*) (see Table 2.2) and copolymer **XXIV**(*4*) on irradiation (I=200 mW/cm^2) followed by relaxation (**a**) at room temperature and (**b**) at a temperature above T_g (copolymers **4–6**: T= 70 °C; copolymer **XXIV** :T=90 °C) [108]

a) First of all, one should mention the considerable difference between the values of Δn_{ind} one and the same polymer irradiated at different temperatures. As can be seen from Fig. 2.35a and b, at the same period of irradiation, the Δn_{ind} values in the LC phase are one order of the magnitude higher than those of Δn_{ind} in the glassy state. Moreover the limited values of photoinduced birefringence are achieved twice as fast as in the LC state (at 70 °C). That means the photoinduced process is much more effective in the visco-elastic (LC state) than in the hard glassy polymer matrix.

b) After switching the laser beam off, a small decrease in the primary induced birefringence takes place and the stable values of birefringence (Δn_{stab}) are preserved in the glassy state. These stable values of photoinduced birefringence are "frozen" in the LC film and remain constant for a prolonged time (for several months). The appropriate UV-Vis absorbance spectra (Fig. 2.27b) of one of the copolymers studied show the change of *cis-trans*-isomer ratio after relaxation. The extent of relaxation at room temperature strongly depends on the glass transition temperature. The Δn_{stab} is achieved slightly faster for more rigid polymers (polymethylacrylates) which have a higher glass transition temperature (compare, for example, curves 1 and 4 in Fig. 2.35a).

Another situation is realized for LC polymers above their glass transition temperature. In this case (Fig. 2.35b), the photoinduced birefringence drastically falls to zero after switching the laser beam off. The influence of temperature on the photoinduced birefringence is shown in Fig. 2.36. In the glassy state the Δn_{stab} values of all copolymers are only slightly dependent on the temperature. However, above the glass temperature a considerable increase in the induced

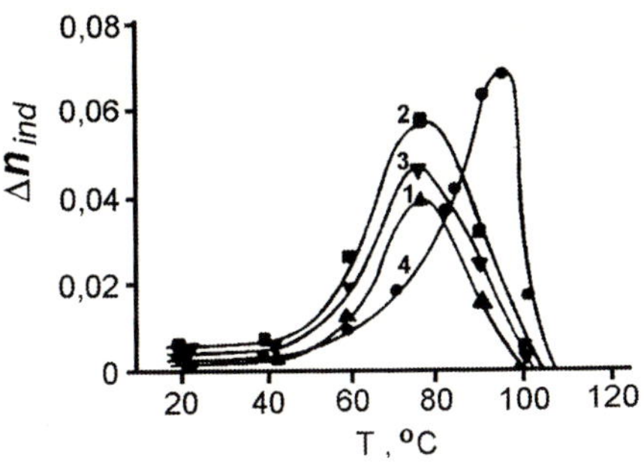

Fig. 2.36. Temperature dependence of the Δn_{ind} of copolymers **5**(*1*), **6**(*2*), **4**(*3*), (see Table 2.2) and copolymers **XXIV** (*4*); irradiation time 60 s [108]

birefringence to values of one order of magnitude higher takes place. A maximum of inducing efficiency is achieved for all copolymers at temperatures of about 30 – 40 °C above the glass transition temperature. So, for example, after one minute of irradiation of copolymer 6 (Table 2.2) (Fig. 2.36, curve 2), a maximum value of $\Delta n_{ind} = 0.06$ is achieved at 80 °C.

A further rise in the temperature close to the clearing temperature causes a sharp decrease in birefringence and in the isotropic melt the photoinduced birefringence cannot be observed.

Such an extreme temperature dependence of Δn_{ind} is probably explained by the following factors. With increasing temperature, the rate of the direct and back reactions of *trans-cis* isomerization also increases which favors a more effective and rapid process of orientational diffusion of dye-fragments to a state in which the latter absorb the "writing irradiation" to a lesser degree. In addition, an increase in the temperature which enhances the side and the main chain mobility also favors an increase in Δn_{ind} values at temperatures exceeding the glass transition temperature.

On the other hand, a further approach to the clearing temperature is accompanied by a lowering of the order parameter and close to the clearing temperature the photoinduced birefringence vanishes.

c) The mesophase type of LC polymers are also likely to affect the photoinduced birefringence. Although this question remains to be resolved because of the difficulty of carrying out the correct comparison of the polymers differing only in the mesophase type, other parameters being equal (such as chemical structure of the main chain, spacer length, dye concentration in copolymer, molar-mass characteristics etc.), nevertheless, one can affirm that a more-ordered type of mesophase greatly prevents the reorientation of mesogenic groups thus showing low values of birefringence. In other words the less-ordered structure of comb-shaped polymers favours higher values of Δn_{ind}.

Some data, shown in the evidence for the statement mentioned above, can be found by comparing the curves presented in Fig. 2.35. As can be seen from this figure, the values of the photoinduced birefringence for copolymers forming a smectic phase (Fig. 2.35, curves 1, 2) rapidly (40 s in the glassy state and 20 s in the visco-elastic state) achieve maximum constant values; at the same time in the nematic phase of copolymer 4 (Table 2.2) (Fig. 2.35, curve 3) the values of Δn_{ind} continue to increase with 60 s and even 600 s of irradiation time to much higher values of Δn_{ind}. Of course, the influence of nematic and smectic structural organization of media on the photoinduced birefringence should be considered as a preliminary result, nevertheless, a contribution of the supramolecular structure of the polymer to the photoinduced processes is beyond any doubt. The different values of order parameter S, calculated from the dichroism of the appropriate absorbance bands of the mesogenic fragments, [35] show strong evidence for a role of the supramolecular organization of the polymeric matrix which undergoes irradiation. So, if the initial order parameter S of the planary-oriented LC film (see polymer 3 in Table 2.2) is equal to 0.4, then after the irradiation, the S value decreases to 0.16 [110, 112]. At the same time, if the initial polymer is in the amorphous state (where S = 0) then after irradiation, we can generate the order parameter appearance with value of S = 0.2 [112].

Highly similar results were obtained for an acrylic copolymer containing the same comonomer units of azo dye (30 mol %) and selectively labeled with ^{13}C in the cyan group of the phenylbenzoate mesogen [114].

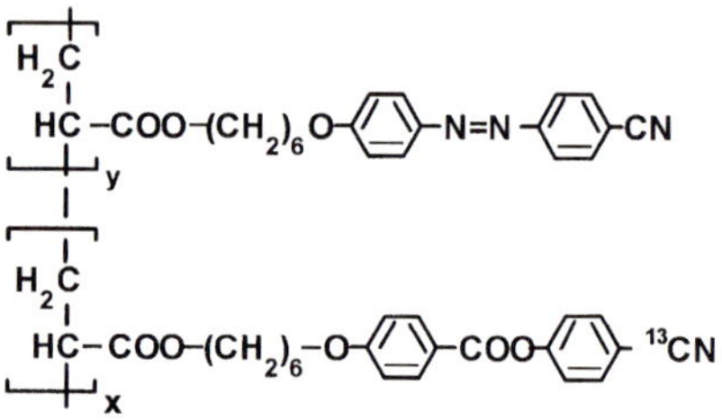

Cop. XXV

$T_g = 32°$; $T_{N\rightarrow I} = 129$ °C

This copolymer was oriented uniform homogeneously, using a combination of magnetic field and surface effects (sample thickness was 75 μm). The initial sample had a high mesogenic group orientation (S=0.6). After irradiation of this nematic copolymer (the xenon lamp with band-pass filter; λ=320–480 nm; time = 30 min.; the power density on the sample $\sim$ 1 mW/cm^2) the order parameter (measured by ^{13}C NMR) was reduced by approximately 25% (S=0.4), which is in good agreement with our data [110, 112].

Above, we discussed the main features of kinetics of the photooptical processes, the influence of the temperature, and the supramolecular structure of the polymer on the photoinduced birefringence. Let us consider some details of these phenomena. In this case our consideration will be limited only by the kinetic regularities of the dynamics of photoinduced birefringence increase at temperatures below the glass transition temperature.

Analysis of the kinetic curves of photoinduced birefringence [115] has shown that the kinetics of Δn_{ind} can be described by a superposition of two processes corresponding to the fast and the slow birefringence increase – Δn^f_{ind} and Δn^s_{ind}, respectively. In this case the general expression for the photoinduced birefringence can be described as follows:

$$\Delta n_{ind}(t) = \Delta n^f_{ind} + \Delta n^s_{ind} \tag{2.10}$$

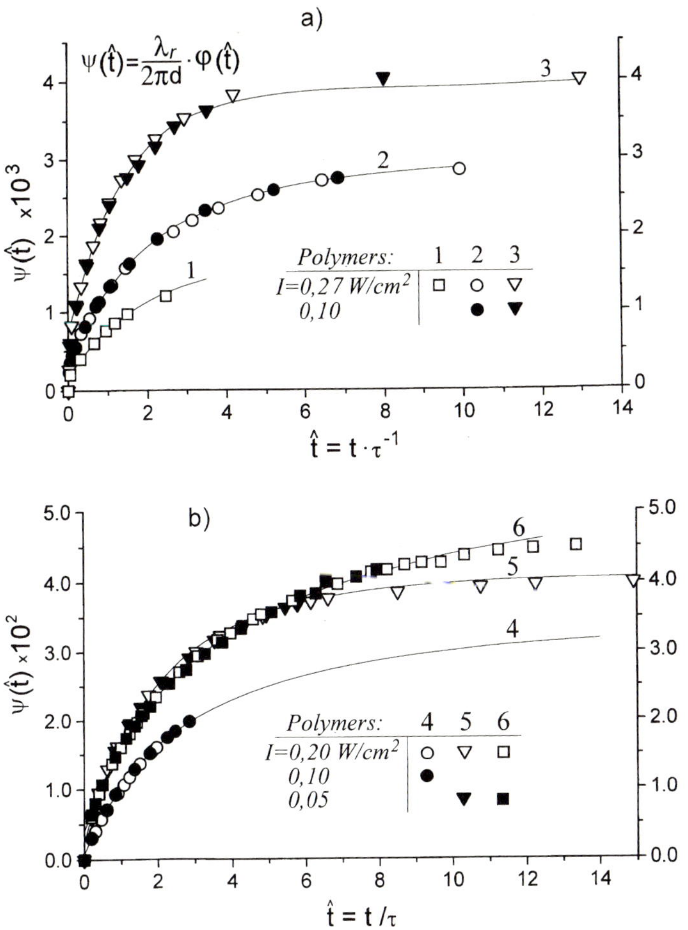

Fig. 2.37a-b. Time dependence of the normalized phase shift $\psi(\hat{t}) = \frac{\lambda_r}{2\pi d} \cdot \varphi(\hat{t})$ for smectic (**a**) and amorphous polymers (**b**). Numbers of the curves correspond to the sample numbers in the Table 2.12. Film thickness: 100(*1*), 40(*2*), 10(*3*), 45(*4*) and 25 μm (*5*, *6*). The *points* correspond to experimental data; all *curves* were calculated according to Eq. (15), taking into account only the slow process ($\tau_f = 0$) [115]

The time corresponding to the fast and the slow birefringence increase we will designate further as τ_f and τ_s respectively (where $\tau_s \gg \tau_f$).

As can be seen from Figs. 2.35 and 2.37 the fast process proceeds a few seconds (around 2–3s) after switching the laser on. The contribution of the fast component to the common value of Δn_{ind} is around 5–10% for amorphous polymers (copolymer 4–6, Table 2.12) and 10–20% for the smectic LC polymers (copolymers 1–3, Table 2.12). It should be noted that the fast process is also

Table 2.12. Photo-induced birefringence Δn_{ind} and constant C of LC (see Table 2.2) and amorphous (see Table 2.3) copolymers containing different concentration of azo dye (λ_r = 632.8 nm; $\Delta n^s_{ind} = \Delta n_{ind}$ at $t \to \infty$) [115]

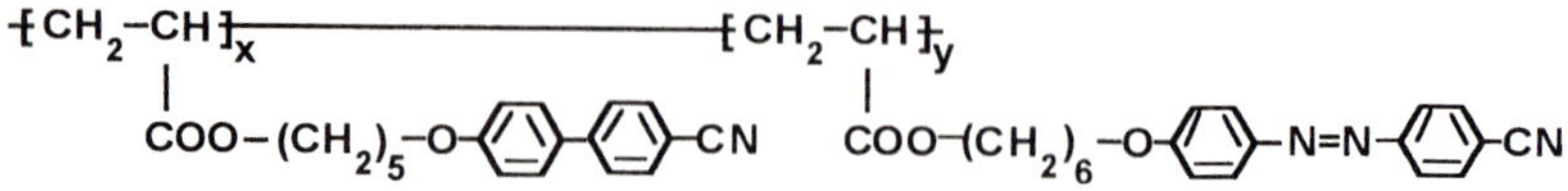

Copolymer	1	2	3
y, mol%	10	20	40
Δn_{ind}	1.8×10^{-3}	3×10^{-3}	3.8×10^{-3}
C*, J/cm^2	70 ± 7	20± 2	9–10
T_g, °C	38	38	38

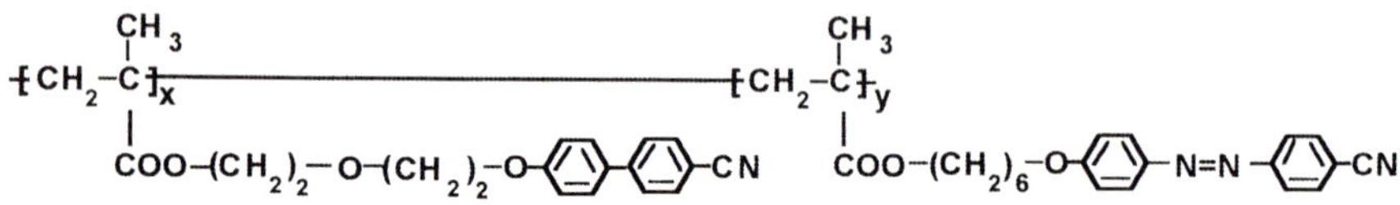

Copolymer	4	5	6
y, mol%	10	20	40
Δn_{ind}	3.4×10^{-2}	4.1×10^{-2}	5.5×10^{-2}
C*, J/cm^2	280± 10	41 ± 2	19 ± 1
T_g, °C	66	68	62

*Values of C were estimated at $\lambda = 488$ nm

observed after switching a laser beam off, i.e. in the process of the photoinduced birefringence relaxation after stopping irradiation. On short exposure, when the irradiation time is less than τ_f long-term recording is not realized. In this case, switching a laser beam off, after the short time exposure, leads to a decrease in Δn_{ind} to zero, as in the situation which is characteristic for Δn_{ind} relaxation in the mesophase at 70 °C (see Fig. 2.35b). The absence of long-term recording on short exposure was also described in Ref. [116].

It should be pointed out that the fast process is characteristic not only for the photoinduced birefringence. The existence of the fast component in the photoinduced birefringence was also observed in studies of photoinduced dichroism [117]. In these cases, the fast component is typical for the kinetics of the photoinduced dichroism independently of whether the azo dye is incorporated in LC polymer [117] or is dissolved in solution [118] (azo dye glycerin solutions; viscosity of such systems is very high $\sim 10^8 - 10^{14}$ Pa · s).

There is no detailed information relating to the fast component of photoinduced birefringence. Probably, the fast process is determined by the abnormally high mobility of some dye fragments partially located at boundary surfaces or defects of the side chain packing. Owing to high mobility of some dye fragments, they most probably do not participate in the preservation of long-term recording, although they can participate in the recording process.

The increase in photoinduced birefringence corresponding to the slow process $[\Delta n^s_{ind}(t)]$, as in photoinduced dichroism [117–118] can be good described by a monoexponental equation [115]:

$$\Delta n^s_{ind}(t) = \left(\Delta n^s_{ind}\right)_{max} \cdot \left[1 - \exp\left(-t \cdot \tau_s^{-1}(I)\right)\right] \tag{2.11}$$

where I -is the "writing" beam intensity; $\left(\Delta n^s_{ind}\right)_{max}$ is the maximum value of Δn_{ind} at $t \to \infty$ and τ_s is the characteristic time for which the photoinduced birefringence Δn_{ind} reaches the value of $(1 - e^{-1})\left(\Delta n^s_{ind}\right)_{max}$. It should be noted that the values of the photoinduced birefringence practically do not depend on the "writing" beam intensity at $t \to \infty$, i.e. $\left(\Delta n^s_{ind}\right)_{max} \approx$ const [115].

The product of τ_s and intensity of irradiation I, is also equal to the constant value C for the slow process

$$\tau_s(I) \cdot I = \text{const} = C \tag{2.12}$$

Taking into account Eq. (2.12) for the description of the kinetics of photoinduced changes in the sample, a parameter for the exposure H was introduced:

$$H = I \cdot t \tag{2.13}$$

Using Eqs. (2.12) and (2.13) one can write:

$$t \cdot \tau^{-1} = \hat{t} = I \cdot t \cdot C^{-1} = HC^{-1} \tag{2.14}$$

The parameters H and C are more convenient for describing the experimental data so far as they enable the results obtained for varying intensity of the "writing beam" to be presented in the form of the common kinetic curve of Δn_{ind} dependence as a function of the normalized time $\hat{t}$ (Fig 2.37a,b)

For the calculation of the curves we used Eqs. (2.15) and (2.16) taking into account absorption of the "writing" beam by the polymeric film [115]. As shown earlier [112, 115], the phase shift between the vertical and horizontal components of an electric vector of the "reading" laser beam $(\vec{E}^{II}_x$ and $(\vec{E}^{II}_y$ in Fig. 2.29), induced by the slow component changes Δn^s_{ind}, in general, can be expressed as:

$$\varphi(t) = \frac{2\pi}{\lambda_r} \int_0^d \Delta n^s_{ind}(z, t)\, dz \tag{2.15}$$

where λ_r is the wave length of the "reading" beam; d is the film thickness; Δn_{ind} is the photoinduced birefringence at the wave length λ_r.

Based on the local changes of the photoinduced birefringence as well as monoexponental kinetics of birefringence increase (for the slow component - Δn^s_{ind}) in the films of LC polymers, Eq. (2.16), connecting the phase shift with the photoinduced birefringence was obtained:

$$\varphi(t) = (2\pi/\lambda_r) \cdot \left(\Delta n^s_{ind}\right)_{max} \cdot d \cdot \left\{1/(\alpha_w \cdot d) \cdot \sum_{k=1}^{\infty} \left(-t/\tau(0)\right)^k \cdot a_k\right\} \tag{2.16}$$

where $a_k = [\exp(-k \cdot \alpha_w \cdot d) - 1]/k \cdot k!$; $\tau(0)$ – is the time corresponding to the Δn_{ind} appearance in the thin superficial layer of the sample; α_w is the absorption coefficient of the polymer film at the specific wave length of the "writing" beam, λ_w.

It should be pointed out that for the practical application of the photooptical phenomena based on the photoinduced birefringence, it is necessary to have materials providing the maximum rate of recording. In this case, for a comparison of the different materials according to this criteria, one can use such parameters as the rate of the phase shift $\varphi(t)$, which is induced by the "writing beam" at the small exposure H of the laser beam action:

$$\frac{\partial\varphi}{\partial H}(H \to 0) = \frac{1}{I(0)}\frac{\partial\varphi}{\partial t}(t \to 0) = \frac{2\pi(\Delta n^s_{ind})_{max}}{\lambda_r \cdot \alpha_w \cdot C}[1 - \exp(-\alpha_w \cdot d)] \qquad (2.17)$$

As mentioned above, in this case, only the slow component of the photostimulated process should be taken into account.

As can be seen from Figure 2.37a and b, the monoexponental approximation (2.11) is in a good agreement with the kinetics of the birefringence increase of the slow component Δn_{ind} in the ordered and the amorphous films of the comb-shaped polymers. It is important to stress that the kinetics of this process can be described by using only two constants – Δn^s_{ind} and C, the values of which for some polymers investigated by us are presented in Table 2.12 [115].

As mentioned above, the more ordered phase hinders a reorientation of the mesogenic side groups with the result that the photoinduced birefringence Δn_{ind} for amorphous polymers has a more higher value (Table 2.12, copolymers 4–6). On the other hand, the appearance of the photoinduced birefringence in the smectic polymers produced by the "writing" beam (spreading along the optical axis of the film; the vector $\vec{E}^I_y$ is perpendicular to the director, see Fig. 2.29) shows strong evidence for the high mobility of side chain fragments even in the ordered phases.

The influence of the ordering degree of the polymer matrix on an increased time of the photoinduced birefringence τ_s or (which is one and the same) on the constant $C(\tau_s \sim C)$ at the constant value of the "writing" beam intensity is more complicated. In our opinion the following factors influence the time τ_s:

1. The photochromic side chain fragments and also the volume embraced by the mesogenic groups (side chain groups not containing dye) undergo reorientation due to the *trans-cis* isomerization of the dye fragments. We will call this volume the "sphere of influence" of the dye fragment (see below, Fig. 2.39).

2. The rate of the *trans/cis* isomerization and probably the value of the photostationary *trans/cis* isomer ratio.

These factors depend on the molecular and supramolecular structure of polymer and a number of other parameters such as temperature, phase state of polymer etc in a complicated manner. This hinders the elucidation of the correlations between the τ_s and different type of ordering of polymer systems.

Let us consider, in general, the influence of the factors indicated in the points 1 and 2 on the time τ_s. Based upon the values of constant C (Table 2.12), one can

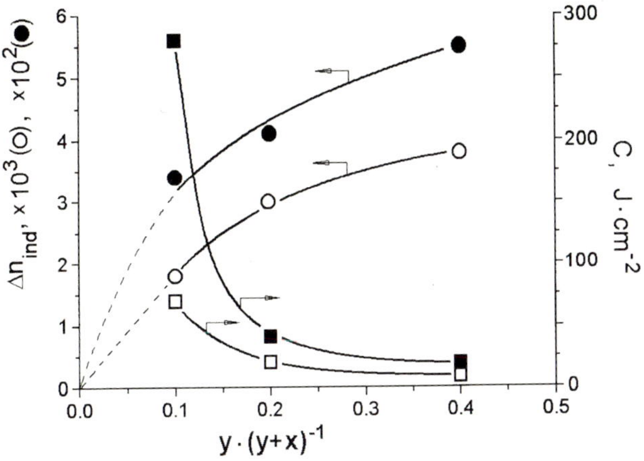

Fig. 2.38. Photoinduced birefringence, Δn_{ind} and constant C as a function of copolymer composition for amorphous (*filled symbols*) and smectic (*open symbols*) polymers [115]

suppose that the "sphere of influence" and the rate of *trans-cis* isomerization in the smectic polymers are less than in the amorphous ones at the one and the same dye concentration. For this reason the difference between the constants C for the smectic and amorphous polymers is not so considerable as for the Δn_{ind}. An increase in the "sphere of influence" should lead to the slowing down of the fixation process of the stationary values of birefringence because, in this case, more mesogens are involved in the reorientation process by an active center. At the same time, an increase in the *trans/cis* isomerization rate (for example, from an increase of writing laser beam intensity) results in a more rapid growth of birefringence, but does not change the value of Δn_{ind}. However, the rate of *trans/cis* isomerization and the time of photostationary equilibrium fixation depends not on the "writing" beam intensity alone. The molecular structure of the mesophase and the degree of ordering of the polymer also influence the above-mentioned values.

The difference between the "sphere of influence" of the amorphous and smectic polymers also follows from the data relating to concentration dependence of Δn_{ind} and C (Fig. 2.38). At low concentrations of dye in polymer y, when the neighboring "spheres of influence" are not overlapped (Fig. 2.38) one can expect that $\Delta n_{ind} \sim y$ (these dependences are shown by the dotted line in Fig. 2.38). An increase in dye concentration leads to the photoinduced birefringence increasing and constant C decreasing due to the decrease in the number of the mesogenic fragments assigned to one active center. As can be seen from this Figure, the mutual influence of dye molecules on each other in the smectic polymers appears at $y \sim 0.1 - 0.2$. In amorphous polymers at these concentrations, the rapid growth of constant C is observed, but the character of Δn_{ind} change is approximately the same as that at higher dye concentration.

For this reason the contribution of the fast component Δn^f_{ind} in the birefringence is higher for smectic polymers than for amorphous ones.

As was pointed out above, long-term recording is not realized with a short action of writing laser beam (exposure $H = I \cdot t < 0.1C$). This case was described in Ref. [116] (see copolymer **XXV**) where the kinetics of the back relaxation of birefringence was reported. According to our evaluation, the exposure used in [116] was insufficient for long-term recording despite the high power density of the laser beam (up to 10 W/cm^2). An estimation of the constant C is accordance with the results of photoinduced dichroism kinetics [117] also gives the values similar to the ones which are presented in Table 2.12. According to Ref. [116] the time of the stationary dischroism fixation τ is equal to 1300 s (with a "writ ing" beam intensity $I \sim 10W/cm^2$) which corresponds to constant $C = I \cdot t = 13J/cm^2$. It is very close to the data of Table 2.12 (see copolymer 3, 6). Taking into account the difference in the chemical structure of these components, different dye concentration (40 and 50 %), different degree of ordering (a nematic copolymer was studied in [116]) as well as the inevitable experimental errors, the close values of constant C may be considered as satisfactory.

Based on the above-mentioned results, one can conclude that the parameter C basically depends on the side chain structure and the azo dye content of the polymer, especially at low small concentrations. At the same time, at higher dye concentrations (more than 10 mol%), the degree of the ordering (supramolecular structure) unlike Δn_{ind} affects the parameter C to a lesser degree.

On the basis of the optical experiments, one can assume the following scheme of side group reorientation in comb-shaped copolymers.

Figure 2.39 shows the isotropic arrangement of the side mesogenic and photochromic groups in the amorphous state and the planar orientation of the photochromic copolymer. (The main chains are omitted for simplicity). On the left, the disposition of individual dye fragments is shown. Its transition moment coincides with the long axis of the azobenzene fragments; α –is the angle between the electric vector of the laser beam and the long axis of the mesogen.

We can distinguish at least three steps in the orientational photoinduced process.

The first step is the absorption of linearly polarized light by azo dye fragments. This absorption is proportional to the square of the cosine of the angle α. The smaller the angle α,the greater is the probability that the appropriate photochromic group will be excited. It is clear that planar orientation is the most favourable one for absorption of the light (in this case the angle α tends to 0). Thus this step may be designated as the angular-dependent absorption.

The second step corresponds to the repeated *trans-cis* and *cis-trans* isomerization processes. As we mentioned above, the geometry of the dye molecules is changed due to the *trans-cis* isomerization and the rod-like *trans*-isomers are transformed into the bent-shaped (crooked) *cis*-isomers. Such transformations lead to the local changes of an orientation of neighboring non-photochromic side groups. In other words, the perturbation ("loosening") of side group packing takes place. This process is probably more effective in the less ordered medium, than it is in the amorphous state. The orientational order in the LC state restricts the reorientation of dye fragments. Nevertheless a certain additional free volume for reorientation appears.

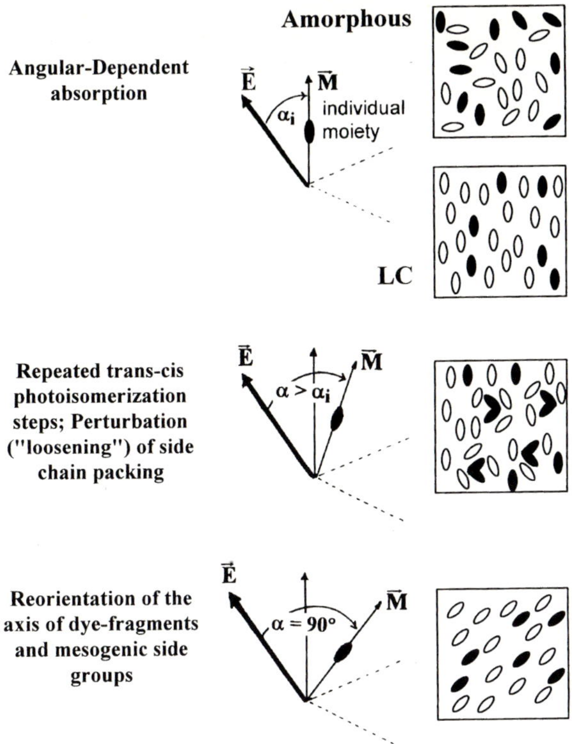

Fig. 2.39. Schematic representation of reorientation of mesogenic groups in the amorphous and planar oriented polymer films

The final stage of the photooptical effect is the reorientation of the photochromic moieties which cause a continuous reorientation of non-photochromic but rod-like side groups. The azo fragments are reoriented in such a way that the long axis of the dye fragments are perpendicular to the electric vector $\vec{E}$. The probability that an individual azobenzene unit exists will again decrease if the angle α becomes greater and vice versa.

Thus this process results in the orientation of the mesogenic side groups in the amorphous polymer and reorientation of the mesogenic side groups in the LC polymer, this is accompanied by a drastic change in the optical properties of the solid polymer films. The key condition of such photooptical reorientation is the existence of strongly anisodimetric mesogenic fragments as the side branches of comb-shaped polymers.

At elevated temperatures (above T_g the photoinduced processes of the structural transformations of the mesogenic groups dictated by their high mobility and high main chain flexibility occur more effectively. It follows that the greater cooperation of this process involves more and more neighboring non-photochromic groups in the orientational transformations (Fig. 2.39). As a result,

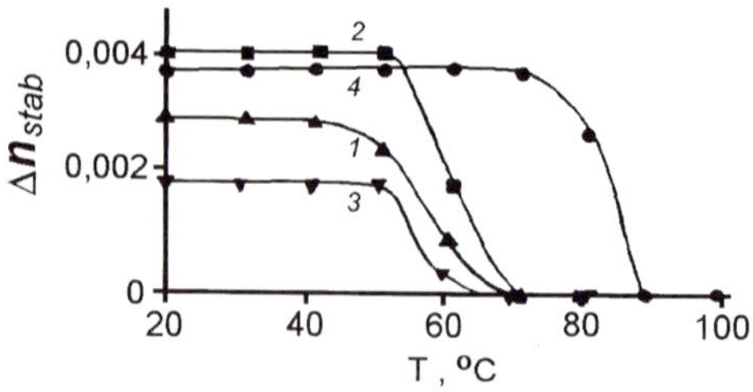

Fig. 2.40. Temperature dependence of the Δn_{stab} of copolymers **5**(*1*), **6**(*2*), **4**(*3*), (see Table 2.2) and copolymers **XXIV** (*4*)

the birefringence above glass transition temperatures increases to values of one order of the magnitude higher than the values of Δn_{ind} in the glassy state.

It should be pointed out that the orientation degree of mesogenic fragments should also be determined by a number of factors among which a free volume is a decisive one. In turn, a free volume depends on the chemical structure of all the macromolecular fragments such as the main chain, its flexibility, length of the spacers as well as the molecular structure of the side mesogenic fragments.

One of the important parameters, which allows us to judge the efficiency of the photoinduced changes in the material, is the photoinduced birefringence preserved in the polymer film after stopping the illumination and following relaxation. This parameter corresponds to the stable birefringence Δn_{stab}.

The maximum values of Δn_{stab} were obtained for comb-shaped polymers irradiated in the amorphous state. For example, the value of Δn_{stab}^{max} for polymer 19 (Table 2.3) was 0.11, which is close to the value of Δn, which is achieved by the homeotropic orientation of LC polymers under the action of electric and magnetic fields. The lower values of Δn_{stab} are characteristic of LC polymers; on average, these values for polymers from Table 2.2 vary between 0.05 and 0.08. This means, as indicated above, that an ordering of the mesogenic groups, realized in the LC phase restricts reorientation of the mesogenic azo-containing monomeric units. However there are no "sterically-hampered" factors which could restrict the reorientation process in amorphous (isotropic) polymer films treated by laser irradiation.

The above-mentioned process leading to the appearance of photoinduced birefringence may be used for recording information and optical data storage. In this case, the question about the possibility of reversible erasure of the recorded information naturally arises. There exist at least two methods of erasure.

As can be seen from Fig. 2.40 the stable fixed birefringence Δn_{stab} practically does not change in the polymer films in the glassy state below T_g. The increase of the temperature, as is seen from the same Figure, leads to the falling of birefringence practically to 0. This characteristic of birefringence changing "prompts" the erasure method of the recorded information, which lies in the heating of the polymer film to above the glass transition temperature. Further, in the case of LC polymers, repeated orientation is necessary for successive recording; in the case of amorphisable polymers, heating the sample above T_{cl} with successive quenching is also necessary for repeated recording.

However, as shown by us in Ref. [16] more convenient and prospective methods of erasure exists. This method consists of the irradiation of LC film with

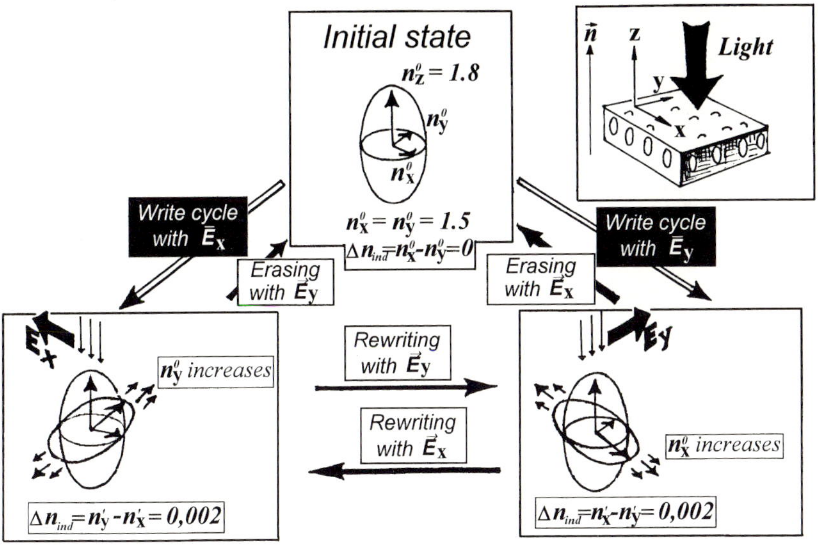

Fig. 2.41. Schemes of the variation of the refractive induces in the write and erase cycles on the homeotropically oriented film of LC polymer

recorded information by a laser beam with its polarization plane opposite to the initial polarization plane of the writing beam. This very simple method permits us to influence effectively the ratio of the values of birefringence in three directions –X, Y and Z. Analysis of the optical data has allowed us to suggest the following picture of the variation of all refractive indices during irradiation.

Figure 2.41 shows the ellipsoids of the refractive indices of LC monodomain polymer film before and after irradiation. As can be seen from the Figure, the initial homeotropically-oriented film presents a uniaxial positive monodomain sample. The initial refractive indices in X and Y directions are equal to each other ($n_x^o = n_y^o$); however the refractive index in Z-direction (n_z^o) is greater than (n_x^o and n_y^o). According to the above-mentioned data (see Figs. 2.28 and 2.29) the increase in the refractive index of the film in the direction perpendicular to the electric vector of the "writing" laser beam after the irradiation of the film is always observed. If the recording is performed by a laser beam with the electric vector polarized in the X-direction ($\vec{E}_x$), an increase in the refractive index n_y^o is observed (see the left side of Fig. 2.41); if the recording is performed by a laser beam with the electric vector polarized in the Y-direction ($\vec{E}_y$) the refractive index n_x^o increases (see the right side of the same Figure). In both the cases, photoinduced birefringence equal to the difference between the new photoinduced refractive indices n_x' and n_y' of the polymer film appears. In other words, the polymer film becomes biaxial and photoinduced birefringence is equal to $\Delta n_{ind} = n_x' - n_y'$. This process corresponds to the recording (writing) cycle.

In order to erase the written image one needs to return to the initial state where $\Delta n_{ind} = 0$. The direction of the electric vector of the "writing" beam should be changed and this beam will act as the "erasing" one. The cycles of

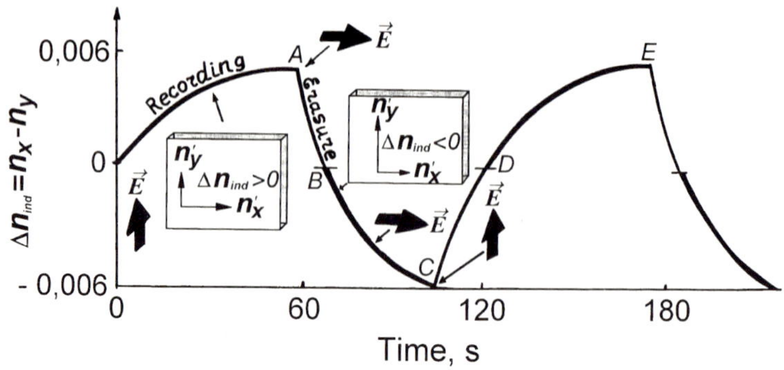

Fig. 2.42. Scheme of reversible optical photorecording showing the photoinduced birefringence as a function of time

erasure are designated by the dark arrows showing how the biaxial ellipsoids (on the left and on the right of the Figure) can be transformed into the initial uniaxial ellipsoid shown in the center of the Figure. Moreover, this scheme shows both the erasure process (the diagonal arrows) and the rerecording (rewriting) process (the horizontal arrows). It can be seen how one can record and erase the image passing from the left part of the figure to the right one and vice versa, by means of changing the electric vector direction of "writing" and "erasing" laser beam.

A generalized diagram of the experimental cyclic process of the information reversible recording on the homeotropically-oriented polymer film is given in Fig. 2.42. Curve OA shows the photoinduced birefringence increase during the irradiation process, i.e. in the recording process. When the polarization plane of the "writing" beam is parallel to Y-coordinate, it means that the refractive index, n'_x, becomes greater than the refractive index, n'_y. (It is shown by the coordinate arrows). When at point A a maximum value of $\Delta n_{ind} = n'_x - n'_y$ is reached a polarization plane (i.e. electric vector E of the laser beam) is turned at 90°. After that, the erasure process proceeds. At point B the image is completely erased ($\Delta n_{ind} = 0$). However, if the laser illumination of the film is continued (with the same direction of the polarization plane) the rise in Δn_{ind} (curve BC) is observed again (in this case $n'_y > n'_x$) and recording of information takes place; however the value of birefringence changes the sign ($n'_x - n'_y < 0$).

By changing the direction of the polarization plane of the laser beam again at point C) one can repeat the recording process. It is important to emphasize that, in this way, one may obtain any values of birefringence lying in the range from Δn_{max} to Δn_{min} which allows the whole "scale of greyness" to be realized.

An example of recording of the test pattern is shown in Fig. 2.30. Another example of recording was obtained from the color standard slide (24 × 36 mm) which was used as a mask for the photo preparation (Fig. 2.43). As can be seen from Fig. 2.43 the microphoto has the good quality and reproduces the half-tints of the image.

To conclude this section, some comparative characteristics of LC materials used for thermooptical and photooptical recording are listed in Table 2.13. As

Fig. 2.43. Photograph of black and white image obtained with a LC polymer

can be seen from the Table, the photooptical recording is characterized by a higher storage capability and higher spatial resolution than with thermooptical recording. In addition, the possibility of reversible recording and a very simple method of erasure in combination with high cyclicity of the LC photochromic

Table 2.13. Comparison of some important characteristics of LC comb-shaped polymers with cyanobiphenyl mesogenic groups for the thermooptical and photooptical recording [21, 43, 115]

Parameter	Thermoooptical recording (Homopolymer+2%dye)	Photooptical recording Dye-containing copolymer
Power density of laser beam, W/cm^2	$10^2 - 10^3$	$10^{-1} - 10^{-2}$
Resolution, lines/mm	200	1000
Storage capability, bit/cm^2	5.10^6	10^9
Write/erase cycles	tested to 10^3	tested to 4×10^2
Bit stability	> 7 years	tested to 3 years

polymer permits us to consider these systems as rather promising ones for practical applications.

2.7 Summary and Outlook

Thus comb-shaped polymers containing mesogenic and photochromic azobenzene groups[10] can be considered as structurally and functionally-integrated systems representing striking examples of molecular and supramolecular organization. The light-induced photochemical transformations occurring in the monomeric units of copolymers (and homopolymers) at the first stage leads to considerable configurational and conformational changes of photosensitive fragments due to *trans-cis* isomerization as it takes place in azobenzene derivatives. These processes undoubtedly occur on the molecular level and result from the redistribution of the electrons in photosensitive fragments of the macromolecules.

Further irradiation of the polymer is inevitably accompanied by photostructural changes in the polymer system. In view of the strong anisotropy of the chromophore groups, the irradiation of polymers by polarized (and even nonpolarized) light induces an orientation of chromophors in the direction orthogonal to the direction of the preferential absorption of incident light. However, due to the high anisotropy of the neighboring side groups united in the microensembles (microdomains), characterized by the orientational order in the direction of the long axes of the side chains a transition from the rod-shaped to the bent-shaped configuration of chromophore groups must lead to a disturbance of the initial supramolecular structure. The photoactivated irradiation, inducing the *trans-cis* isomerization of photoanisotropic side groups, stimulates a reorientation of the neighboring monomer units, which can proceed like a cooperative process.

If the first initiating photochemical process is carried out very quickly, the second process proceeds as a time-dependent and diffusion-controlled process. The periodical change of the electric vector of the activated irradiation results in the cyclic macroscopic reorientation of the microdomains and the reconstruction of the supramolecular structure. Probably, the local electric field induced by the orientational ordering of the polar anisotropic fragments is also essential for the processes of such reorientation. It is noteworthy that this process proceeds not because of a change in molecular ensemble as a whole (as it occurs due to an electric field) but because of a photoisomerization mechanism when the photochromic molecule, after the "bending" –transformation (induced by *trans-cis* isomerization), performs the specific function of an initiator of the repacking of the mesogenic fragments and rebuilding of their surroundings. So, in this case,

[10] We considered here the concrete comb-shaped polymer systems containing only the azobenzene derivatives as the photochromic side groups; however, all these considerations may also be used for other polymer systems with another molecular structure and a similar mechanism of photoinduced transformation.

the light acts as the "regulated external field" permitting us to regulate the supramolecular structure of the polymer. We hope that all the above-mentioned results will stimulate further investigations into thermo-, electro- and photo-induced phenomena directed towards the elucidation of the correlations between the molecular and the supramolecular structure of such functionalized polymer systems.

In the near future, such studies may lead to an elaboration of new electro-, thermo- and photochemoactive supramolecular devices, the active principle of which will be based on the photoinduced changes of the structure and the optical properties of polymer materials.

On the other hand, the investigation and establishment of the regularities between molecular and supramolecular structure of the functionalized polymers and their optical, photochemical, elctrooptical and thermooptical properties would be an essential contribution to the elaboration of the fundamental problems of electro- and photooptical polymer media.

Acknowledgements. The authors are greatly indebted to their colleagues from the Chemistry Department of the Humboldt University in Berlin – Drs. J. Stumpe, Läsker and Th. Fischer, as well as Dr. R. Ruhmann from the Center of Macromolecular Chemistry in Berlin for joint experimental work and valuable discussions. The authors are also grateful to Mrs. T. Zabolotskaya for assistance in the manuscript preparation.

This work was supported in part by the Russian Foundation of Fundamental Research (grant 96-03-33820), and the Scientific Program "The Russian Universities".

2.8 References

1. Chemla DS, Zyss J (eds) (1987) Non-linear optical properties of organic molecules and crystals, vol 1-2. Academic Press, Orlando
2. Ulmann A (1991) An introduction to ultrathin organic films from Langmuir-Blodgett to self-assembly. Academic Press, New York
3. Prasad PN, Donald RU (eds) (1988) Non-linear optical and electrooptical polymers. Plenum, New York
4. Williams DJ (ed) (1983) Non-linear optical properties of organic and polymeric materials. Am Chem Soc, Washington, ACS Symposium Series, v 233
5. Arakelian SM, Chilingarian YuS (1984) Non-linear optics of liquid crystals. Nauka, Moscow
6. Blinov LM, Chigrinov VS (1994) Electrooptic effects in liquid crystal materials. Springer, Berlin Heidelberg New York
7. Shibaev VP, Plate NA (1984) Adv Polym Sci 60/61: 173
8. Finklelmann H, Rehage G (1984) Adv Polym Sci 60/61: 99
9. Plate NA, Shibaev VP (1980) Comb-shaped polymers and liquid crystals. Nauka, Moscow (in Russian); Plate NA, Shibaev VP (1987) Comb-shaped polymers and liquid crystals. Plenum, New York
10. Shibaev VP (1988) In: Plate NA (ed) Liquid-crystal polymers. Chimia, Moscow p190 (in Russian); (1993) Plenum, New York, p193
11. McArdle CB (1989) In: McArdle CB (ed) Side chain liquid crystal polymers, Blackie, Glasgow London, p357
12. Wendorff JH, Eich M (1989) Mol Cryst Liq Cryst 169: 133
13. Eich M, Reck B, Ringsdorf H, Wendorff J (1986) Proc SPIE-Int Soc Opt Eng 682: 93
14. Plate NA, Talrose RV, Shibaev VP (1984) Pure and Appl Chem 36: 403

15. Plate NA, Talrose RV, Shibaev VP (1986) In: Pethrick RL (ed) Polymer Yearbook Harwood Academic Publishers, London, p 195
16. Shibaev VP, Yakovlev IV, Kostromin SG, Ivanov SA, Zverkova TI (1990) Vysokomolek Soed A32: 1552 (in Russian); Translated in Polymer Sci USSR A32: 1478(1990)
17. Shibaev VP, Kostromin SG, Plate NA, Ivanov SA, Vetrov VYu, Yakovlev IA (1983) Polymer Commun 24: 364
18. Ivanov SA, Yakovlev IA, Vetrov VYu, Kostromin SG, Shibaev VP (1983) Pis'ma v ZhTF 9: 1349 (in Russian)
19. Haase W (1988) In: McArdle CB (ed) Side chain liquid crystal polymers. Blackie, Glasgow London, p 309
20. Shibaev VP, Belayev SV (1990) Vysokomolek Soed A32: 2266 (in Russian); Translated in Polymer Sci USSR A32: 2384 (1990)
21. Shibaev VP (1994) Mol Cryst Liq Cryst 243: 201
22. Freidzon YaS, Shibaev VP, Plate NA (1974) Proceedings of the III all-union conference on liquid crystals, Ivanovo, p 214 (in Russian); Shibaev VP, Plate NA (1977) Vysokomolek Soed A19: 923 (in Russian); Translated in Polymer Sci.USSR A19: 1065 (1978)
23. Kreuder W, Ringsdorf H (1983) Markromol Chem., Rapid Commun 4: 807
24. Hermann-Schonherr O, Wendorff J, Ringsdorf H, Tschirner P, (1987) Makromol Chem., Rapid Commun 7: 791
25. Engel M, Hisgen B, Keller R, Kreuder W, Reck B, Ringsdorf H, Schmidt HW, Tschirner P (1985) Pure and Appl Chem 57: 1009
26. Shibaev VP, Freidzon YaS, Kostromin SG (1994) In: Shibaev VP, Lui Lam (eds) Liquid crystalline and mesomorphic polymers. Springer, Berlin Heidelberg New York, p 77
27. Shibaev VP, Kostromin SG, Plate NA (1982) European Polymer J 18: 651
28. Kostromin SG (1982) Dissertation for the degree of Candidate of Sciences, Moscow State University, Moscow (in Russian)
29. Kostromin SG, Ngo Duy Cuong, Garina ES, Shibaev VP (1990) Mol Cryst Liq Cryst 193: 177
30. Pham Ahn Tuan, Kostromin SG, Shibaev VP (1992) Polym Bull 29: 49; (1993) 30: 249
31. Kostromin SG, Pham Ahn Tuan, Shibaev VP (1994) Vysokomolek Soed B36: 1807 (in Russian)
32. Plate NA, Talrose RV, Kostromin SG, Shibaev VP, Kresse H (1981) Abstracts of 27th Intern Symp on Macromolecules (IUPAC) Strasbourg, France, vol 2, p 978
33. Schmidt HW (1989) Angew Chem Int Ed Adv Mater 28: 940
34. Ringsdorf H, Urban C, Knoll W, Sawodny M (1992) Makromol Chem 193: 1235
35. Belayev SV, Zverkova TI, Panarin YuP, Kostromin SG, Shibaev VP (1986) Vysokomolek Soed 28B: 789 (in Russian)
36. McArdle CB (ed) (1992) Applied photochromic polymer systems. Blackie, Glasgow London
37. Wegner G (1993) Mol Cryst Liq Cryst 235: 1
38. Grebenkin MF, Ivashenko AV (1989) Liquid-crystalline materials. Chimia, Moscow, p 82 (in Russian)
39. Barachevskii VA (ed) (1987) Properties of photosensitive materials and their application in holography. Nauka, Leningrad (in Russian)
40. Kartuschanskii (ed) (1988) Perspectives and possibilities of unsilver photography. Chimia, Leningrad (in Russian)
41. Sekkat Z, Dumont M (1991) Appl Phys B53: 121; (1993) Synthetic Metals 54: 373; (1992) Appl Phys B54: 486
42. Bâlzani V, Scandola F (1991) Supramolecular photochemistry. Ellis Harwood, Chichester
43. Shibaev VP, Kostromin SG, Ivanov SA, Läsker L, Fischer Th, Stumpe J (1994) Proceedings of international conference on liquid crystal polymers, 6–9 September, 1994, Beijing, China, p5
44. Yang-Kyoo Han, Dong-Yu Kim, Yong Ha Kim (1992) J Polymer Sci-Polym Chem A30: 1177
45. Nazeni A, Willams G, Attard G, Karasz F (1992) Polymer Adv Technol 3: 157
46. Ringsdorf H, Schmidt HW, Baur G, Kifer R, Windshed (1986) Liquid Cryst 1: 319
47. Ruhmann R, Zschuppe V, Dittmer M, Wolff D (1992) Makromol Chem 193: 3073
48. Reck B, Ringsdorf H (1985) Makromol Chem Rapid Commun 6: 291
49. Eich M, Wendorff J, Reck B, Ringsdorf H (1987) Makromol Chem Rapid Commun 8: 59

50. Reck B, Ringsdorf H (1986) Makromol Chem Rapid Commun 7: 389
51. Wada T, Zhang G, Sasabe H (1993) Mol Cryst Liq Cryst 224: 1
52. Natansohn A, Rochon P, Gosselin J, Xie S (1992) Macromolecules 25: 2268
53. Natansohn A, Rochon P, Brown D (1993) Polymer Preprints 34: 739
54. Natansohn A, Rochon P (1993) Preprints of 3rd Pacific polymer conference, December 10–14, Gold-Coast, Australia, p 723
55. Yitzchaik S, Berkovic G, Krongauz V (1990) Chem Mater 2: 162; (1990) Macromolecules 23: 3539
56. Stone PJ, Miller LS, Walton DJ (1993) Mol Cryst Liq Cryst 235: 209
57. Yokoyama S, Kakimoto M, Imai Y (1993) Mol Cryst Liq Cryst 227: 295
58. Krongauz V (1992) In: McArdle CB (ed) (1993) Applied photochromic polymer systems Blackie, Glasgow London, p 121
59. Matsuoka M (1993) Mol Cryst Liq Cryst 224: 85
60. Blinov LM (1983) Electro-optical and magneto-optical properties of liquid crystals Wiley, Chichester
61. Talrose RV, Kostromin SG, Shibaev VP, Plate NA, Kresse H, Sauer C, Demus D (1981) Makromol Chem., Rapid Commun 2: 305
62. Shibaev VP, Kostromin SG, Talrose RV, Plate NA (1981) Dokl Acad Nauk SSSR, 259: 1147 (in Russian)
63. Talrose RV, Shibaev VP, Plate NA (1983) Vysokomolek Soed., A25: 2467 (in Russian)
64. Ringsdorf H, Schmidt HW (1984) Makromol Chem 185: 1327
65. Goozner R, Finkelmann H (1985) Makromol Chem 186: 2407
66. Coles HJ (1985) Faraday Disc Chem Soc 79: 201
67. Talrose RV, Plate NA (1988) In: Plate NA (ed) Liquid-crystal polymers Chimia, Moscow, p 296 (in Russian); (1993) Plenum, New York London, p 303 (English translation)
68. Shibaev VP, Talrose RV, Korobeinikova IA, Plate NA (1989) Liquid Cryst 4: 467
69. De Gennes PG (1974) The physics of liquid crystals Clarendon Press, Oxford
70. Blinov LM (1978) Electro-optical and magneto-optical properties of liquid crystals Nauka, Moscow, p 6 (in Russian)
71. Casagrande C, Fabre P, Veyssie M, Weill G (1984) Mol Cryst Liq Cryst 113: 193
72. Korobeinikova IA, Talrose RV, Shibaev VP, Plate NA (1987) Vysokomolek Soed A29: 1037 (in Russian)
73. Barmatov EB, Stroganov LB, Talrose RV, Shibaev VP, Plate NA (1993) Vysokomolek Soed A35: 162; Translated in Polymer Sci A35: 183 (1993)
74. Ringsdorf H, Zentel R (1982) Makromol Chem 183: 1245
75. Kiefer R, Baur G (1989) Liquid Cryst 5: 1497
76. Shibaev VP, Barmatov EB, Stroganov LB (1993) In: Buka A (ed) Modern topics in liquid crystals. World Scientific, Singapore New Jersey London Hong Kong, p 76
77. Van der Meer B, Potma F, Dekker A, de Jeu W (1982) Mol Phys 45: 1227
78. Coles H, Bankcraft (1993) Mol Cryst Liq Cryst 237: 97
79. Paskhovsky E, Litvina TG, Kostromin SG, Shibaev VP (1992) J Phys II, France 2: 1577
80. Scarp K, Langerwall ST, Stebler B (1980) Mol Cryst Liq Cryst 60: 215
81. Shibaev VP (1982) Abstracts of the first all-union symposium on liquid crystal polymers, December 14–16, Chernogolovka (Moscow region) p 5 (in Russian)
82. Shibaev VP, Kostromin SG, Plate NA, Ivanov SA, Vetrov VYu, Yakovlev IA (1985) In: Blumstein A (ed) Polymeric liquid crystals. Plenum, New York, p 345
83. Shibaev VP, Yakovlev IA, Kostromin SG, Ivanov SA (1990) In: Abstracts of 33-rd IUPAC International symposium on macromolecules, July 8–13, 1990, Montreal, Canada, section 1.6.4
84. Yakovlev IA, Kostromin SG, Ivanov SA, Shibaev VP (1990) In: Abstracts of All-Union Conference on problems of optical data storage, October 4–12, 1990, Telavi, p 110 (in Russian)
85. Coles HJ, Simon R (1985) In: Blumstein A (ed) Polymeric liquid crystals. Plenum, New York p 351
86. Coles HJ, Simon R (1985) In: Chapoy (ed) Recent advances in liquid crystal polymers. Elsevier Appl Sci Publishing, London, p 322

87. MCArdle CB, Clark M, Haws C, Wiltshire M, Parker A, Nestor G, Gray G, Lacey D, Toyne K (1987) Liquid Cryst., 2: 537
88. Nakamura T, Ueno T, Tani C (1989) Mol Cryst Liq Cryst 169: 167
89. Tomilin MG (1988) In: Kartuschanskii AL (ed) Perspectives and possibilities of unsilver photography Chimia, Leningrad, p 194 (in Russian)
90. Kakichashvili SD (1972) Optica and Spectroscopia 33: 171; 33: 214 (in Russian)
91. Nikolova L, Todorov T (1984) Opt Acta 31: 579
92. Todorov T, Nikolova L, Tomova N (1984) Appl Opt 23: 4309; 23: 4588
93. Shatalin ID (1989) Optica and Spectroscopia 66: 362 (in Russian)
94. Anderle K, Wendorff JH (1994) Mol Cryst Liq Cryst 243: 51
95. Eich M, Wendorff JH (1987) Makromol Chem., Rapid Commun 8: 467
96. Wendorff JH, Eich M (1990) J Opt Soc of America 7: 1428
97. Ortler R, Brauchle Ch, Miller A, Riepl G (1989) Makromol Chem., Rapid Common 10: 189
98. Ikeda T, Kurihara S, Karanjit D, Tazuke S (1990) Macromolecules 23: 3938
99. Ikeda T, Horiuchi S, Karanjit D, Kurihara S, Tazuke S (1990) Macromolecules 23: 42
100. Ikeda T, Hasegawa S, Sasaki T, Miyamoto T, Mao-Peng Lin, Tazuke S (1991) Makromol Chem 192: 215
101. Ikeda T, Miyamoto T, Sasaki T, Kurihara S, Tazuke S (1991) Mol Cryst Liq Cryst 188: 235
102. Wagenblast G, Beck K, Etzbach K (1993) Polymer Preprints 34: 699
103. Brown GH (1971) Photochromism, techniques of chemistry, vol III Wiley-Interscience, New York
104. Rau H (1973) Angew Chem 85: 248
105. Guillet J (1985) Polymer Photophysics and photochemistry. Cambridge University Press, Cambridge
106. Horie K, Mita I (1989) Advances in Polymer Sci 88: 77
107. Shibaev VP, Kostromin SG, Ivanov SA, Fischer Th, Läsker L, Stumpe J (1994) Abstracts of 35-th International symposium on Macromolecules (IUPAC), July 11–15, Akron, USA, p 457
108. Ivanov SA, Yakovlev IA, Kostromin SG, Shibaev VP, Läsker L, Kreysig D, Stumpe J (1991) Makromol Chem., Rapid Commun 12: 709
109. Shibaev VP (1992) Abstracts of 14th International liquid crystal conference, vol II, June 21–26, Pisa, Italy, p833
110. Läsker L, Fischer Th, Stumpe J, Kostromin SG, Ivanov SA, Shibaev VP, Ruhmann R (1994) Mol Cryst Liq Cryst 246: 347
111. Stumpe J, Müller L, Kreysig D, Hauck G, Koswig HD, Ruhmann R, Rübner J (1991) Makromol Chem., Rapid Commun 12: 709
112. Läsker L, Fischer Th, Stumpe J, Kostromin SG, Ivanov SA, Shibaev VP, Ruhmann R (1996) Mol Cryst Liq Cryst (in press)
113. Weigert F (1920) Ann Phys 63: 681
114. Spiess HW, Wiesher V, Schmidt-Rohr K, Boeffel C, Pawelzik V (1991) Makromol Chem., Rapid Commun 12: 457
115. Shibaev VP Ivanov SA, Kostromin SG (1996) Vysokomolek Soed (in press)
116. Wiesner U, Antonietti M, Boeffel C, Spiess HW (1990) Makromol Chem 191: 2133
117. Wiesner U, Reynolds N, Boeffel C, Spiess HW (1992) Liquid Cryst 11: 251
118. Neporent BS, Stolbova OV (1993) Optica and Spectroscopia 14: 624 (in Russian)

CHAPTER 3

Cyclic Liquid-Crystalline Siloxanes as Optical Recording Materials

F.H. Kreuzer, Ch. Bräuchle, A. Miller, and A. Petri

3.1 Introduction

Organic materials for optical data storage applications is a fascinating and growing field of research [1]. Of special interest are liquid-crystalline polymers (LCPs), especially side chain polymers [2]. They have been proposed as optical recording media by many authors [3–6].

The mechanism of the storage process can be generally understood as a variation of the birefringence by changing the director of calamitic molecules and/or the degree of order of the liquid-crystalline system. In addition changes in the optical density and photochromic effects may play an important role [7–11]. Besides the effects of electric or magnetic fields – as in low molar mass liquid crystal technology – there are two additional procedures for achieving such variations:

a) induction of phase transition by thermal heating
b) reorientation by photochemical *cis-trans* rearrangements.

These techniques work as writing or erasing processes but the storage process itself requires the conservation of the effects introduced into the material or to the aligning layer at the walls of a liquid-crystalline (LC)-cell [12]. That means that the decay of the information after writing should be zero in the ideal case.

The stabilization of the information contained in the liquid crystalline phase can be done by wall effects if the liquid crystal is positioned between glass plates (especially in the case of low molar mass materials [13]), or by viscous effects below the glass transition temperature. Additionally the information can be fixed by a polymerization [14–16] or crosslinking reaction.

Most of the materials studied in the past were either polymeric or monomeric compounds. Oligomers, especially cyclics combine the advantages of polymeric systems i.e. the possibility of connection of different mesogens in one molecule, glass transition temperatures T_g comparable to polymers and the advantages of low molar mass compounds i.e. moderate viscosities at temperatures above T_g.

The cyclic liquid-crystalline (CLC) siloxanes reported as information recording media in this chapter may be generally classified as side chain oligomers/ polymers.

3.2
Cyclic Liquid-Crystalline Siloxanes

3.2.1
General Synthetic Aspects

The synthesis of cyclic siloxanes with mesogenic side groups is similar to the linear species [17–19]. Some peculiar aspects have to be discussed with respect to the application properties of the oligomeric materials. One of the most important

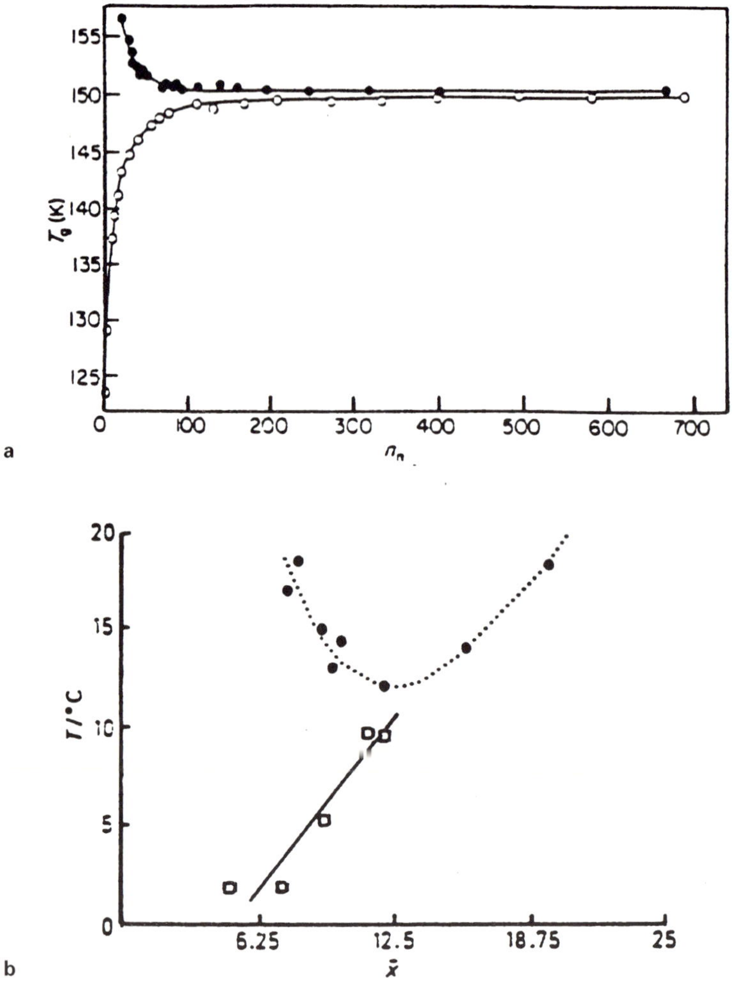

Fig. 3.1a,b. Glass transition temperature T_g as function of the degree of polymerization; **a** dimethylsiloxanes (○ linear, ● cyclic [22]; **b** liquid-crystalline siloxanes containing the cyanobiphenyl mesogenic groups (□ linear, ● cyclic) [20]

points is the low and well-defined degree of polymerization (DP) of the available cyclics. DPs from 4–7 are the most favoured species, however ring systems with DPs up to 24 have been studied recently [20]. As cyclics are free of end groups the depression of T_g by trimethylsilyl endgroups is eliminated, but entropic effects may also play an important role [21]. So cyclics with DPs < 100 exhibit glass transition temperatures higher than the corresponding linear siloxanes [22]. This effect is also valid for CLC siloxanes with DPs < 10. As a consequence CLC-siloxanes exhibit relatively high glass transition temperatures at moderate bulk viscosities (Fig. 3.1).

Although the cyclics as starting materials are very well-defined compounds, the resulting products are mostly mixtures of many but predictable compounds due to statistical effects. These statistical aspects are important in the case of co-oligomers with low DPs. In long-chain copolymers the statistical distribution yields inhomogeneities along the siloxane backbone, but short-chain systems or cyclics with 4–10 siloxane units actually represent a collection of chemically different substituted molecules. The amount of the different species in the case of a copolymer is determined by the binomial distribution.

Assuming two mesogens M1 and M2 of the same reactivity attached to a cyclic siloxane with n siloxane units:

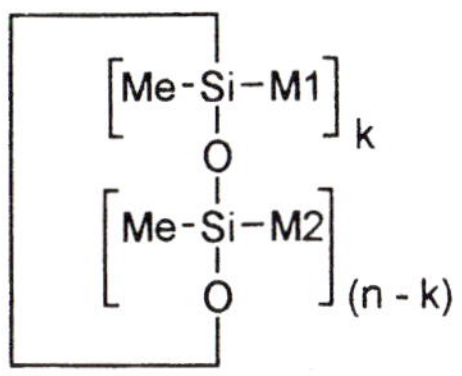

where k is the number of mesogens M1 attached to the cyclic siloxane with n Si-atoms, n-k the number of mesogens M2. The probability p(n, k) of finding k

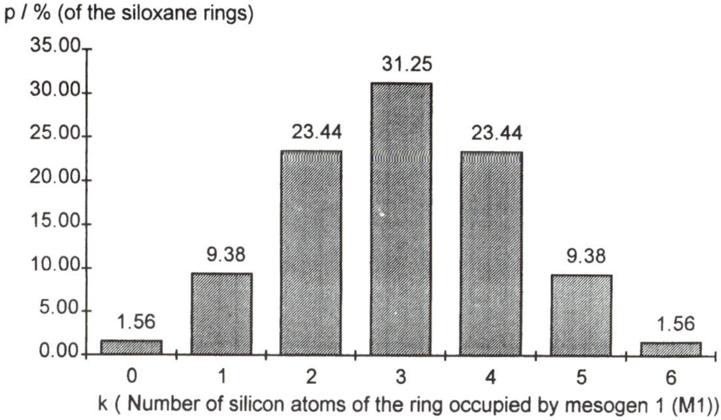

Fig. 3.2. Product distribution of a hexameric copolymer containing two mesogens M1 and M2 with the molar ratio 1:1

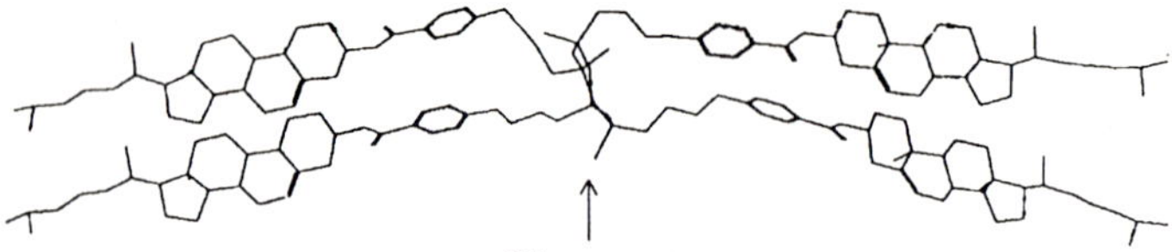

Fig. 3.3. Bundle structure of a cyclic liquid-crystalline siloxane with 4 Si-O-units

mesogens M1 attached to the siloxane ring with n Si-atoms results from the binomial distribution [23]:

$$p(n,k) = \frac{n!}{k! \times (n-k)!} pM1^{k} \times (1 - pM1)^{(n-k)} \times 100 \qquad (3.1)$$

where p is the % of cyclic siloxane molecules substituted by k mesogenic groups M1; pM1 mole fraction of M1 in the mixture of both mesogens and k is changed from 0 to n.

Figure 3.2 shows that the regular hexamer appears in the overall composition of the cooligomer mixture only in a concentration of 31.25%.

Nevertheless, the phase behavior of CLC-siloxanes is similar to the corresponding linear systems. With elongated mesogens the calamitic structures were observed. Bundle or bowl like structures are the most probable ones [24–31] (Fig. 3.3).

3.2.2 Synthesis

3.2.2.1 *Synthesis of Siloxanes*

The cyclic siloxanes with Si-H groups can be prepared by standard methods [32].

3.2.2.2 *Connection of Mesogens to the Siloxane by Hydrosilylation*

The following considerations are valid for CLC oligomers/polymers as well as for linear systems. Most of the observations were made on linear siloxanes but they are also valid for cyclics. The key reaction for the synthesis of CLC siloxanes is the hydrosilylation reaction. Therefore some special aspects of this reaction should be mentioned. The hydrosilylation is an addition of a SiH bond to a double bond, catalyzed by Pt and other transition metal complexes or peroxides and has been well known since the 1950s [33].

Starting materials are Si-H functional siloxanes and mesogenic compounds with terminally unsaturated spacers. As the catalyst, mostly Pt-compounds are used such as $H_2PtCl_6 \cdot H_2O$, platinum(II)-olefin-complexes and platinum(0) complexes with phosphines [34].

Due to the importance of this reaction in the chemistry of LC siloxane side chain polymers, some of the peculiarities of the addition of Si-H-groups to olefins should be mentioned[1].

The hydrosilylation usually yields β-adducts so the silicon unit is connected to the ω-position of the olefin:

$$\equiv \mathrm{Si} - \mathrm{H} + \mathrm{CH_2} = \mathrm{CH} - \mathrm{R} \overset{(\mathrm{Cat.})}{\longrightarrow} \equiv \mathrm{Si} - \mathrm{CH_2} - \mathrm{CH_2} - \mathrm{R}$$
$$\beta - \text{adduct}$$

In additional to the desired reaction, several side reactions are observed:

a) α-Addition

$$\equiv \mathrm{Si} - \mathrm{H} + \mathrm{CH_2} = \mathrm{CH} - \mathrm{R} \overset{(\mathrm{Cat.})}{\longrightarrow} \equiv \mathrm{Si} - \mathrm{CH(CH_3)} - \mathrm{R}$$
$$\alpha - \text{adduct}$$

b) Isomerization

$$\mathrm{CH_2} = \mathrm{CH} - \mathrm{CH_2} - \mathrm{R} \overset{(\mathrm{Cat.})}{\longrightarrow} \mathrm{H_3C} - \mathrm{CH} = \mathrm{CH} - \mathrm{R}$$

c) Hydrogenation

$$\equiv \mathrm{Si} - \mathrm{H} + \mathrm{CH_2} = \mathrm{CH} - \mathrm{R} \overset{(\mathrm{Cat.})}{\longrightarrow} \mathrm{H_3C} - \mathrm{CH_2} - \mathrm{R}$$

d) Splitting of propene

Allylic compounds containing an atom with a free electron pair in the γ -position undergo a splitting reaction studied in detail by Speier et al. [35]:

$$\equiv \mathrm{Si} - \mathrm{H} + \mathrm{CH_2} = \mathrm{CH} - \mathrm{CH_2} - \mathrm{R} \overset{(\mathrm{Cat.})}{\longrightarrow} \equiv \mathrm{Si} - \mathrm{R} + \mathrm{CH_2} = \mathrm{CH} - \mathrm{CH_3}$$
$$\mathrm{R} = -\mathrm{OR'}, -\mathrm{Cl}, -\mathrm{O_2CR'}, \ldots.$$

The loss of significant amounts of propylene is one of the most important side reactions of allyloxy groups containing mesogens and can lead to a loss up to 30% of the C3-spacer [36, 37].

The polymers resulting from the hydrosilylation reaction contain remarkable amounts of mesogens not connected with the siloxane. These residues not only result from the applied excess of mesogens but also from side products of hydrogenation, isomerization, or splitting reactions.

The standard methods for purification are chromatographic procedures, re-precipitation and the purification with supercritical carbon dioxide [38].

3.2.3 Synthesis of Crosslinkable Materials

Photochemically crosslinkable materials are needed for permanent optical storage applications.

[1] Synthesis of side chain liquid crystal polysiloxanes was considered in detail by G. Gray in the book "Side Chain Liquid Crystal Polymers" (Ed. C. McArdle), Blackie, Glasgow London, 1989.

CLC-siloxanes bearing photoreactive groups can be synthesized by using the trimethylsilyl group as the protecting group to introduce methacrylic functions [39, 40].

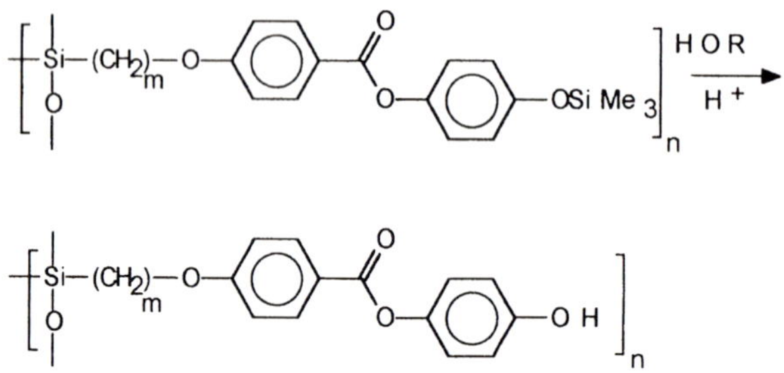

Subsequent esterification by an acid chloride in the presence of an HCl acceptor or by an appropriate acid anhydride yields the desired ester.

The application properties of cyclic materials can also be tuned for adjustment e.g. of viscosity and clearing temperature by adding polymerizable monomeric additives.

3.3 Optical Recording with Cyclic Liquid-Crystalline Siloxanes

3.3.1 Permanent Recording

3.3.1.1 *Photopolymerization*

For permanent recording of a pattern, CLC-siloxanes can be used exhibiting smectic, nematic or cholesteric phases. Equivalent to the cholesteric behavior are compounds which exhibit S_{A^*} and S_{C^*} phases due to their dependence of reflectivity on temperature. The degree of reflectance of the cholesteric, S_{A^*} and S_{C^*} phases is theoretically 100% for circularly polarized light. The wavelength of reflection (WLR) is defined by the pitch of the helical structure and can be tuned simply by the concentration of the chiral compound responsible for the induction of a right- or left-handed structure. In cholesteric liquid crystalline siloxanes (ChLC siloxanes) mostly cholesterol derivatives or dihydrocholesterol derivatives are used. CLC siloxanes containing these compounds reflect left-handed circularly polarized light. For special purposes right-handed reflecting materials may be necessary. LC siloxanes with these properties have been obtained using doristerol derivatives [41].

The photocrosslinking process works via polymerization of methacrylic groups which are introduced into the molecule directly or by a polymeranalogous reaction [42]. Two typical generations of patterns are possible: patterns consisting of variations of the order parameter of the LC material or of

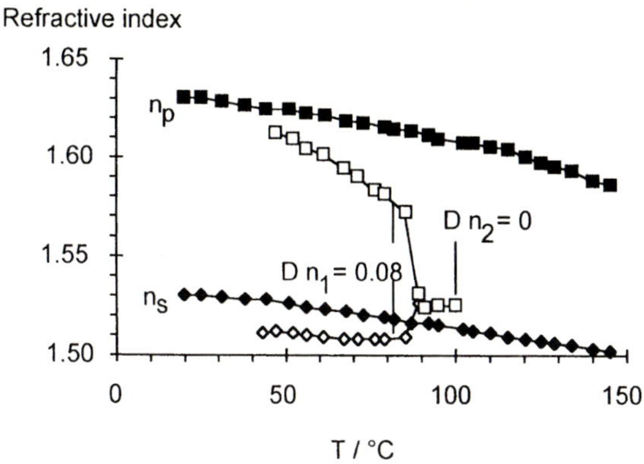

Fig. 3.4. Temperature dependence of refractive index of a noncrosslinked (□◇) and a crosslinked material (■◆). The pattern obtained exhibits a Δn of 0.08 surrounded by isotropic material with Δn=0. After crosslinking the temperature dependence of Δn is very low

variations of the wavelength of the reflected light. Crosslinking has a strong effect on the change of the index of refraction Δn as demonstrated in Fig. 3.4.

Figure 3.5 shows the processing steps to fix the order parameter and therefore Δn at different temperatures by photocrosslinking. In the first step, the illuminated areas are photocrosslinked at a temperature T_1 below the clearing isotropization temperature T_1. In the second step the temperature is raised to $T_2 > T_1$ in order to achieve an isotropic distribution in the nonphotocrosslinked areas. Finally, these areas are fixed by photocrosslinking in the third step. If only for the first step a photoinitiator is used and for the second step a thermally activated "radical source", the second irradiation step can be substituted by heating the sample. Heating can be done either by a hot stage or by a laser beam.

Operating in the region below the clearing temperature, patterns with different Δn values can be written in. The same procedure was applied to obtain patterns of different reflectivity or different colors of reflection with cholesterics. The different effects which can be used for optical recording with photocrosslinkable CLC siloxanes are summarized in Table 3.1.

Using cholesteric materials the patterns obtained can be detected by the brilliant colors of cholesteric reflection as well as by the polarization properties of cholesteric layers. Since the reflectivity is 100% for circularly polarized light, the signal to noise ratio for the read-out of the stored pattern is very high.

Table 3.1. Examples for possible combinations of different effects for optical recording with photocrosslinkable CLS Siloxanes; Δn_B is the Δn of the background; λ_{R_B} is the λ of reflection of the background which should give no signal with the applied detector; λ_{R_1}, λ_{R_2} ...... should give signals with the applied detector

Phase	Orientation	Effect responsible for detection of pattern	Background
nematic, smectic	planar	$\Delta n \neq 0$ $\Delta n_1 \neq \Delta n_B$	$\Delta n_B = 0$ (isotropic) $\Delta n_B \neq \Delta n_1$
	homeotropic	scattering	pseudoisotropic
cholesteric, S_{A^*}, S_{C^*}	planar	λ_R $\lambda_{R_1} > \lambda_{R_2} > \lambda_{R_1}$	isotropic $\lambda_{R_1} < \lambda_{R_B}$ or $< \lambda R_2$

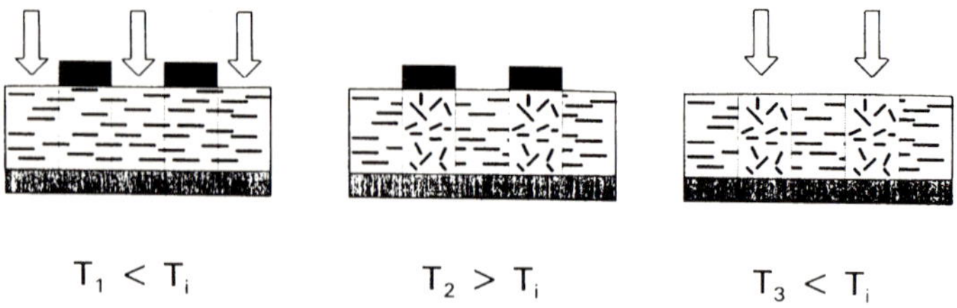

Fig. 3.5. Processing steps to obtain Δn patterns in CLC siloxanes

3.3.1.2
Photobleaching

3.3.1.2.1
In the Presence of Benzophenone

In the case of cholesterics, an additional mechanism for writing was investigated [43]. This process is similar to the photobleaching process and works with the aid of a photoreactive substance e.g. benzophenone. This is, of course, only one example showing a mechanism of irreversible destruction of the LC-material during the recording process. The material used for these experiments exhibits a cholesteric phase and contains 10% (w/w) benzophenone coupled to the siloxane ring via a C3-spacer.

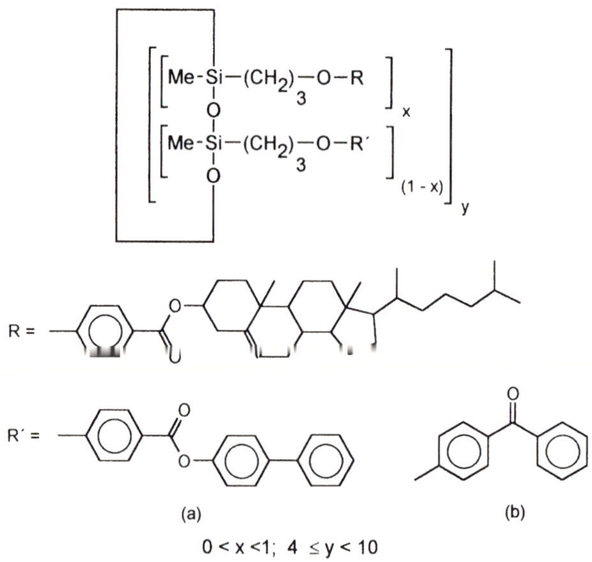

The reading wavelength detecting the reflectivity of the sample is adaptable to any wavelength by varying parameter x in the cholesteric siloxane. For the following experiments the composition was $x = 0.4$ and as R'(b) was used. The

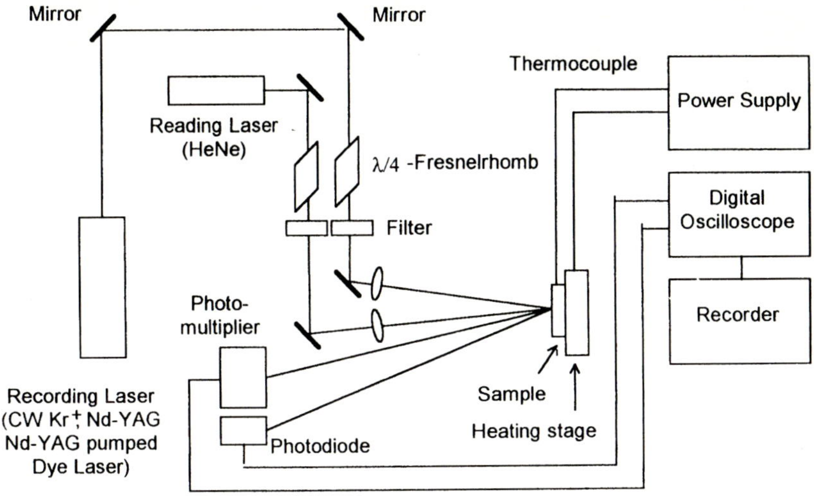

Fig. 3.6. Experimental set-up for investigation of writing and reading characteristics. The Kr^+ laser was used for writing and reading experiments in the CW mode. In the pulsed experiments the Nd-YAG was used for writing and the He-Ne laser for reading. For CW experiments, a frequency doubled Nd-YAG laser can also be used

doped siloxane shows a reflectivity of at least 80% for properly oriented samples and circularly polarized light.

The photochemical process produces a change of reflectivity ΔR of 80% dependent on the energy of the incident detection beam. This promises a high signal to noise ratio for data encoding. The well-known photoreaction of benzophenone in polymer matrices [44, 45] at an exciting wavelength of 380.7 nm above the threshold intensity of $7 \times 10^{-9} W/\mu m^2$ at a saturation energy of 7×10^{-10} $J/\mu m^2$ cleared the material at the point of irradiation. A general set-up for these experiments is shown in Fig. 3.6.

It is possible that, besides the attack of the photochemically generated ketyl radical from benzophenone, the bleaching process may be done by a Photo-Fries rearrangement of the phenylbenzoate group.

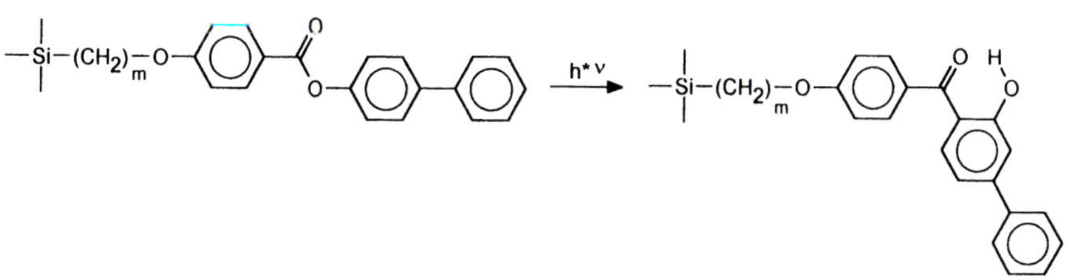

The elongated shape of the mesogenic side group is lost by the Fries reaction and converted to a bent one. The bent molecule is not mesogenic furthermore and therefore the LC properties of the compound are lost.

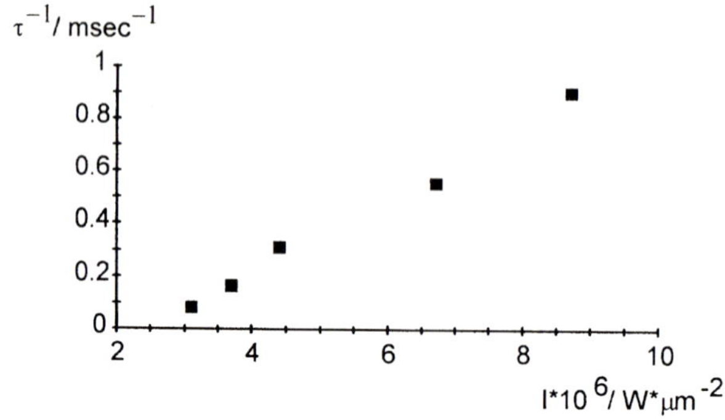

Fig. 3.7. Dependence of response time on writing intensity

3.3.1.2.2
In the Presence of Carbon Black

In order to sensitize the CLC-siloxane over a broad range of the spectrum from UV to the near infrared carbon black as a sensitizer was used. In moderate concentrations (0.5%–1%(w/w)) the reflectivity of the cholesteric materials is not influenced too much. The carbon black used for the experiments had a BET surface of 150 m^2/g. The samples had a thickness of 20μm – 40μm and were oriented on glass or polyethylene surfaces heating the liquid crystal close to its clearing point of 170 °C. The reflectivity change ΔR can be produced by a CW laser beam (power density up to $9 \times 10^{-6} W/\mu m^2$ above a threshold intensity of $2.86 \times 10^{-6} W/\mu m^2$.

The increase of ΔR as a function of time takes place with time constants τ in the range of milliseconds. The time constants τ depend on the laser intensity as shown in Fig. 3.7.

The time constants and threshold intensities are independent of the recording wavelength from 530 nm-752 nm, indicating that the system is also sensitive in the near infrared region. ΔR can also be achieved by using only one laser pulse of 10 ns so data rates of 100 MHz may be possible (Fig. 3.8). In this case a threshold energy of $2.54 \times 10^{-9} J/\mu m^2$ was found.

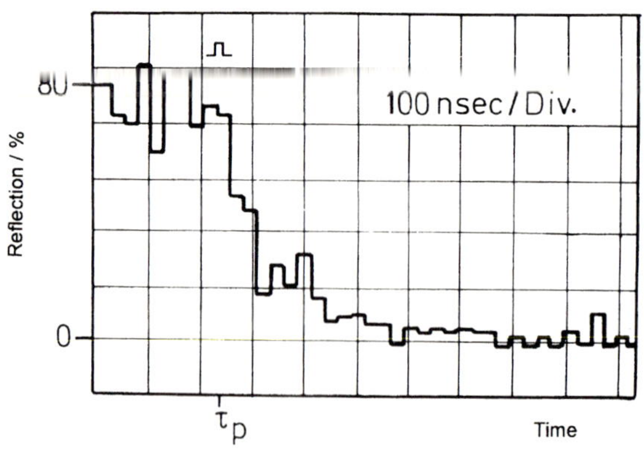

Fig. 3.8. Reflectivity of CLC siloxane, doped with carbon black at λ=632.5 nm during writing with a 10 ns laser pulse at λ=680 nm. At τ_p the recording laser pulse starts

The rise time of ΔR of the written signal is only dependent on the pulse duration. The size of the written spots does not exceed the size of the recording laser beam ($\geq 200\mu m$). Using a semiconductor laser with $\lambda \approx 800$ nm for writing and reading a diffraction limited storage density of 1.33×10^8 bits/cm^2 can be estimated [46].

Below this threshold energy of 2.54×10^{-9} J/μm^2 recording processes are also possible but with a reduced difference in the intensity of the reflected light ΔR max.

Information can be written in with energies down to 0.3×10^{-9} J/μm^2 without total destruction of the illuminated area. The writing process is limited to the surface and after heating above the glass transition temperature the material recovers its light reflecting properties.

The undoped CLC siloxane exhibits a threshold energy of 3×10^{-7} J/μm^2, a hundred times higher than the doped material. From the fact that external heating did not reduce the threshold energy, it was concluded that the writing process is accompanied by a real photochemical reaction. Similar observations were made with doped materials.

3.3.2 Reversible Recording

3.3.2.1 *Reversible Thermal Recording*

As mentioned before by applying a writing energy below the threshold energy reversible recording is possible. If carbon black is used only as an absorber and concentrated in a separate layer, irradiation is possible from the back side of the sample and the photoreaction is generally avoided. In the case of direct irradiation, the applied energy should not exceed the threshold energy of the undoped material. By this technique reversible writing and erasing is possible.

Any kind of IR absorbing substrate e.g. glasses of the type KG3 (Schott) work as well.

3.3.2.2 *Recording by cis trans Isomerization*

A quite different mechanism of optical signal storage uses the *cis-trans* isomerization of dye stuffs especially of azo compounds. Many LC materials containing these groups have been synthesized previously [47, 48] and the isomerization process was investigated [49, 50, 51][2].

Advantages of CLC siloxanes in contrast to the linear species may be attributed to a relatively high glass transition temperature connected with a moderate molecular weight. Most of the investigations were made using cholesteric phases albeit nematic and smectic systems work as well.

---

[2]These LC materials are considered in Chapter 2 of this book.

In the case of cholesterics two different mechanisms for the reading process exist. The first one is indicated by double refraction and uses polarized light. It works in the transmission mode and is applied to nematics and smectics as well. The second one works in the reflection mode and is a peculiar effect of cholesterics.

3.3.2.2.1
Materials

The cyclic siloxanes used for these experiments had the following structure and composition [52]:

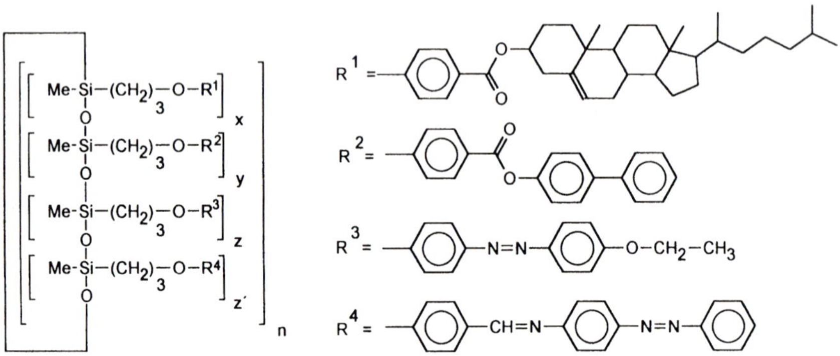

$$x + y + z + +z' = 1;\ n = 4$$

As mentioned in Sect. 3.2.1, the chemical formula represents only an idealized molecule. The real materials consist of a set of cyclic molecules with different substitution patterns. So it is clear that most of the molecules of Table 3.2 are free of photoreactive molecules if only concentrations of 5% or 10% of the azo dye are used (Fig. 3.9).

3.3.2.2.2
Determination of Holographic Material Parameters

The determination of hologram efficiency and sensitivity was carried out using the following experimental set-up: (Fig. 3.10)

Table 3.2. CLC Siloxanes containing azo groups (λ_R-is the wavelength of reflection of the cholesteric film)

Nr.	n	x	y	z	z′	T_g, °C	T_{cl}°C,	λ_R, nm
1	4	0.4	0.5	0	0.1	50	192	790
2	4	0.45	0.5	0.05	0	49	206	695

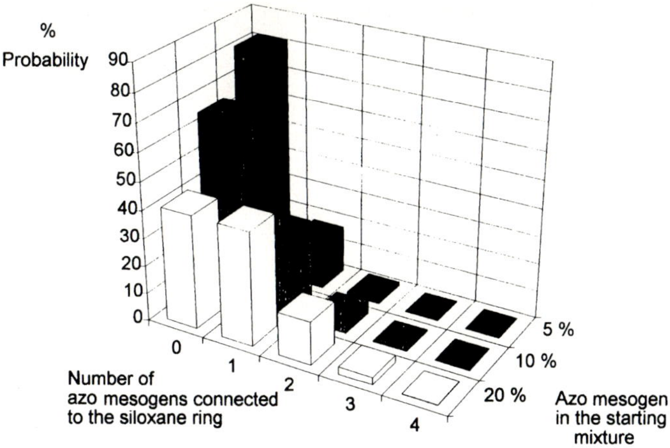

Fig. 3.9. Using 5, 10 and 20 mol% azo dye, the resulting LC-siloxane rings are connected to an extent of 19% 34% and 59% with the azo mesogens

The coherent beams O (object beam) and R (reference beam) form a sinusoidal interference pattern in the CLC siloxane sample. The fringe distance Λ is given by

$$\Lambda = \lambda_w / 2 \cdot n \cdot \sin q \tag{3.2}$$

With the index of refraction n of the sample and all other parameters as indicated in Fig. 3.10. The hologram efficiency results from the ratio of intensities of the diffracted beam I_d and the reading beam I_o:

$$\eta = I_d / I_o \tag{3.3}$$

The growth of the hologram $\eta = f(t)$ shows a quadratic time dependence:

$$\eta = (S \cdot I_w \cdot t)^2 \tag{3.4}$$

S is the sensitivity of the system and I_w is the writing intensity. For long irradiation times this equation is no longer valid because saturation effects dominate.

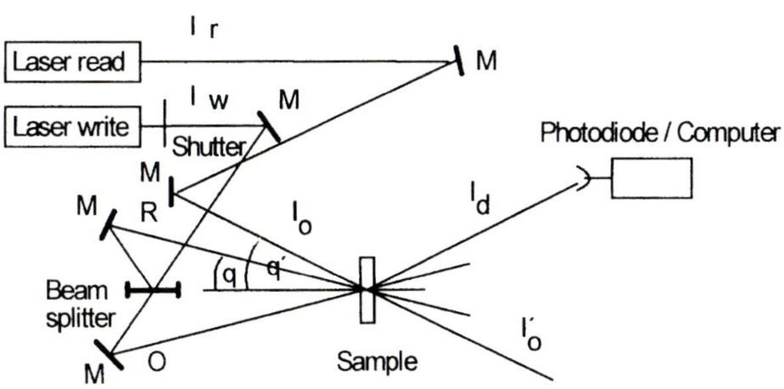

Fig. 3.10. Experimental set-up for the measurement of efficiency and sensitivity by holography (*M*: mirror, λ_w = 413 nm, 476 nm, λ_r = 633 nm) [52]

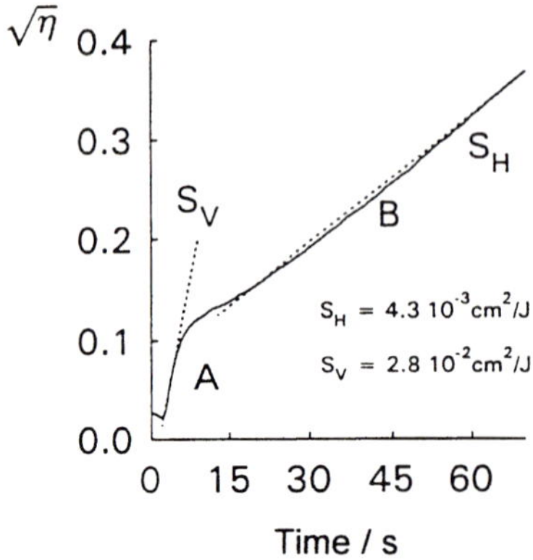

Fig. 3.11. Beginning of a hologram growth of a cholesteric cyclic siloxane. For linearizing $\sqrt{\eta}$ is plotted versus time [53]

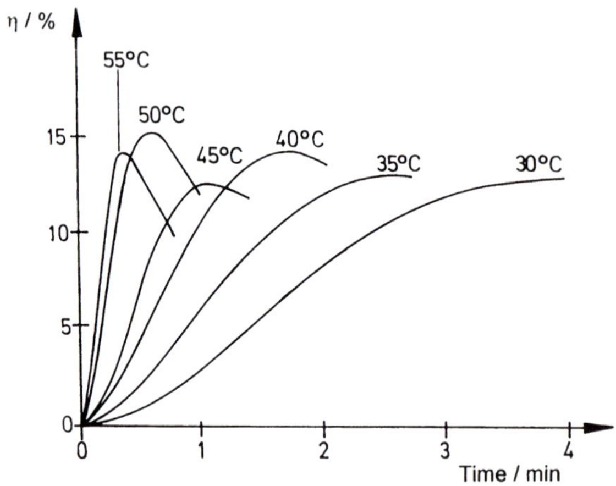

Fig. 3.12. Typical hologram growth curves at different temperatures. I_w = 0.9 W/cm^2, λ_w = 413 nm, θ=3.5°, sample thickness d=25 μm [52]

A typical hologram growth curve for CLC siloxanes shows a very fast increase of the hologram at the beginning. It is concluded that this part of the reaction should be attributed to the *cis-trans* isomerization of the azo dye stuff itself followed by the slower rearrangement of the LC-matrix.

Two different sensitivities S_V and S_H can be measured as shown in Fig. 3.11. The sensitivity $S_V = 2.8 \times 10^{-2}[cm^2/J]$at the beginning of the photo reaction in part A is nearly one order of magnitude higher than in part B ($S_H = 4.3 \times 10^{-3}[cm^2/J]$).

As shown in Fig. 3.12 the hologram growth curve is temperature dependent. For higher temperatures, i.e. lower viscosities, the rearrangement of the CLC-siloxane matrix can be achieved with higher sensitivity. However, close to T_g the hologram is no longer stable and decreases after longer irradiation.

3.3.2.2.3
Writing with Polarized Light

The photochemically induced reorientation of the *cis-trans* isomerizing molecules can be achieved in two different ways, by polarized or unpolarized light. If the plane of polarization of the incident light and of the transition dipole moment of the photoisomerizing molecule are parallel, a process of rearrangements-the photoinduced rotational diffusion [54, 55] - starts.

A simple two-dimensional sketch illustrates this fundamental process:

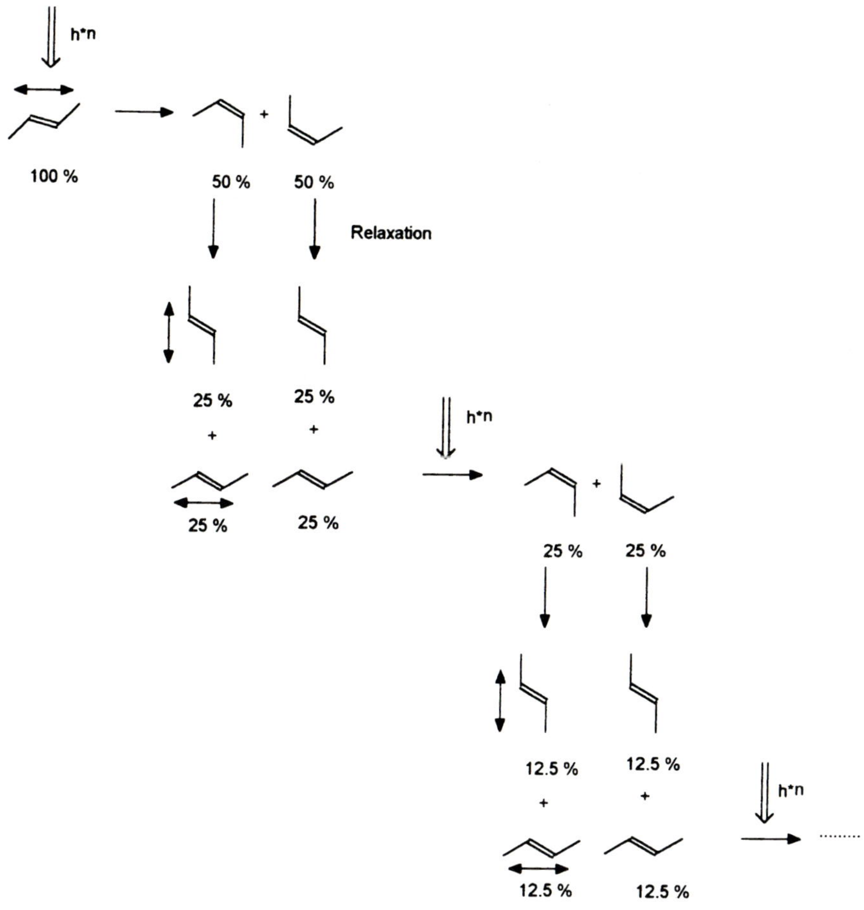

The relaxation processes work statistically so the final direction of the molecules is the direction with the lowest excitation probability. If the light propagates in the x-direction, the movement of the molecules can be described as summarized in Table 3.3.

The photoinduced reorientation yields two different results depending on whether linearly polarized (100) or unpolarized (110) light is used. In the case of linearly polarized light the x-position should also be populated but normally only an y/z-population is observed.

From the light-induced rotational diffusion, deformations of the cholesteric structure result are accompanied by interesting visible effects [56]. For the experiments described in this chapter cholesteric materials with a long reflection

Table 3.3. Possible orientations of mesogens induced by polarized irradiation. The light is propagating in the x-direction

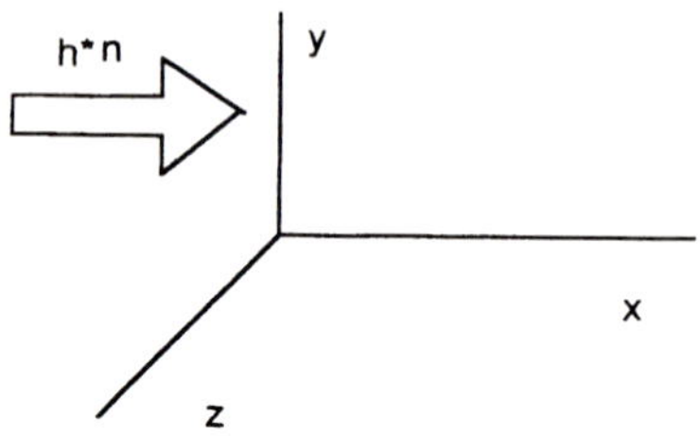

Direction of the E-vector			Resulting direction of the mesogenic groups		
X	Y	Z	X	Y	Z
0	0	1	(1)	1	0
0	1	0	(1)	0	1
0	1	1	1	0	0

wavelength were used. These materials were synthesized with dihydrocholesterol instead of cholesterol.

Figure 3.13 shows the effect of formation of new reflection bands by illumination with linearly polarized light. These new bands are typically higher order Bragg reflection bands. They can be used, e.g. for data storage in the visible, if the

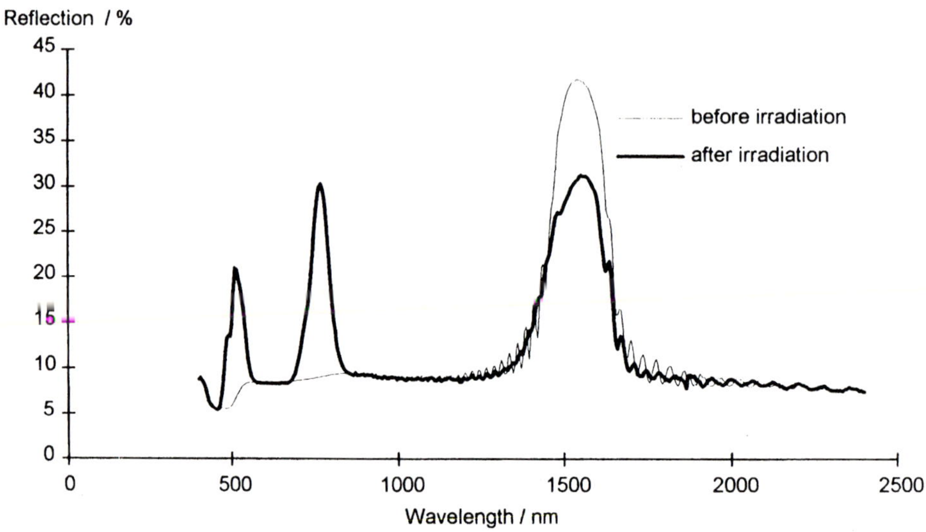

Fig. 3.13. Higher order Bragg reflection bands generated by irradiation of a cholesteric LC-siloxane layer with linearly polarized light (Hg-lamp, intensity 200 mW/cm^2, irradiation time: 5 min, transmission, crossed polarizers)

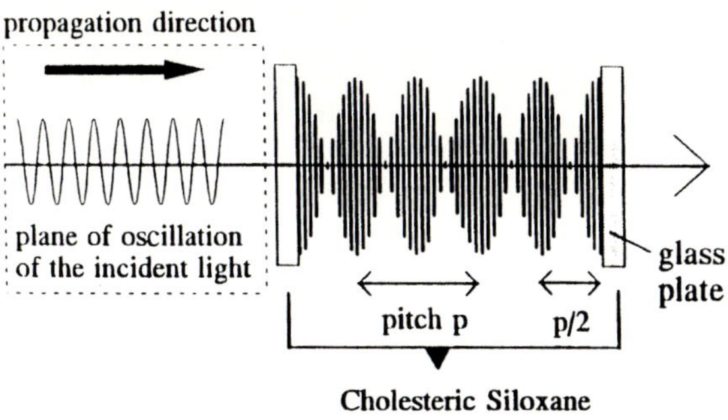

Fig. 3.14. The interaction of linearly polarized light with azo dyes in cholesteric liquid-crystalline siloxanes [56]

first order reflection band is in the near infrared. The stored pattern appears in the colors of the new reflection bands if the reflection mode is used for read-out.

The observation of new reflection bands is based on photoselection and can be explained with Fig. 3.14. This figure illustrates the situation for ChLC siloxanes. For linearly polarized light layers with strong absorption (long bars) and no absorption (short bars) alternate with the periodicity given by a spacing of half of the pitch p/2. Thus photoinduced reorientation will start to change the index of refraction with the periodicity p/2 which gives rise to a new reflection grating and higher orders of reflection.

Deformations of cholesteric structures were achieved also by magnetic fields [57–59] and calculated for different types of deformation (nonpolar, polar, aperiodic) [60]. The reported effects are in principle the same as with light but with the significant difference that magnetic or electric fields do not generate these effects below the glass transition temperature and cannot be used for the storage of pattern in cholesteric samples.

A more detailed investigation shows that the reflection grating, produced as shown in Fig. 3.14, polarizes the reflected light linearly in the new reflection

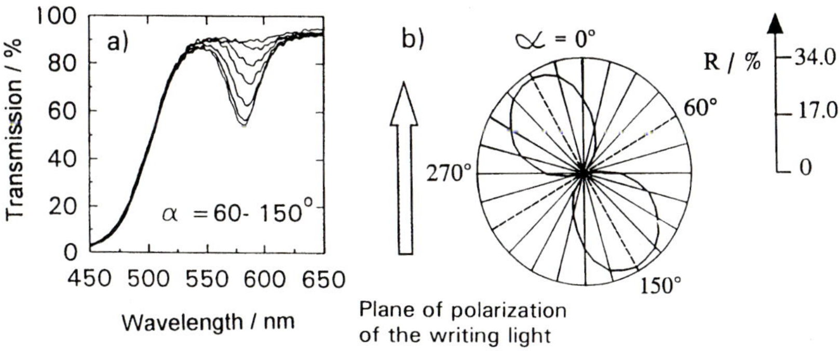

Fig. 3.15a,b. Angle dependence of the intensity of an illuminated sample of a CLC-siloxane: **a** transmission as a function of wavelength showing the 2nd order reflection at approx. 580 nm (t = 45 s, I = 250 mW×cm-2, λ_w = 413 nm, stepsize $\Delta\alpha$ = 15°) **b** polar diagram of the reflected intensity calculated from the transmission spectra

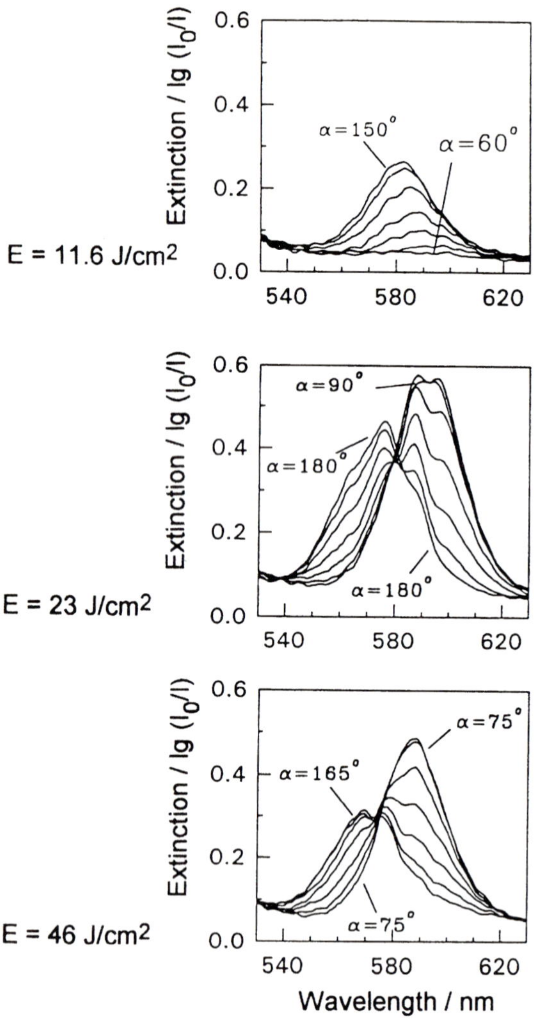

Fig. 3.16. Development of the spectral shape of the 2nd order reflection band after prolonged illumination with linearly polarized light

bands. This is demonstrated by absorption spectra with linearly polarized light. In Fig. 3.15 the 2nd order reflection band of a CLC siloxane was measured by rotating the plane of the linearly polarized light for each spectrum in steps of 15°.

From a practical point of view, the spots generated by irradiation with linearly polarized light therefore contain not only a response of reflection or no reflection, but also include information on the angle of the plane of the reflected and polarized light different from the writing light. This effect can be used to increase the storage density or the signal to noise ratio.

The spectral shape of the generated reflection band depends on the applied energy as is shown in Figs. 3.16 and 3.17. For higher energies a complicated polarization dependence develops in which both wings of the new reflection band seem to be perpendicular to each other. This behavior is not fully understood yet.

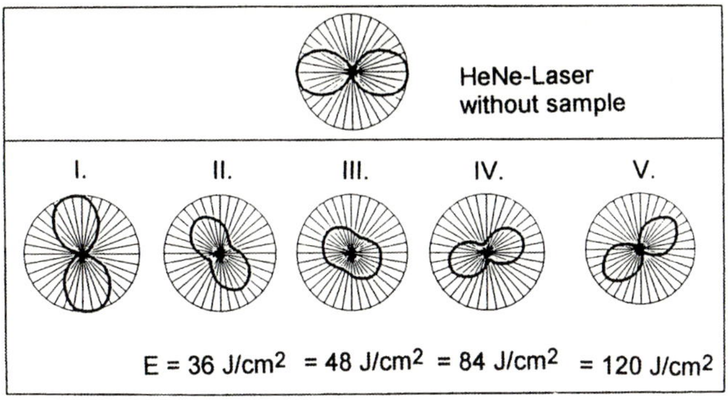

Fig. 3.17. Polarization of a HeNe-Laser without a sample and after irradiation of samples with different energies

The spots generated in a ChLC siloxane have a Gaussian profile resulting from the profile of the writing beam and thermal flux in the sample. From this, it is clear that the center of the spot represents an area containing a higher energy of irradiation than the outer parts with the result that the planes of polarization are nearly perpendicular and the wavelength of detection is different by about 15 nm. Figure 3.18 represents only a scheme and the intensities were calculated by a $\sin^2$-function valid for the absorption probability (Fig. 3.18).

3.3.2.2.4
Writing with Unpolarized Light

Table 3.3 results shows that writing with unpolarized or circularly polarized light (110) yields a homeotropic orientation [61, 62]. If unpolarized light is used for

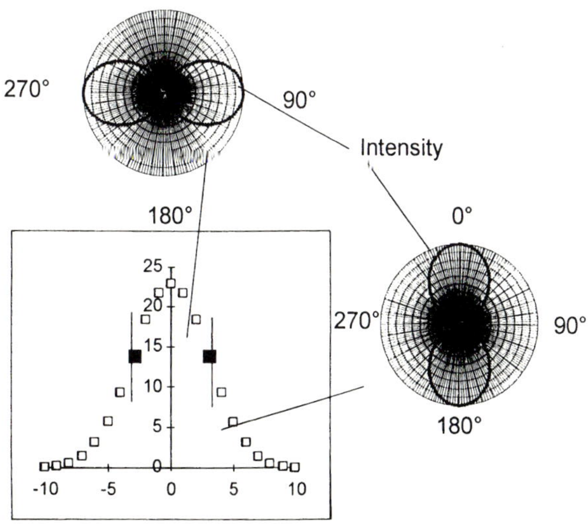

Fig. 3.18. Schematic illustration of a written dot in ChLC siloxane with the location of different planes of polarization at two WLR

Table 3.4. Materials used for experiments with unpolarized light

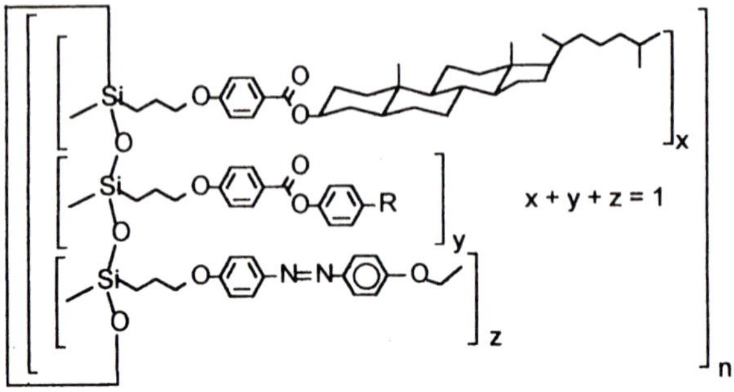

	R	x	y	z	n
I:	Phenyl	0.4	0.4	0.2	4
II:	CN	0	0.7	0.3	4

I:	g	54	n*	193	i			
II:	g	27	s_A	137	n	165	i	

irradiation, the mesogenic groups exhibit the lowest transition dipole moment when they are oriented parallel to the propagation direction of the incident beam.

The beginning of the reorientation perpendicular to the planar orientation is accompanied by a significant scattering effect. The illuminated parts look black in transmission. This scattering is reduced step by step and the illuminated parts become clear, transparent and pseudoisotropic.

The materials used for these experiments had the following structure (Table 3.4).

The LC siloxanes were sandwiched between two polyimide-coated glass plates. The thickness of the cells was about 10μm. To produce oriented samples, the cells were heated well above the glass transition temperature of the liquid crystal. Then the two glass plates were sheared. This results in well-oriented samples of I (cholesteric phase with the helix axis perpendicular to the cell surface) and II (planar uniaxial orientation, smectic layers perpendicular to the cell surface). Unoriented turbid samples were prepared by heating the cells to the isotropic state followed by subsequent rapid cooling without shearing.

The samples were irradiated with unpolarized light of a 200 W Hg-lamp (HBO 200 W). The spectral composition of the light was restricted from 350 nm to 900 nm (4 cm water cuvette as heat filter). The integral light intensity at the sample surface was 200 nW/cm^2, but only the part of the spectrum which is absorbed by the azo dye contributes to the reorientation. The irradiation took 5 hours to 15 hours.

The orientation of the samples before and after irradiation was determined by measuring the birefringence (Fig. 3.19).

Perpendicular irradiation of oriented or unoriented samples of I or II with unpolarized light results in a homeotropic orientation of the mesogenic groups and the azo dyes. The samples appear dark between crossed polarizers in-

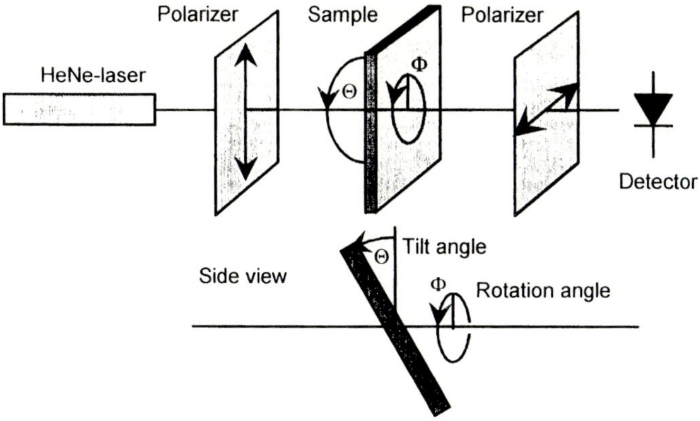

Fig. 3.19. Experimental set-up for the determination of the orientation of the samples. The sample is placed between two polarizers and roated around the measuring beam (angle Φ), while the tilt angle θ (angle between beam axis and cell-normal) is held constant

dependent of the rotating angle if the viewing direction is parallel to the irradiating beam, but appear bright, if they are tilted. In the cholesteric sample I the reflection band vanishes too. Oblique irradiation results in the orientation sketched in Fig 3.20, the azo dyes and the mesogenic groups are oriented parallel to the irradiating beam. (Figs. 3.21–3.23).

Irradiation of a turbid sample of I results in addition to the homeotropic orientation, in a drastic decrease in light scattering (from 20% before irradiation to 1% after irradiation), and an increase in transmission (from 65% before irradiation to 92% after irradiation, 99% if corrected for reflection losses at the glass surfaces).

The reorientation sketched in Fig. 3.20 is in agreement with experiments done with linearly polarized light. The dye molecules isomerize from *trans* to *cis* and from *cis* to *trans*. During this process they move statistically until they reach an

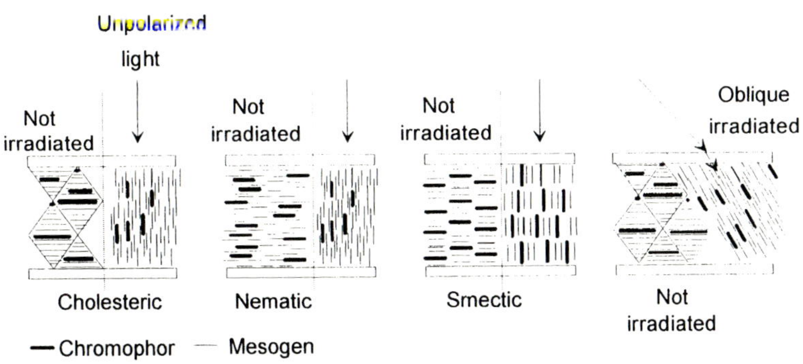

Fig. 3.20. Sketch of the reorientation by irradiation with unpolarized light

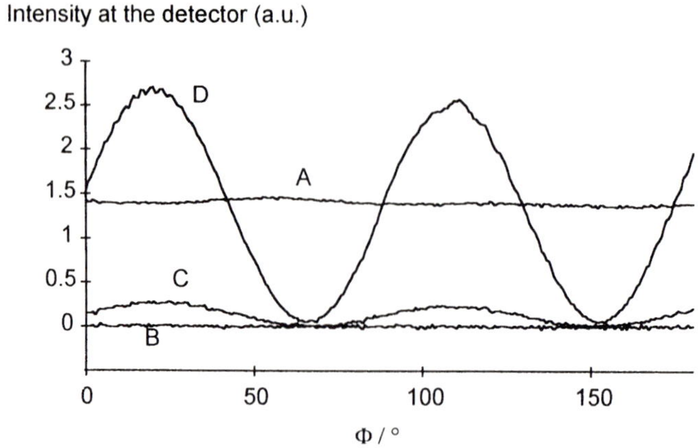

Fig. 3.21. Transmission during rotation (angle Φ) of an oriented sample of I between crossed polarizers before and after irradiation for different tilt angles θ (see set up of Fig. 3.19);
A: before irradiation $\theta = 0°$
B, C, D: after irradition $\theta = 0°$, 20° and 40° respectively

orientation where they cannot be excited: their transition moment M is now perpendicular to the polarization P of the exciting light (photo selection). For unpolarized light the only orientation with M perpendicular P is the one where M is parallel to the direction of the irradiating beam.

This reorientation takes place below the glass transition temperature of the liquid crystal. The light-induced changes are frozen in at this temperature.

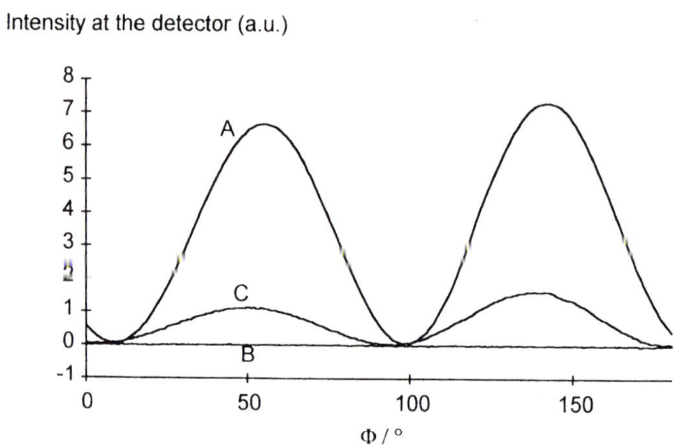

Fig. 3.22. Transmission between crossed polarizers (a.u.) of an oriented sample of II before and after irradiation for different tilt angles θ (see set up Fig. 3.19); *A:* before irradiation θ=0 °C *B,C:* after irradiation $\theta = 0$ °C and 20 °C, respectively
A: before irradiation $\theta = 0°$
B, C: after irradition $\theta = 0°$ and 20°, respectively

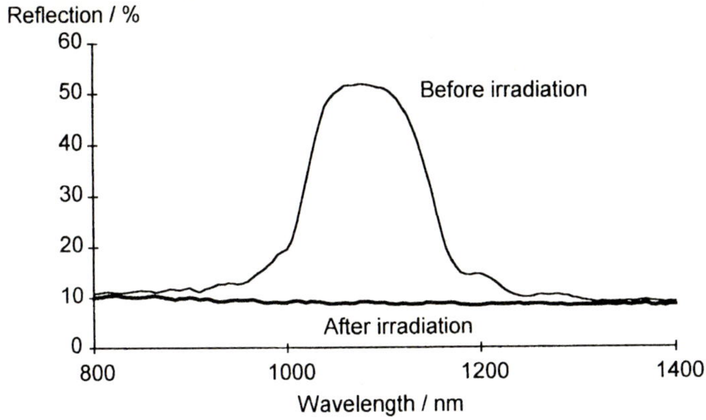

Fig. 3.23. Reflection (unpolarized light) of an oriented sample of I before and after irradiation

3.3.2.2.5
Erasing

All patterns stored in noncrosslinkable materials can be erased by heating the samples above the glass transition temperature (50 °C–60 °C). Up to this temperature the information is stable for years. The decay rate of a written information depends on the T_g (Fig. 3.24).

The heating process works in a hotstage as well as with laser light (Fig. 3.25).

Erasing with laser light may cause some peculiarities due to the small areas of the heated spots. Therefore the quality and uniformity of the erased areas are the better the larger the heated part of the sample.

3.3.2.2.6
Sensitivity

As demonstrated in Fig. 3.21 the materials show two types of sensitivities: The first one, attributed to the *cis-trans* isomerization does not yield stable patterns.

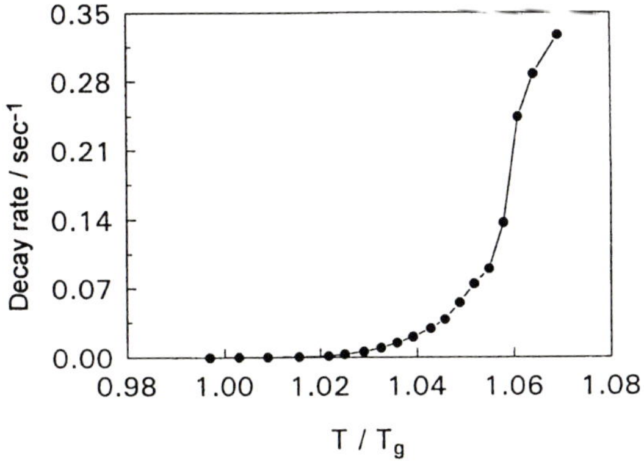

Fig. 3.24. Decay rate as a function of the relative temperature (T/T_g)

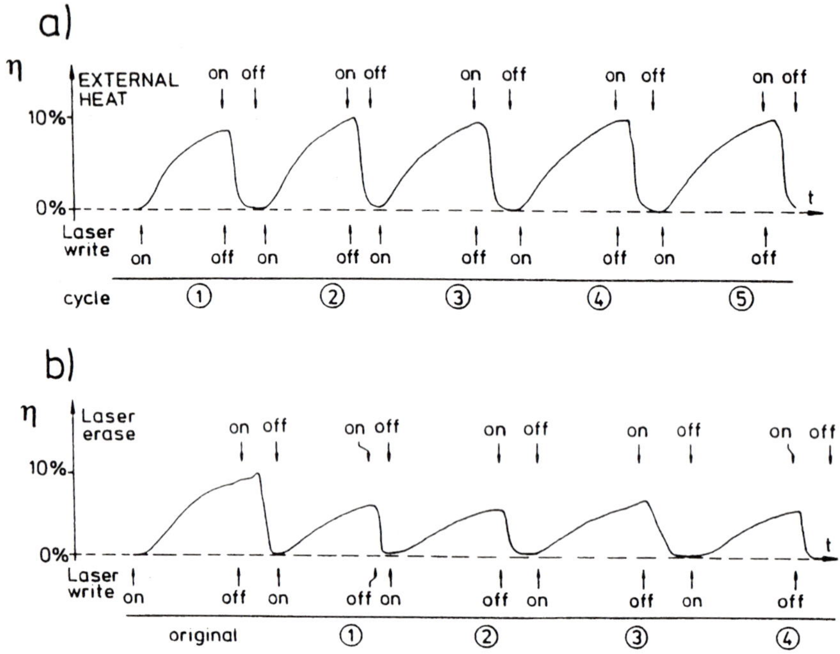

Fig. 3.25a,b. Writing and erasing cycles by heating in a hot stage (**a**) and by laser light (**b**)

The second one is accompanied with a rearrangement of the liquid-crystalline matrix. This process is relatively insensitive. To improve the sensitivity, the writing process can be done at elevated temperatures but below T_g (Fig. 3.26).

3.3.2.2.7
Resolution

Depending on the sample preparation resolutions in μm range are possible (Fig. 3.27).

Instead of the yellow azo-containing systems, stilbene derivatives yield coloured images [63].

3.3.2.3
Recording of SHG-Patterns

Recording of patterns exhibiting areas with the ability to generate second harmonic signals which are embedded in non active surroundings can be obtained by applying light and simultaneously an electric field to a ChLC siloxane sample containing an appropriate azo dye [64]. This poling effect works even below the glass transition temperature of the sample and offers us the opportunity of creating nonlinear optical wave guides.

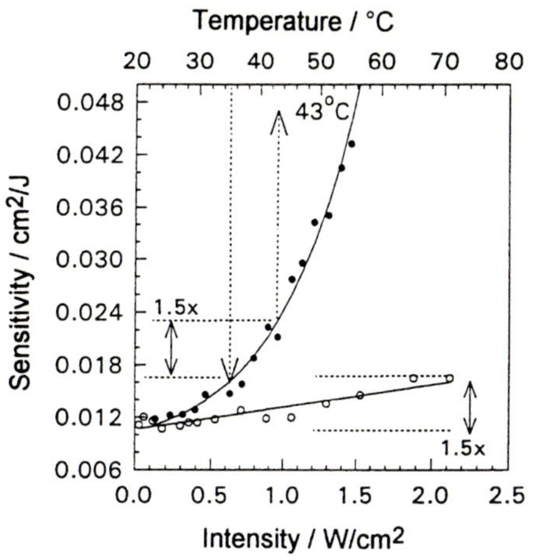

Fig. 3.26. Influence of temperature (•) and intensity at (35 °C) of the writing beam (o) on the sensitivity

Materials which can be patterned by this method contain azo dyes of the type:

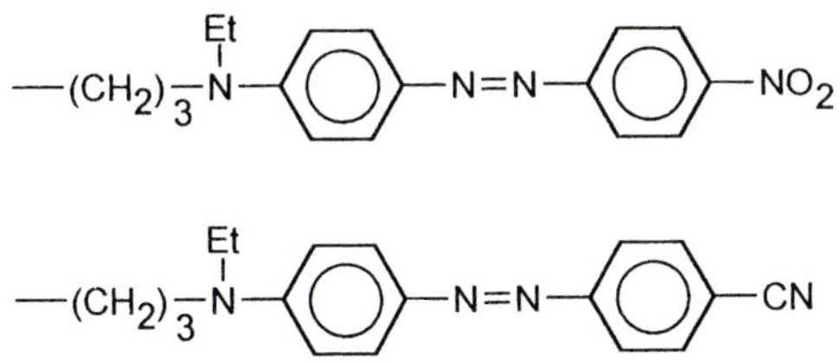

The patterning process converts the planar-oriented, cholesteric arrangement of the molecules by irradiation with circularly or nonpolarized light to a homeotropic one. The electric field induces a preferential orientation of the azodipoles. This results in a non centrosymmetric structure which is a pre-

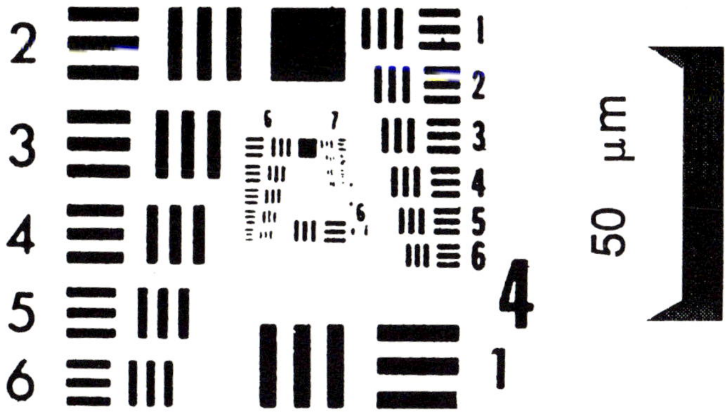

Fig. 3.27. Image of the USAF test pattern. Microphotograph between polarizers, written with 10 J/cm^{-2}

requisite for second harmonic generation. The applied electric fields are in the order of 22 MV/m, the light intensities in the order of 150 mW/cm^2 and the irradiation time 60 min–120 min.

3.4 References

1. Feringa BL, Jager WF, de Lange B (1993) Tetrahedron 49: 8267
2. Nakamura T, Ueno T, Tani C (1989) Mol Cryst Liq Cryst 169: 167
3. Shibaev VP, Kostromin SG, Platé NA, Ivanov SA, Vetrov VYu, Yakovliev IA (1983) Polymer Commun 24: 364
4. Platé NA, Talroze RV, Shibaev VP (1984) Pure Appl Chem 56: 403
5. Eich M, Wendorff JH, Reck B, Ringsdorf H (1987) Makromol Chem., Rapid Commun 8: 59
6. Eich M, Wendorff JH (1987) Makromol Chem., Rapid Commun 8: 467
7. Cabrera I, Krongauz V, Ringsdorf H (1987) Angew Chem 99: 1204
8. Cabrera I, Dittrich A, Ringsdorf H (1991) Angew Chem 103: 106
9. Natarajan LV, Bunning TJ, Tondiglia V, Patnaik S, Pachter R, Crane RL, Adams WW (1992) Proc of the Sec Int Symp on Chem of funct Dyes Kobe, Kinki Chem Soc 2: 405
10. Natarajan LV, Tondiglia V, Bunning TJ, Crane RL, Adams WW (1992) Adv Mat Opt Elect 1: 293
11. Natarajan LV, Bunning TJ, Klei EH, Crane RL, Adams WW (1991) Macromolecules 24: 6554
12. Ikeda T, Suzuki Y, Seki T, Hosoki A, Aoki K (1988) Langmuir 4: 1214
13. Sasaki T, Ichimura K (1994) J Amer Chem Soc 116: 625
14. Shannon PJ, Armstrong World Industries, Inc (1984) USP 4637896
15. Tsutsui T, Tanaka R (1980) Polymer 21: 1351
16. Shannon PJ, Gibbons WM, Sun ST (1994) Nature 368: 532
17. Kreuzer F-H, Gawhary M, Winkler R, Finkelmann H, Consortium für elektrochemische Industrie GmbH (1981) EP 60 355 (1982) Chem Abstr 98: 10013u
18. Riepl G, Kreuzer F-H, Miller A, Consortium für elektrochemische Industrie GmbH (1989) EP 333 022 A2, Chem Abstr 112: 67360s
19. Spes P, Hessling M, Kreuzer F-H, Consortium für elektrochemische Industrie GmbH (1991) DE 394 0148 A1, Chem Abstr 115: 123530y (1991)
20. Richards RDC, Hawthorne WD, Hill JS, White MS, Lacey D, Semlyen JA, Gray GW, Kendrick TC (1990) J Chem Soc Chem Commun 95
21. Clarson SJ, Semlyen JA, Dodgson K (1985) Polymer 26: 930
22. Clarson SJ, Semlyen JA, Dodgson K (1990) Am Chem Soc Polym Prepr 31: 563
23. Kreuzer F-H, Maurer R, Spes P (1991) Makromol Chem Makromol Symp 50: 215
24. Bunning TJ, Klei HE, Samulski ET, Crane RL, Linville RJ (1991) Liq Cryst 10: 445
25. Pachter R, Bunning TJ, Adams WW (1991) Comp Polym Sci 1: 170
26. Bunning TJ, McNamee SG, Klei HE, Samulski ET, Ober CK, Adams WW (1992) Amer Chem Soc Polym Prepr 33: 315
27. Pachter R, Bunning TJ, Socci EP, Farmer BL, Crane RL, Adams WW (1992) Amer Chem Soc Polym Prepr 33: 671
28. Socci EP, Farmer BL, Bunning TJ, Pachter R, Adams WW (1993) Liq Cryst 13: 811
29. Pachter R, Bunning TJ, Crane RL, Adams WW (1993) Makromol Chem Theory Simul 2: 337
30. Bunning TJ, Vezie DL, Lloyds PF, Haaland PD, Thomas EL, Adams WW (1994) Liq Cryst 16: 769
31. Gresham KD, McHugh CM, Bunning TJ, Crane RL, Klei HE, Samulski ET (1994) J Pol Sci Part A, 32: 2039
32. Noll W. Chemistry and Technology of Silicones, Academic, New york (1968)
33. Speier JL, Webster JA, Barnes GH (1957) J Amer Chem Soc 79: 974
34. Ojima I, The hyrosilylation reaction in: The Chemistry of Organic Silicon Compounds, Patai EdS, Rappoport Z, 1479, (John Wiley & Sons 1989)
35. Ryan JW, Menzie GK, Speier JL (1960) J Amer Chem Soc 82: 3601

36. Kreuzer F-H, Andrejewski D, Haas W, Häberle N, Riepl G, Spes P (1991) Mol Cryst Liq Cryst 199: 345
37. Yu JM, Teyssié D, Boileau S (1992) Polym Bull 28: 435
38. Krishnamurthy S, Chen SH (1989) Makromol Chem 190: 1407
39. Andrejewski D, Gohary M, Luckas HJ, Winkler R, Kreuzer F-H (1990) Consortium für elektrochemische Industrie GmbH, EP 358,208 A2, Chem Abstr 113: 68893a (1990)
40. Kreuzer F-H, Maurer R, Stohrer J (1992) Abstr 14th Int Liq Cryst Conf Pisa 2: 936
41. Weitzel P, Kreuzer F-H, Maurer M, Consortium für elektrochemische Industrie (1992) DOS 4234845 A1
42. Andrejewski D, Gohary M, Luckas HJ, Winkler R, Kreuzer F-H, Consortium für elektrochemische Industrie GmbH (1990) EP 358,208A2, Chem Abstr 113: 68893a (1990)
43. Pinsl J, Bräuchle Ch, Kreuzer F-H (1987) J Mol Electronics 3: 9
44. Wagner PJ (1976) Top Curr Chem 66: 1
45. Bräuchle Ch, Burland DM, Bjorklund GC (1981) J Phys Chem 85: 125
46. Bell AE (1984) SPIE Proc 382: 2
47. Eich M, Wendorff JH, Reck B, Ringsdorf H (1987) Makromol Chem Rapid Commun 8: 59
48. Eich M, Wendorff JH (1987) Makromol Chem Rapid Commun 8: 467
49. Anderle K, Birenheide R, Werner MAJ, Wendorff JH (1991) Liq Cryst 9: 691
50. Stumpe J, Müller L, Kreysig D, Hauck G, Koswig HD, Rhmann R, Rübner J (1991) Makromol Chem Rapid Commun 12: 81
51. lvanov S, Yakovlev l, Kostromin S, Shibaev V (1991) Makromol Chem., Rapid Commun 12: 709
52. Ortler R, Bräuchle Ch, Miller A, Riepl G (1989) Makromol Chem., Rapid Commun 10: 189
53. Petri A, Thesis, Ludwig-Maximilians-Universität München, München 1993
54. Michl J, Thulstrup EW (1986) Spectroscopy with polarized light, Verlag Chemie, Weinheim, p 213
55. Anderle K, Birenheide R, Eich M, Wendorff JH (1989) Makromol Chem., Rapid Commun 10: 477
56. Petri A, Bräuchle Ch, Leigeber H, Miller A, Weitzel H-P, Kreuzer F-H (1993) Liq Cryst 15: 113
57. Frank FC (1958) Discuss Faraday Soc 25: 1
58. De Gennes PG (1968) Solid State Commun 6: 163
59. Mayer RB (1968) Appl Phys Lett 14: 208
60. Dreher R(1973) Solid State Commun 12: 519
61. Miller A, Kreuzer F-H, Leigeber H (1992) DOS DE 42 06 089 A 1
62. Miller A, Leigeber H, Weitzel H-P, Petri A, Bräuchle Ch (1992) Abstracts 21st Freiburger Arbeitstagung Flüssigkristalle
63. Boldt P, Leupold J, Weitzel HP, Leigeber H (1994) GPat Appl P4409207
64. Anneser H, Feiner F, Petri A, Bräuchle Ch, Leigeber H, Weitzel H-P, Kreuzer F-H, Haak O, Boldt P (1993) Adv Mater 5: 556

CHAPTER 4

Photoregulation of Liquid Crystal Alignment by Photochromic Molecules and Polymeric Thin Films

K. Ichimura

4.1 Introduction

Alignment of liquid crystal molecules is a key technology for applying the mesophases to display devices. Liquid crystal molecules are uniformly oriented by the action of substrate surfaces in display cells which are driven by the application of electric field to switch the director of liquid crystals. The alignment is essentially governed by the nature of substrate surfaces [1]. The modes of the alignment are roughly classified into two categories; homeotropic and planar states. In the former case, the long molecular axis of liquid crystals is arranged perpendicular to the substrate surface whereas liquid crystal molecules are oriented parallel to the surface in the latter. Homogeneous alignment refers to the unidirectional orientation of the molecular axis in the planar mode and displays birefringence with excellent optical quality. The conventional liquid crystal display devices work by switching of the alignment between homeotropic and homogeneous modes by the action of an applied electric field.

A great number of procedures for liquid crystal alignment have been demonstrated, and the alignment mechanism has also been extensively discussed [1]. The most familiar way to achieve homeotropic alignment involves the treatment of substrate surfaces with amphiphilic compounds or long alkyl silylating reagents. A representative method to give homogeneous alignment is based on a unidirectional rubbing treatment of a polymer thin film including a polyimide covering substrate surfaces. These phenomena suggest that the liquid crystal alignment is determined by the orientational nature of the substrate surface and therefore indicate that the alignment is capable of being regulated by external stimulation other than electric fields if the surface structure is responsive to the external stimuli.

There have been a lot of reports stating that mesophase changes are caused by the photoreaction of certain compounds which are similar to liquid crystals in shape and size and dissolved in mesophases [2]. The mesophase changes are induced by photoreacted molecules which act as impurities to destroy the mesophase because of the drastic change in the molecular structure. For example, when a cholesteric liquid crystal is contaminated with azobenzene, reversible modification of the cholesteric pitch is brought about by alternate illumination with UV and visible light for the photoisomerizations [3]. Adiabatic phase changes between nematic and isotropic phases [4, 5] as well as between smectic and nematic phases [6] are triggered by the photoisomerization of azobenzenes.

It should be stressed that the photocontrol of liquid crystal alignment mentioned in this chapter is quite different from this kind of photoinduced mesophase change [6]. What takes place here is the alteration of the orientational direction of low-mass liquid crystals without any mesophase modification. In order to realize reversible alignment alteration in a photochemical way, photochromic molecules was employed because they give rise to reversible alignment alterations induced by the interchange of the molecular structure upon alternate excitation with actinic light of two different wavelength regions. This type of photoresponsive liquid crystal cells was reported for the first time in 1988 [7]. In one of the typical cases, the photoinduced alignment modification is caused by a surface covered with photochromic units with a surface density of one unit per 1 nm^2. This situation led one to call the surface a "command surface" because each photochromic unit results in the reversible alignment alteration of about 10^4 liquid crystal molecules showing a remarkable molecular amplification.

There are two types of command surfaces for photochemical liquid crystal switching. The first one is prepared by chemical modifications of substrate surfaces with photochromic molecules. Surface-selective modification can be achieved by using silica plates or polymer thin films. The second one consists of coating or adsorption of polymers substituted with photochromic units preferentially at the side chains on substrate surfaces.

The surface-assisted alignment photoregulation driven by the command surfaces is classified into two groups, depending on the mode of the director change. The first involves the alignment alteration between the homeotropic state and planar or homogeneous state (Fig. 4.1). This type of liquid crystal switching resembles that in the conventional electrically driven liquid crystal cells. The second one is achieved by the action of linearly polarized light with surface photochromic molecules which bring about azimuthal reorientation of liquid crystal molecules in a planar state (Fig. 4.2). This type of molecular reorientation is quite unique since it cannot be realized by the application of an electric field.

This chapter deals with recent research trends of the newborn photoresponsive liquid crystal systems driven by the action of photoactive molecular and polymeric films covering substrate surfaces.

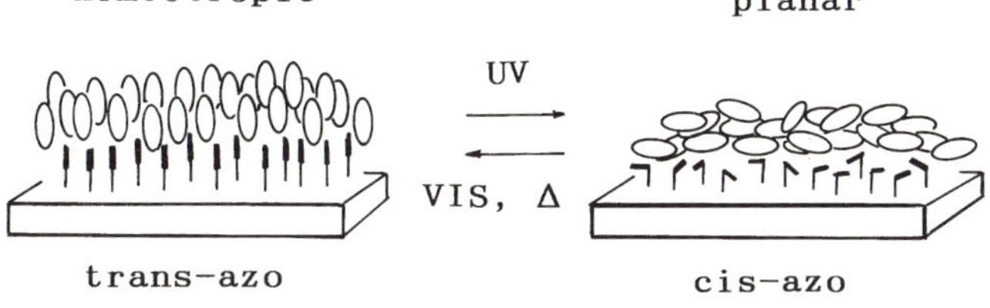

Fig. 4.1. Schematic representation of surface-assisted alignment regulation between homeotropic and planar modes triggered by surface photochromic units

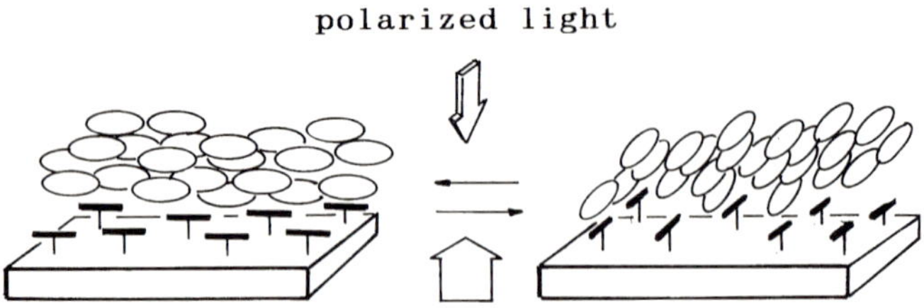

Fig. 4.2. Illustrative representation of photo-induced in-plane alignment regulation

4.2
Alignment Regulation by Surface-Modified Silica

4.2.1
Surface Modification of Silica and Cell Fabrication

Silylation is of very practical significance for chemical modification of silica surfaces with organic molecular layers through covalent bonding(s) [8]. Surface structures of silica treated with silylating reagents have been studies extensively in connection with composite polymeric materials reinforced by glass fibers [9]. The structures and configurations of organic residues attached to silica surface are markedly influenced by the structure of silylating residues. Although single molecular layers are formed by monoalkoxy silylating reagents, it has been mentioned that trialkoxysilyl reagents afford a polysiloxane network after hydrolysis to form complexed structure. On the other hand, a sophisticated method was developed by Sagiv to cover the inorganic surfaces with the use of long-chain alkylsilyltrichloride by chemisorption which brings about the formation of closely packed monomolecular layers resembling a Langmuir-Blodgett (LB) membrane [10].

Modification of silica surfaces has been frequently carried out with use of photochromic derivatives bearing alkoxysilyl residues to prepare the command surfaces [7, 11]. Although the structural elucidation of the surfaces modified with photochromic units is not yet fully understood, this technique provides a convenient procedure for making the surface photoactive. A general procedure for preparing photochromic silylating reagents I consists of the synthesis of a photochromic carboxylic acid and the subsequent condensation with a 3-aminopropyl derivative of mono-, di- or triethoxysilane (Fig. 4.3). Because of Si–O bond formation between the alkoxylsilylating reagents and silica surface takes place under harsher reaction conditions than silylation with the use of alkyltrichlorosilane, heat treatment is required to fix the photochromic units covalently. A typical experimental procedure involves dipping a silica plate in 1 wt % ethanol solution of the corresponding silylating reagent, followed by heating the plate at 120 °C for 20 min. The plate is subsequently subjected to cleaning by means of

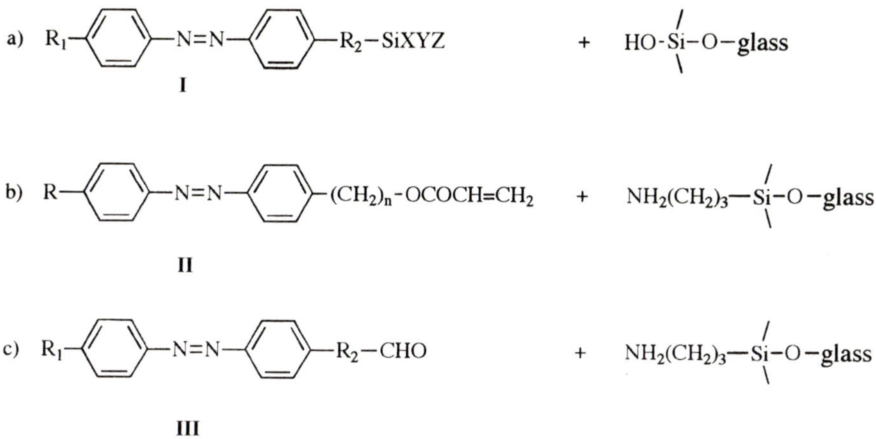

Fig. 4.3. Methods of introduction of azo-chromophores to a silica surface

ultrasonication in a good solvent such as dichloromethane. An alternative route to command surfaces involves 3-aminopropylation of the silica surface, followed by attaching photochromic units through amido bond formation [11], Michael addition with acryloylated photochromic derivatives II [12] and Schiff base formation with photochromic aldehydes III [13, 14].

The amount of photochromic residues attached to the silica is estimated spectroscopically under the assumption that the absorption coefficient in a solution is not markedly altered by the chemisorption. It should be noted that this convenient method is rather qualitative because photochromic moieties like azobenzenes sometimes form aggregations and stand perpendicular to surfaces more or less. A more adequate way of estimating the surface concentration of photochromic units is based on detaching the units by alkaline treatment to measure the amount of the chromophore in a solution phase spectroscopically [11]. The value estimated by the latter is not far from those obtained by the former method. Generally, trialkoxysily derivatives give the largest number of photochromic residues per unit area among silylating reagents bearing mono, di- and trialkoxysilyl residues [11]. The average number of photochromic residues can be readily controlled by modifying the silica surface in the presence of an alkyl silylating reagent as a two-dimensional diluent [11].

The most conventional way of making photoresponsive liquid crystal cells is to put a nematic liquid crystal between two surface-modified substrates. A hybrid cell is occasionally fabricated with a photochromic plate and a substrate plate which brings about homogeneous or homeotropic alignment caused by conventional surface treatment technique as exemplified above. Photoinduced alignment alteration is carried out conveniently by monitoring linearly polarized light such as an He-Ne laser beam passed through a liquid crystal cell during the course of photoirradiation for photochromism at the substrate surface.

4.2.2
Factors Affecting Alignment Regulation

4.2.2.1
Structure of Azobenzene Derivatives

Because of their good availability and acceptable photofatigue resistance, azobenzene derivatives have been most extensively employed for surface photofunctionlization. One of the characteristic features of this chromophore involves the rodshaped molecular structure which can act in fact as a mesogen. On *trans-cis* photoisomerization, the chromophore is subjected to a drastic molecular

Table 4.1. Structural effect of azobenzene units on photoalignment regulation

Chemical structure of $R_1-C_6H_4-N{=}N-C_6H_4-R_2$ — glass		Alignment[a]	Photoregulation[b] Liquid crystals	
R_1	R_2		DON-103	RO-571
H	$O(CH_2)_5CONH(CH_2)_3Si(O-)_3-$	p	–	
Cl	$O(CH_2)_5CONH(CH_2)_3Si(O-)_3-$	p	–	
CH_3O	$O(CH_2)_5CONH(CH_2)_3Si(O-)_3-$	p	–	+
$CH_3(CH_2)_5O$	$O(CH_2)_5CONH(CH_2)_3Si(O-)_3-$	p	–	
NC	$O(CH_2)_5CONH(CH_2)_3Si(O-)_3-$	p	–	–
NCPhOCO	$O(CH_2)_5CONH(CH_2)_3Si(O-)_3-$	p	–	–
$CH_3(CH_2)_5$	$OCH_2CONH(CH_2)_3Si(O-)_3-$	h	+	–
$CH_3(CH_2)_5$	$O(CH_2)_2CONH(CH_2)_3Si(O-)_3-$	h	+	
$CH_3(CH_2)_5$	$O(CH_2)_3CONH(CH_2)_3Si(O-)_3-$	h	+	
$CH_3(CH_2)_5$	$O(CH_2)_5CONH(CH_2)_3Si(O-)_3-$	h	+	–
$CH_3(CH_2)_5$	$O(CH_2)_5CONH(CH_2)_3Si(CH_3)(O-)_2-$	p	–	+
$CH_3(CH_2)_5$	$O(CH_2)_5CONH(CH_2)_3Si(CH_3)_2O-$	p	–	–
$CH_3(CH_2)_5$	$O(CH_2)_{10}CONH(CH_2)_3Si(CH_3)_2O-$	h	+	
$CH_3(CH_2)_7$	$O(CH_2)_5CONH(CH_2)_3Si(O-)_3-$	h	+	
cyclohexyl	$O(CH_2)_5CONH(CH_2)_3Si(O-)_3-$	h	+	
H	$OCO(CH_2)_2NH(CH_2)_3Si(O-)_3-$	p	–	
CH_3	$OCO(CH_2)_2NH(CH_2)_3Si(O-)_3-$	p	–	
CH_3O	$OCO(CH_2)_2NH(CH_2)_3Si(O-)_3-$	h	+	
$CH_3(CH_2)_5$	$OCO(CH_2)_2NH(CH_2)_3Si(O-)_3-$	h	+	
$CH_3(CH_2)_5$	$O(CH_2)_2O-C_6H_4-CH{=}N(CH_2)_3Si(O-)_3-$	h	+	
$CH_3(CH_2)_5$	$O(CH_2)_5O-C_6H_4-CH{=}N(CH_2)_3Si(O-)_3-$	h	+	
$CH_3(CH_2)_5$	$O(CH_2)_{10}O-C_6H_4-CH{=}N(CH_2)_3Si(O-)_3-$	h	+	

[a] p: planar, h: homeotropic. [b] +: positive photoregulation, –: no photoregulation

DON-103 (I)

RO-571 (II)

EXP-CIL (III)

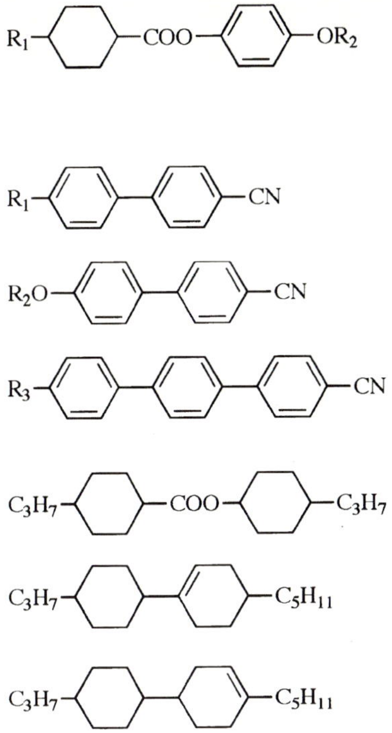

Fig. 4.4. Chemical structures of some nematic liquid crystals used for photoregulation

shape alteration between the rodlike and V-shaped structure, accompanied by the marked polarity change[1]. The photoinduced alignment regulation depends markedly on the substituents on the azobenzene chromophore having the general formula I.

As compiled in Table 4.1, a substituent at the *para*-position (R_1) plays an important role in photoalignment behavior [11]. The substituent effect on the photoalignment is also dramatically dependent on the nature of liquid crystals. In the case of a cyclohexanecarboxylate-type nematic liquid crystal (DON-103) (Fig. 4.4), hydrophilic substituents including chloro, cyano and methoxy groups are not appropriate for the present purpose since the homeotropic alignment should be caused by the *trans*-isomer of the surface azobenzenes before photoirradiation to achieve the alignment photocontrol. Azobenzene molecules without any substituent are not photoregulative. These groups induce a planar mode and consequently no alignment alteration after UV irradiation for the *trans*- to *cis*-photoisomerization. Hydrophobic substituents like hexyl, octyl and cyclohexyl residues give rise to homeotropic alignment and results successfully in the alignment conversion into the planar mode upon UV illumination. In contrast, a cyanobiphenyl-type liquid crystal (RO-571) demonstrates quite dif-

[1] *tran-cis* isomerization azobenzene is also considered in Chapter 2 of this book.

ferent behavior [15], as seen in Table 4.1. Since this type of liquid crystal is apt to align homeotropically, it is inevitable to choose a suitable azobenzene surface-modifier and/or an optimal mixing ratio of the azobenzene and a photochemically inert silylating reagent [15]. These indicate that the molecular interactions between liquid crystal molecules and surface azobenzene units play a crucial role in the photochemical alignment regulation.

Contradictory results were obtained for the methoxy substituent. When the azobenzene unit with the hydrophilic residue is introduced by Michael addition of the corresponding acryloyl derivative II ($R = OCH_3$, $n = 0$) to an aminated glass surface, the surface gives a homeotropic alignment which can be converted into the planar mode by UV irradiation [12]. This suggests that the spacer effect should be also taken into account on the alignment. In fact, as seen in Table 4.1, a longer spacer such as a decamethylene unit seems to be favourable for photoregulation; the photoregulativity of the modified silica surface is dependent on the spacer length when an azobenzene unit is introduced with use of a derivative bearing a monoethoxysilyl group. Furthermore, as discussed later, even an azobenzene without a substituent becomes photoregulative for the liquid crystal alignment if the chromophore is introduced into the side chain of poly(vinyl alcohol) (PVA) through a longer spacer.

Simultaneous modification of the silica surface with photochemically inert alkyl silylating reagents occasionally affects the photoregulativity [16]. A typical example is the effect of dodecylsilylating reagent. Although a surface modified with nonsubstituted azobenzene

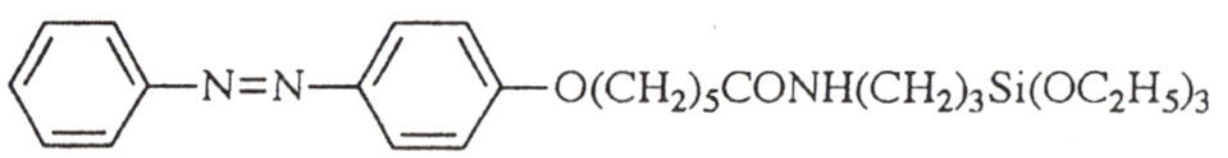

(IV)

IV fails to regulate the liquid alignment, as stated above, co-modification with dodecyltriethoxysilane makes the alignment homeotropic and the silica plates photoactive for the alignment alteration. An appropriate choice of the molar ratio of the azo-silylating reagent and the dodecylsilylating reagent is a crucial factor for achieving photoregulation. If the shorter octyl residue is employed for the simultaneous surface modification, the alignment remains planar and is not altered by UV irradiation in every molar ratio. Furthermore, octadecyltriethoxysilane as a co-modifier results in a homeotropic alignment which, however is not converted into planar mode.

Attempts have been made to regulate a cholesteric mesophase of cholesterol derivatives and a smectic mesophase by means of a command surface incorporating *p*-hexylazobenzene units. No alignment alteration is observed in either case. The failure to reorient the cholesteric mesophase is possibly because the size and shape of the cholesterol derivatives are quite different from those of the azobenzene and so cause no effective side-by-side molecular interaction on the substrate surface which is necessary condition for the alignment control.

In smectic phases, the molecular movement of liquid crystals is suppressed so that the reorientation transfer from molecule to molecule is highly inhibited and there is no photoinduced alignment modification. This fact seems to be rather useful from a practical viewpoint. Although cells filled with nematic liquid crystals have a potential value to photorecording systems, one of the problems is to stabilize photoimages written in the alignment change because the thermal reversion of the *trans*-isomer erases the planar alignment on storage spontaneously. Thus, there is a proposal to obtain persistent photoimages by freezing them in a smectic mesophase with a rigid ordered structure according to the following procedure [15]. A cell fabricated with glass plates modified with a mixture of *p*-hexylazobenzene silylating reagent and ethyltriethoxysilane is filled with K-24 which exhibits a smectic mesophase between 21.5 °C and 33.5 °C ($T_{S \to N}$) and a nematic mesophase between 33.5 °C and 40.5 °C ($T_{N \to I}$) and warmed above $T_{N \to I}$ and subsequently cooled down to display a homeotropic alignment. No alignment transformation is induced upon UV irradiation when the cell is kept below $T_{S \to N}$. On the other hand, UV exposure at 37 °C above $T_{S \to N}$ to produce a nematic phase and subsequent cooling down below $T_{S \to N}$ bring about stable parallel alignment even though the *cis*-isomer on the surface is converted into the *trans*-isomer. This memory effect of the smectic mesophase affords the principle for permanent storage of erasable photoimages. Images by UV irradiation can form latent photoimages which can be developed by warming up above $T_{S \to N}$ and fixed by rapid cooling down.

4.2.2.2
Photochromic Units Other Than Azobenzene Derivatives

With regard to elucidating the mechanism and to extending the concept of the photoalignment, exploration of photochromic units capable of regulating the liquid crystal alignment is an important subject. As suggested by the illustration shown in Fig. 4.1, the drastic molecular shape-change based on the geometrical isomerization may play an essential role in the alignment regulation. Stilbene photoisomerizes quite similarly in the shape and size to that of azobenzene. In fact, a stilbene substituted with octyloxy residue (5) at the *p*-position induces the liquid crystal alignment alteration upon the reversible photoisomerization [13].

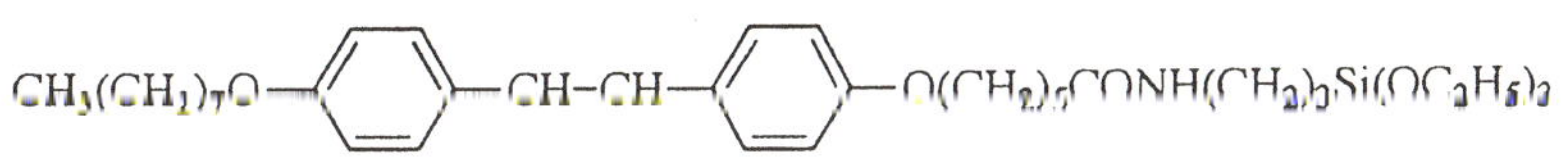

(V)

Because the chromophores possess absorption bands at shorter wavelength region, alternate irradiation was made by light of 313 nm for the *trans*-to-*cis* isomerization and by light of 254 nm for the reverse isomerization.

Other representative photoactive residues displaying geometrical isomerization like *p*-dimethylcinnamylidene **VI** and retinylideneamino groups **VII** are introduced onto an aminated silica surface through Schiff's base formation of the corresponding aldehyde [13].

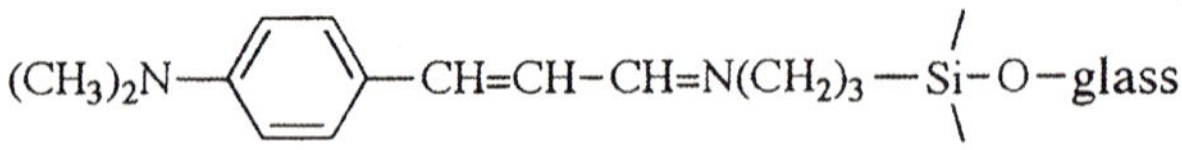
(VI)

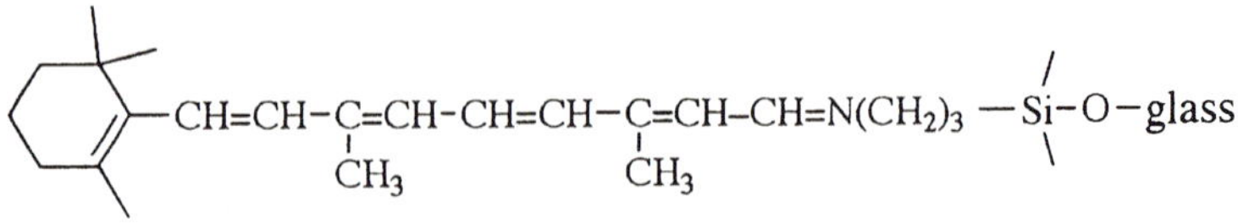
(VII)

These are all inactive for the photoregulation, however, since initial states before photoirradiation afforded a planar mode. A surface modified with 5-nitrospiropyran **VIII** does not allow us to convert the planar mode into a homeotropic one upon UV irradiation for the same reason.

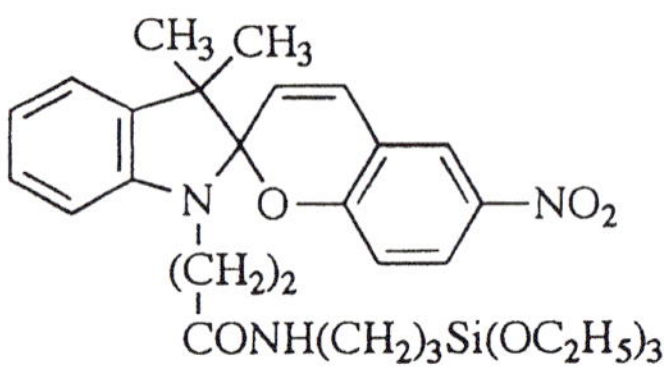
(VIII)

It is worthy of mention that the possibility of these photochromic units inducing an alignment alteration cannot be entirely excluded. As revealed above, although a non-substituted azobenzene moiety is not able to control the liquid crystal reorientation, suitable molecular design and appropriate choice of a co-modifier may make the chromophore photoregulative. An attempt has been made to regulate liquid crystal alignment by a molecular layer of a novel photochromic compound, hemithioindigo **IX** which displays extremely high photo-fatigue resistance in a diluted solution.

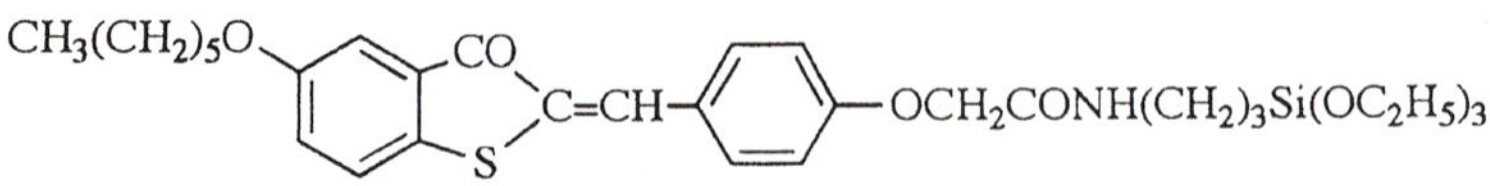
(IX)

Although a surface modified with the chromophore bearing hexyl residue resulted in homeotropic alignment, UV irradiation does not bring about the planar mode [13]. In this case, photoirradiation leads to a monotonous disappearance of the chromophore without isomerization possibly due to intermolecular photoreaction. Photoregulativity of chalcones X is also affected by substituent.

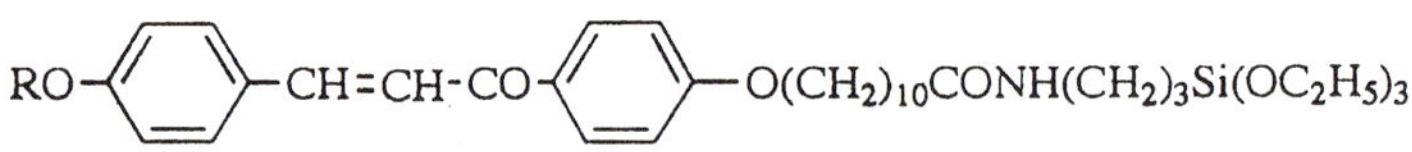
(X)

These results are compiled in Table 4.2.

The photochromism of α-hydrazono-β-ketoesters **XI** which have hardly been investigated, involves the geometrical photoisomerization around a C=N bond, accompanied by switching of a hydrogen bonding upon alternate irradiation with UV and visible light (Fig. 4.5) [17]. Quite similarly to the "azobenzene case", reverse isomerization is also induced thermally. This somewhat complicated structural change brought about by alternate photoirradiation resembles that of

Table 4.2. Structural effect of some photochromic units on photoalignment regulation

Photochromic compound	Alignment[a]	Photoregulation[b]
V	h	+
VI	p	–
VII	h	–
VIII	p	–
IX	h	–
Xa	p	–
Xb	h	+

[a]p: planar, h: homeotropic. [b]+: positive photoregulation, -: no photoregulation.

azobenzenes. Substituting the phenyl ring with a hexyl or methoxy group, the hydrazonoketoester units are introduced onto a silica surface by silylation or Michael addition of acrylate derivatives to an aminated silica (Table 4.3) [18].

The surfaces modified by these methods gives homeotropic alignment except where covered with a methoxy-substituted derivative attached by the Michael addition. Upon UV irradiation, liquid crystal molecules alter their alignment to form the planar mode. In contrast to the surface azobenzenes, the photochemical reverse process caused by visible light irradiation does not result in the recovery of the alignment. Interestingly, the planar mode is stable toward photoirradiation and converted into the homeotropic alignment only by heating the cell. Thus, this type of the liquid crystal cell is driven through a combination of UV irradiation and heat treatment. It should be pointed out that the spectral alteration of

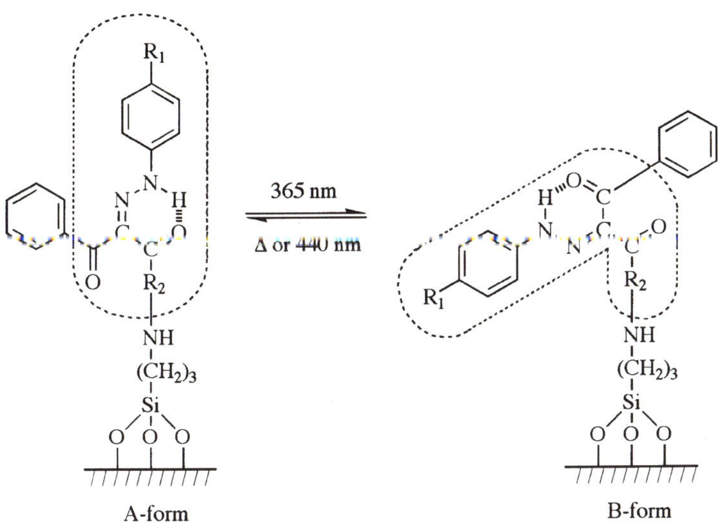

Fig. 4.5. Photochromic reaction of α-hydrazono-β-ketoesters **XI**

Table 4.3. Alignment modes of liquid crystal cells made up with silica glass plates surface-modified with α -hydrazono-β-ketoesters XI (see formula XI in Fig. 4.5)

Structures of XI		Alignment modes[a]			
R_1	R_2	Exposure to			
		Initial	UV	VIS	Heating[b]
C_6H_{13}	$NH(CH_2)_3Si(O-)_3$ – glass	h	h	h	–
C_6H_{13}	$OCH_2CONH(CH_2)_3Si(O-)_3$ – glass	h	p	p	h
C_6H_{13}	$O(CH_2)_5CONH(CH_2)_3Si(O-)_3$ – glass	h	p	p	h
CH_3O	$OCH_2CONH(CH_2)_3Si(O-)_3$ – glass	h	p	p	h
C_6H_{13}	$O(CH_2)_2OCO(CH_2)_2NH(CH_2)_3Si(O-)_3$ – glass	h	p	h	h
C_6H_{13}	$O(CH_2)_6OCO(CH_2)_2NH(CH_2)_3Si(O-)_3$ – glass	h	p	h	h
C_6H_{13}	$O(CH_2)_{11}OCO(CH_2)_2NH(CH_2)_3Si(O-)_3$ – glass	h	p	h	h
CH_3O	$O(CH_2)_2OCO(CH_2)_6NH(CH_2)_3Si(O-)_3$ – glass	r	r	r	–

[a]h: homeotropic, p: planar, r: random. [b]alignment after heating at 65–70 °C.

the photochromic hydrazonoketoesters attached to a silica surface is quite minute with respect to both shifts in the absorption maximum and absorbance. However, such a negligibly small spectral change is dramatically amplified with liquid crystal alignment alteration to help make photoimages visible to the naked eye.

4.2.2.3
Surface Photochromic Reaction and Alignment Alteration

Detailed studies have been made on the relationship between photochromic reaction of azobenzene derivatives on substrate surfaces and the photoinduced alignment alteration. The photoreaction takes place without any disturbance due to the surface binding. An exceptional example was observed for the α-hydrazono-β-ketoester unit [18]. As mentioned above, no photoisomeric form reverses to the original one upon visible light irradiation when the unit is attached to a silica surface although the reverse isomerization takes place readily by visible light irradiation for solution photochromism.

Since the *cis*-isomer of the azobenzene chromophore takes up a larger area per molecule in a closely packed LB membrane than the *trans*-isomer does, no marked photoisomerization from *trans* to the *cis* takes place when azobenzene units are incorporated in the middle of long-alkyl chain of an aliphatic carboxylic acid [19]. The absorption spectrum of the chromophore is drastically modified because of the formation of the J-aggregate. Various types of aggregation have been observed for bilayer membranes of amphiphilic molecules incorporating azobenzenes [20]. Azobenzene derivatives attached relatively densely on a silica surface through silylation demonstrate partial formation of aggregation which disappears when the average density of the chromophores is reduced by co-modification with an inert silylating reagent like ethyltriethoxysilane [11].

The efficiency of the liquid crystal alignment is affected considerably by the average density of the chromophore [11]. There is a critical value for photoregulation; a minimum average density of *p*-hexylazobenzene unit for the alignment regulation is about 1.0 nm^2, and no photoresponce is observed when the average density exceeds 1.2 nm^2. Within the critical density value, the higher the average density of the chromophore, the longer the induction period becomes for the alignment change from the homeotropic to the planar mode so that the sensitivity is reduced. In the relationship between the average density and the induction period upon UV irradiation, the latter is abruptly enhanced at about 0.5 nm^2 average density. This value corresponds to double the minimum occupied area of *trans*-azobenzene amphiphilic molecules of 0.25 nm^2 obtained by the LB technique [21]. This situation suggests that when the effective average density exceeds about 0.5 nm^2 the *trans*-azobenzene units may interact readily with the liquid crystal molecules which can stick to the azobenzene molecular layer because of the reduction of space filling of the surface azobenzene. In fact, Hiltrop and Stegemeyer demonstrated that a loosely packed LB membrane of lecithin molecules transferred onto a substrate plate effectively causes a homeotropic alignment whereas a densely packed LB membrane affords a rather distorted alignment [22]. This phenomenon is interpreted in terms of specific interaction between the amphiphilic molecules and liquid crystal molecules which stick to the loosely packed LB membrane.

Rearrangement of *trans*-azobenzene chromophores on the substrate surface upon contact with a liquid crystal layer is observed spectroscopically [23]. It is quite convenient that the azobenzene chromophore bears a typical dichroic character. The π-π^* transition moment along with the longer molecular axis corresponds to the absorption at the longer wavelength of around 350 nm while an absorption band at a shorter wavelength of about 250 nm is assignable to the transition moment perpendicular to the molecular axis of azobenzene. Thus, a ratio of absorbance at 250 nm to that at 350 nm affords a measure of the orientation of the chromophore; the larger the ratio, the more the chromophore units stand perpendicular to the surface. An unequivocal example was observed for a *p*-hexylazobenzene attached to a silica surface by a decamethylene spacer. The electronic absorption spectrum of the monolayered azobenzene resembles that of the corresponding azobenzene in solution, indicating that the direction of the chromophore units is quite random. When the surface is wetted with a UV-transparent, nematic liquid crystal (EXP-CIL), the absorption spectrum is markedly changed so that the absorbance ratio of the azobenzene is much reduced, showing that the azobenzene moiety sprouted from the surface is reoriented perpendicularly to the surface to give a homeotropic alignment.

The photoisomerization of the surface azobenzene units brings about a considerable molecular shape change which causes the reorientation of liquid crystal molecules surrounding the photoisomerizable groups. As a result, liquid crystal molecules at a boundary region align parallel to the surface, and these reoriented molecules trigger the reorientational relaxation of bulk liquid crystal molecules to afford a planar alignment. There has been some argument as to whether the liquid crystal alignment mode is correlated with the surface energy of the substrates [1]. On the one hand, it has been clearly demonstrated that the surface energy is determined by the nature of monolayered molecules covering solid surfaces. According to the Friedel-Creagh-Kmetz rule which claims that the nematic liquid crystal alignment is determined by the surface energies of the substrate (γ s) and the surface tension of the liquid crystal (γ LC); the alignment is homeotropic for γ s $<$ γ LC and parallel for γ s $>$ γ LC [1]. In fact, surface energies of the azo-modified plates are in some cases altered on photoisomerization of the surface azobenzenes [23]. However, the photoinduced surface energy change is not a prerequisite for the alignment regulation since some azo-modified surfaces bring about no distinct surface energy modification even though the same surface can regulate the alignment photochemically [24]. This implies that the specific interaction owing to the matching of molecular structure of the photochromic unit and the liquid crystal molecule is an essential requirement for photoregulation.

4.2.2.4
Properties of Photoresponsive Liquid Crystal Cells

The photoisomerization takes place very quickly in the picosecond range. Possibly the surface photoisomerization is completed on the same time scale. The subsequent process to relax the liquid crystal alignment has been measured by excitation of a cell with a pulsed UV laser beam for *trans* to *cis* photoisomerization, followed by monitoring the transmitted intensity of a linearly polarized He-Ne laser beam as a probe light [25]. The values of the relaxation time of a nematic liquid crystal residue lies within the range between 50 msec and 300 msec, depending on the cell conditions including cell thickness, the nature of the liquid crystal as well as the average density of surface azobenzene. It should be stressed that this response time is nearly equal to a decay time of the conventional electro-optical nematic cell. This is very reasonable since the reorientation of liquid crystals after turning off an electric field is determined solely by the physical properties of liquid crystals and the nature of the aligning layer covering the cell substrates.

Some evaluations have been made on the photoresponsive cell from a practical point of view [11]. The photoresponse is affected by cell thickness. The change in transmittance of probing light through the cell as a function of irradiation period obeys Eq. (4.1) where polarized light of λ nm wavelength with intensity I_0 is passed through a layer of thickness d of a mesophase layer having birefringence Δn and a crossed polarizer, giving transmitted light intensity I.

$$I = I_0 \sin^2(2\beta)\sin^2(\pi d \Delta n(\alpha)/\lambda) \tag{4.1}$$

Eq. (4.1) was originally derived to elucidate the relationship of tranmittance of a probing light with applied voltage of electrode-driven liquid crystal cells and is modified to another equation by replacing Δn at an applied voltage V with $\Delta n(\alpha)$ at tilt angle α. Here, β stands for an angle contained by the polarization axis of probe light and a homogeneous axis. Since β is constant in this case, the transmittance during the alignment alteration is dependent solely on the change in the tilt angle α. The experimental photoresponse curves are in line with those of theoretical ones unless the thickness of the mesophase layer exceeds about 20 μm [11]. A thicker cell of 100 μm shows no distinct photoresponse.

Although cells fabricated with azo-modified plates respond reversibly upon alternate irradiation with UV and visible light for the *trans* to *cis* and for the reverse isomerization, the cells become inactive for upon photoexcitation after about 2000 irradiation cycles. In general, azobenzenes are lost predominantly by the hydrogen abstraction of the azo group absorbing UV light to form hydrazobenzenes with a quantum yield of about 10^{-4} [26]. The photofatigue of the cells may arise from the degradation of azobenzene upon UV irradiation.

Fine photoimages can be written-in using the photoresponsive cell because the photosensitive molecular layers are firmly bound on substrate surfaces. Two μm lines and spaces are resolved. This situation enables one to make a holographic grating according to the following procedure [6]. Whole areas of a cell surface modified with hexylazobenzene is first exposed to UV light to yield a planar alignment. Subsequently, the cell is illuminated with two crossed beams of an Ar gas laser emitting 488 nm light to record a holographic grating. Although diffraction efficiency is very low (<0.1%), transient formation of a grating is observed.

4.3 Alignment Regulation by Polymer Thin Films

4.3.1 Surface-Modified Polymer Thin Films

Analogous to surface modification of silica plates, a thin film poly(vinyl alcohol) (PVA) spin-coated on a glass plate is subjected to surface-selective introduction of azo-chromophores. Surface-selective modification is performed by esterification with a *p*-hexylazobenzene acid chloride **XII** or acetalization with *p*-hexylazobenzene aldehydes **XIII** in benzene which is a poor solvent for PVA (Fig. 4.3).

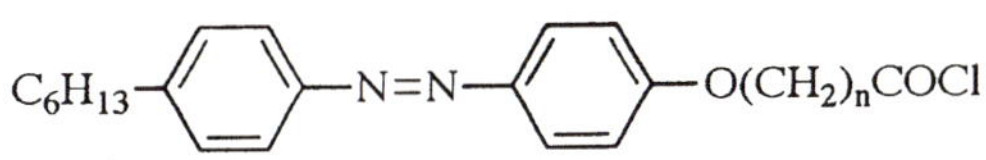

(**XII**)

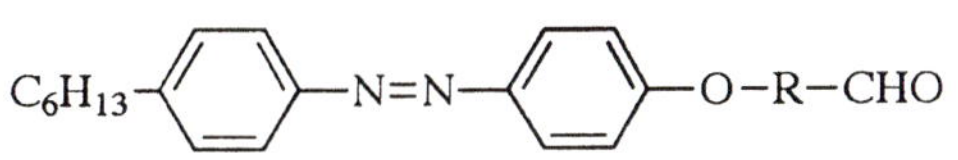

(**XIII**)

The advantage of this procedure is to enable one to achieve the so-called rubbing treatment of the surface-modified films for homogeneous alignment [14]. Although the *trans-cis* photoisomerization of azobenzene units on silica surface brings about the alignment change from a homeotropic into a planar mode, as illustrated in Fig. 4.1, the direction of the alignment is not uniaxial to yield a multi-domain texture which reduces the optical quality of a cell considerably unless the surfaces are subjected to the rubbing treatment. A cell is made by sandwiching a nematic liquid crystal between two plates covered with modified PVA thin film rubbed in the same direction. On UV irradiation, the alignment is converted from a homeotropic into a homogeneous alignment with good optical quality. A photoresponsive twisted nematic (TN) cell can be assembled by setting the rubbing direction of a couple of the substrate plates in a crossed position. This type of technique has potential value since a photoresponsive liquid crystal cell is readily available simply by coating the azo-modified PVA film on any kind of transparent substrates including flexible plastic sheets. Furthermore, guest-host cells are assembled by dissolving dichroic dyes in liquid crystals to visualize photoimages through a single polarizer sheet.

4.3.2 Spin-Coated Polymer Thin Films

An alternative way of preparing command surfaces consists of spin-coating a polymer bearing azobenzene substituents in the side chain [27, 28]. Two kinds of azobenzene polymers have been prepared. The preparation of the first involves esterification of PVA with an azobenzene acid chloride or a carboxylic acid with the aid of picryl chloride as a condensation reagent to give PVA derivatives substituted with azobenzenes (mAznPVA: **XIV**) in which m and n are the carbon numbers of a tail alkyl group and that of a polymethylene spacer, respectively [27].

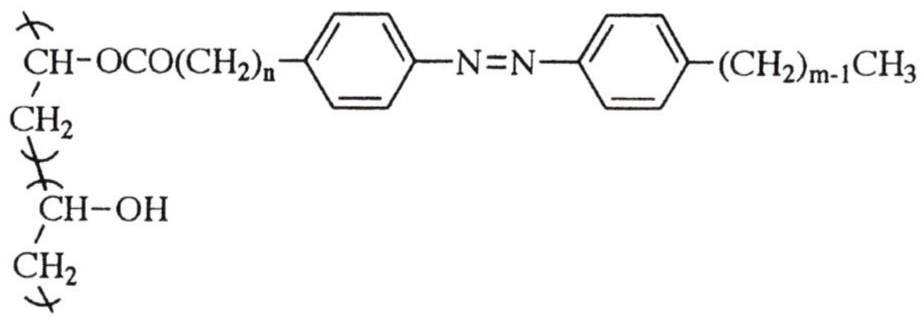

(XIV)

The second type of polymers are synthesized by radical polymerization of acrylated or methacrylated azobenzenes **XVI** [28].

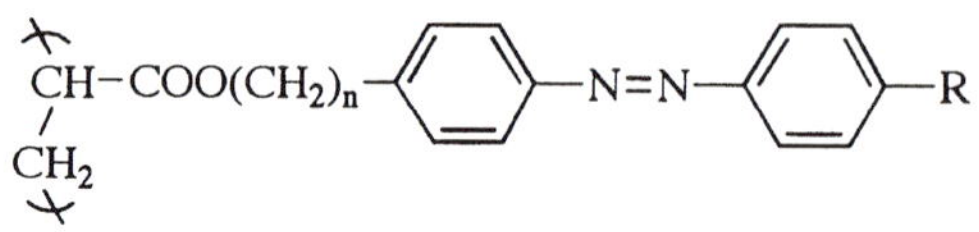

(XV)

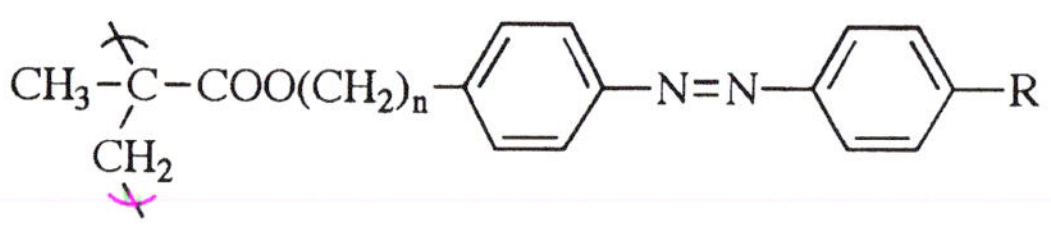

(XVI)

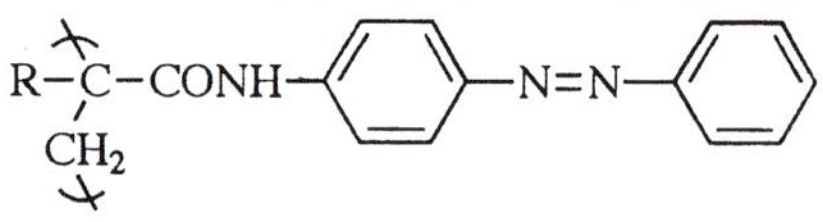

(XVII)

Acrylamide-type polymers **XVII** are also prepared (Table 4.4). While the procedure to cover a substrate plate with these azobenzene polymers is simple and convenient, the relationship between the polymer structure and the photoregulative function is very complicated, as summarized in Table 4.5.

It is noteworthy that the azobenzene chromophores attached in side chains of vinyl polymers sometimes display a blue shift in their electronic absorption band

Table 4.4. Chemical structure and abbreviations of azobenzene polymers

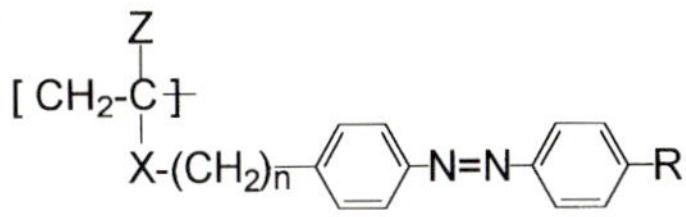

Abbreviations	Structure of polymer			
	Z	X	R	n
HAzPAE	H	COO–	H	0
CIAzPAE	H	COO–	CH_3-	0
C6AzPAE	H	COO–	$CH_3(CH_2)_5-$	0
C10AzPAE	H	COO–	CH_3O-	0
HAzPME	CH_3-	COO–	H	0
C1AzPME	CH_3-	COO–	CH_3-	0
C6AzPME	CH_3-	COO–	$CH_3(CH_2)_5-$	0
C10AzPME	CH_3-	COO–	CH_3O-	0
C6AzC11PME	CH_3-	COO–	$CH_3(CH_2)_5-$	11
HAzPAA	H	CONH–	H	0
HAzPMA	CH_3-	CONH–	H	0

Table 4.5. Photocontrol of liquid crystal alignment by spin-coated films of azobenzene-pendent polymers

Polymer	Aggregation of Azobenzene groups		Alignment of liquid crystals							
	before baking	after baking	DON-103				RO-571			
			Spin-coating		Adsorbed		Spin-coating		Adsorbed	
			Init[a].	Contr[b].	Init.	Contr.	Init.	Contr.	Init.	Contr.
HAzPAE			p	No			p	No		
C1AzPAE			p	No	h	No	p	No	h	No
C6AzPAE			p	No	h	No	p	No	h	No
C10AzPAE			p	No			p	No		
HAzPME	–	+	p	No	h	Yes	p	No	h	No
C1AzPME	–	+	p	No	h	Yes	h	Yes	h	No
C6AzPME	–	++	p	No	h	Yes	h	Yes	h	No
C10AzPME	–	–	p	No			p	No		
C5AzC11PME	++	++	h	Yes						
HAzPAA	–	–	p	No	h	No	p	No	h	No
HAzPMA	–	–	p	No	h	Yes	p	No	h	Yes
1Az6PVA	–	++	h	Yes	h	No				

[a]initial alignment; [b]controllability of alignment; –: no aggregation; +: slight aggregation; ++: prominent aggregation; p: planar alignment; h: homeotropic alignment; Yes: photoregulative; No: no photoregulation.

due to the π, π^*-transition because of aggregations. For example, heat treatment of a thin film of 6Az1PVA (XIV: m = 6, n = 1) results in a spectral change due to the aggregation of the azo-chromophore which reverses again to the monomeric form after one cycle of alternate UV and visible light irradiation for the reversible *trans-cis* isomerization. Similar spectral characteristics have been observed more or less for polyacrylates **XV** as well as polymethacrylates **XVI** possessing azobenzenes in their side chain. As revealed in the preceding sections, liquid crystal alignment photocontrol is achieved by the action of monolayered photochromic thin films. Therefore, coating of thick polymeric films is no necessary condition, and the thickness can be considerably reduced for the photocontrol. For this purpose, adsorption of azo polymers on substrate silica surfaces is useful procedure for photoalignment control [27, 28].

There exists a marked difference in photoresponse pattern of alignment alteration between the PVA-based azo-polymers and the acrylate- or methacrylate-based azo-polymers. A thin film of a PVA substituted with *p*-hexylazobenzene affords a rapid switching of alignment between homeotropic and planar modes whereas poly(methacrylates) bearing azobenzenes in their side chains shows a relatively long induction period before alignment modification from a homeotropic to a planar mode upon UV irradiation. As discussed in the previous section [11], the photoinduced alignment alteration is occurs after a certain degree of the photoisomerization from the *trans-* into the *cis*-form of the azobenzene units which are bound covalently on the silica surface as a molecular layer so that there is a induction period before the photoalignment change. In the case of poly(acrylates) **XV** or poly (methacylates) XVI, the photoisomerization rate of the azo chromophore in the polymers is much slower than the surface azobenzenes because of the restricted free volumes in the polymer matrices below their T_g.

The photoinduced alignment alteration described so far takes place reversibly between bistable states. An extraordinary alignment change can be observed when a thin film of a polymethacrylate substituted with *p*-hexylazobenzene in the side chain through a long spacer (C6AzC11PME, XVI, R = hexyl, n = 11) is used as a command surface [29, 30]. Before irradiation, a nematic liquid crystal (DON-103) shows a marbled texture. UV irradiation brings about the alignment change to form a transient homeotropic mode at an early stage of irradiation, and finally the texture is transformed into the schlieren. This alignment transformation between the marbled and schlieren textures through the homeotropic state takes place reversibly upon alternate illumination with UV and visible light for the azobenzene photoisomerization. The transient homeotropic alignment is stable when the film is not subjected to further irradiation. This phenomenon has been interpreted in terms of morphological modification of the polymer thin film. The methacrylate homopolymer exhibits a blue-shifted π, π^*-transition band owing to the aggregation of the *trans*-chromophore. As mentioned above, a lot of azo polymers display the aggregation of the chromophore in solid films by heat treatment, and the photoisomerization from *trans* to *cis* results in the monomerization of the chromophore which forms no aggregation any longer upon photoinduced reversion into the *trans*-isomer unless the polymers are subjected to annealing. In marked contrast to this type of azopolymers, the aggregation of the azo chromophore of C6AzC11PME is readily formed even at room temperature so that visible light irradiation for the *cis*- to-*trans*- photoisomerization leads at first to the monomeric *trans*-form which is converted into the aggregation upon further illumination.

4.3.3
Polymeric Langmuir-Blodgett (LB) Membranes

Since the command surfaces work in principle at a monomolecular level, the Langmuir-Blodgett (LB) technique yields significant information concerning the nature of photochromic molecular films [23]. So far, a lot of studies have been done on photoresponsive properties of photochromic amphiphiles spread on water/air interfaces (Langmuir membranes; L-membranes) as well as membranes transferred onto substrate (LB membranes) [21, 31, 32]. Because photochromic reactions are accompanied by a considerable change in the molecular structures, L-membranes display marked alteration of surface pressures, reflecting a change in the occupying areas of photoactive moieties. Referring to LB membranes embedding photochromic moieties showing geometrical photoisomerization such as thioindigoes and azobenzenes, photoconversion of *trans*-isomers into the *cis* counterparts is severely suppressed owing to the necessity of expansion of occupying areas of the chromophores while the reverse isomerization from the *cis*- to the *trans*-isomer takes place with great ease since the molecular membranes require the reduction of the occupying areas [19, 33]. This situation does not meet the requirement to afford command surfaces if azobenzene amphiphiles, for example, are transferred onto substrate plates in their *trans*-form since the isomerization is inhibited upon UV irradiation as far as LB membrane deposition is made under applying a moderate surface pressure. Thus one of the necessary conditions for LB membranes as command surfaces to demonstrate reversible photoisomerizations is to ensure free volumes in the membranes.

Polymeric amphiphiles substituted with pendent photochromic units are good candidates to give photoactive LB membranes for command surfaces because occupying areas are determined not by photochromic units, but by polymeric backbone chains which possess larger occupying areas. In fact, reversible photoisomerization has been observed for azobenzene units which are bound to a polyelectrolyte through a coulombic interaction [34]. To obtain command surfaces derived from azo-LB membranes, a series of PVA substituted with azobenzene in the side chain ((see formula XIV for) mAznPVA) have been prepared [23, 35]. Owing to the partial esterification of the polymer with azobenzene derivatives, the polymers have an amphiphilic character based on hydrophilic vinyl alcohol units and hydrophobic azobenzene units substituted with a long-chain alkyl.

Deposited membranes of mAznPVA are photoregulative for liquid crystal alignment. Employment of an appropriate surface pressure during the deposition of a molecular film enables one to control the average density of the azo chromophores on surfaces. A minimum density of the photochromic units for the liquid crystal photoregulation is around one unit per 1 nm^2 [23] and quite in line with the value (1.2 nm^2) obtained by the chemical adsorption method through the silylation [11]. Being similar to other deposited LB films, multilayered azo polymers demonstrate molecular orientation during the dipping procedure to lead to a homogeneous liquid crystal alignment, the direction of which is parallel to the dipping direction [23].

With use of this novel technique, systematic studies have been carried out to elucidate factors of molecular structures of the azo chromophores governing the photochemical liquid crystal switching [23, 35–37]. A spacer length (n) between the azo chromophore and the PVA backbone plays a critical role in the liquid crystal photoregulation. In keen contrast to a spin-coated film of 6Az1PVA having a short spacer, a monolayered LB film of the same polymer demonstrates no photoregulation of LC alignment. Monolayered LB membranes of azo polymers having a relatively long spacer of n = 5 or 10 (6Az5PVA or 6Az10PVA) can control the reversible liquid crystal alignment upon photoirradiation, confirming that a single molecular layer is enough to work as a command surface to regulate liquid crystal switching. The dichroic character of the azobenzene moiety provides conformational information of the photochromic units on the substrate surfaces. Although the electronic absorption spectrum of a single LB layer of 6Az1PVA is not essentially influenced by contact with a liquid crystal layer, monolayered membranes of the other polymers which are active for photoregulation display considerable alteration in the absorption spectra when the films are converted with a liquid crystal layer, suggesting that the azo chromophores of the latter two demonstrate specific molecular interaction with liquid crystal molecules to result in a reorientation perpendicular to a surface. These result support that the working mechanism presented in Fig. 4.1 is reasonable. The photoactivity of a spin-coated film of 6Az1PVA may be ascribed to the random orientation of the polymeric backbone which itself acts as a spacer between the azochromophore and the polymer chain.

The effects of spacer length and tail length on the photoresponse of liquid crystal cells are summarized in Table 4.6. It is noteworthy here that the number of molecular layer plays an important role; although no photochemical alignment alteration is achieved by the monolayered 6Az1PVA membrane, a cumulative layer is positive for photocontrol. The photoregulative properties of the polymeric LB films are influenced not only by the spacer length, n, but also by the tail length denoted as m. Unsubstituted (m = 0) azobenzene bound to PVA backbone through a longer spacer (n = 10) can regulate the liquid crystal switching in contrast to the results in the chemical adsorption method in which unsubstituted azobenzene does not act as a command molecule as shown in Table 4.4. Furthermore, although the rates of the photochemical as well as thermal isomerization are not affected by the tail length, longer tail residues bring about the reduction of the rate of the photoswitching from homeotropic to planar modes; the longer the tail length, the slower becomes the photoresponse. This tail-length effect is particularly remarkable for monolayered membranes. The liquid crystal switching by 10Az10PVA (m = 10, n = 10) membrane is twenty times slower than that by Az10PVA (m = 0, n = 10). This slower response is a reflection of the reorientation of photoisomerized azobenzene chromophores embedded in LB membranes and is a result of the subsequent interaction of the chromophore with liquid crystal molecules after a sufficient relaxation period.

Table 4.6. Structural effect of amphiphilic azobenzene polymers of LB films obtained on the base of PVA on azimuthal alignment photocontrol under an action of linearly polarized UV light

Photoregulativity			
Polymer	Number of layers		
	1	3	9
0Az10PVA	+	r	p
1Az10PVA	+	r	p
4Az10PVA	+	r	p
6Az10PVA	+	r	p
8Az10PVA	+	r	p
10Az10PVA	+	r	p
12Az10PVA	–	r	p
6Az1PVA	–	–	p
6Az5PVA	–	+	p

-: no photoresponse; +: reoriented perpendicular to the polarization plane; r: in sufficiently oriented; p: oriented parallel to the dipping direction and not affected by polarized photoirradiation.

The following deductions are made in conjunction with the working principle of the alignment photocontrol between the homeotropic and planar mode from the results discussed above. Photochromic units with a suitable length of both tail and spacer interact with liquid crystal molecules intimately at a surface to bring about perpendicular conformation of the command molecules to result in homeotropic alignment. The photoisomerization of command molecules takes place rapidly to induce the reorientation of LC molecules surrounding the isomerized command molecules. The rearrangement of liquid crystal molecules at a boundary region is transferred from molecule to molecule to induce the mesophasic alignment change in the whole cell.

One of the significant characteristics of the LB technique involves readiness of the control of the average density of the commanding azobenzene unit on the substrate surface by an appropriate choice of an average density at a water/air interface at which a polymeric membrane is transferred onto a substrate plate [38]. Deposition of a membrane after UV irradiation to achieve the *trans*- to-*cis* isomerization is also a novel technique to reduce the average density of the chromophore since the *cis*-isomer displays a larger occupying area. It has been revealed that the electronic absorption spectra of LB monolayers are markedly dependent upon the average density of the chromophore. The spectral shape and absorption maxima of monolayered 6Az10PVA are roughly in line with those in a solution when the monolayer is transferred at an average density of 1.02 nm^2, indicating unequivocally that the azo chromophores are oriented in a random fashion without any specific interaction among the chromophores. In contrast, remarkable hypsochromic shift of the absorption band for the π, π^* -transition is observed when the film deposition is made at an average density $=$ 0.30 nm^2. This reflects the aggregation of the azobenzenes through the side by side interaction with such a dense packing. Furthermore, the spectroscopic analysis of dichroic nature ofthe azobenzene reveals that the chromophore units stand almost perpendicular to the substrate surface plane. UV irradiation modifies the molecular as well as orientational structure of the surface azobenzenes successively so that the tilt angle of the nematic liquid crystals is regulated as a function of the extent of the photoisomerization [39].

Detailed studies on molecular orientation of LB films of 6Az1PVA have been carried out with aid of FT-IR measurements and agree essentially with the results mentioned above [40]. The frequencies of the CH_2 stretching bands are sensitive to the conformation of the hydrocarbon chains and afford crucial information on the ordered structure of the LB films. The results show that the hydrocarbon chain in a monolayered film is *trans-zigzag* to imply an ordered structure whereas that in multi-monolayered films contains *gauche* conformers, being indicative of disordered structure which is favourable for molecular interactions between the azo chromophores and LC molecules to display the commanding ability.

Concerning the working mechanism of the command surfaces, a novel technique has been used to reveal molecular reorientation of polymeric LB monolayers by measuring the generation of Maxwell displacement current across a single layer of 6Az1PVA spread on a water surface upon photoirradiation to initiate the photoisomerization of azobenzenes incorporated in a monolayer [41]. The displacement current of a monolayered film of the polymer flows on the water surface as a result of three factors; orientational change of polar molecules, changes in the molecular density on the water surface, and changes in the surface potential of water. A major concern in this study is to relate the displacement current with the second factor to obtain information on the orientational change in monolayers on an air/water interface. The displacement-current measurement has been also made on a mixture of the polymer with nematic liquid crystals in order to obtain some information on the reorientational event of molecules at the surface [42].

4.4 Surface-Assisted Azimuthal Alignment Regulation

4.4.1 Photoinduced Molecular Reorientation

When actinic light is linearly polarized, photons are predominantly absorbed by molecules with a transition moment parallel to the polarization plane of the light to result in optical anisotropy more or less [43]. Within a time range where mobility of photoreacted molecules is suppressed, the exposed materials display dichroism (referred to as photodichroism) in their absorption spectra and angular dependence of refractive indices to cause photobirefringence beyond the absorption wavelength ranges. This type of photochemical phenomena induced by polarized light was first observed by Weigert who studied the photoinduced dichroic properties of photoemulsions and solid solutions of photobleachable dyes [44] referred to as the Weigert effect. Before describing the azimuthal reorientation of liquid crystal molecules by means of linearly polarized light, a brief survey will be made of the closely related photoanisotropic phenomena which attract intensive interest because of the potential value in practical applications for optical devices including erasable optical recordings.

Photoinduced anisotropy has been observed for polymeric materials doped with photochromic molecules including spiropyrans [45], azobenzenes [45], a fulgide [46] and anthracenes [47]. Exposure to linearly polarized light results in angular-selective photobleaching or photocoloration. The photodichroism thus formed is rather a transient phenomenon which disappears on prolonged irradiation along with the completion of the photoreaction. The photoanisotropy is destabilized by adding plasticizers in the polymer matrices to enhance the mobility of doped photochromic molecules in the matrices [45]. Since the molecular motion is suppressed by polymeric networks, heating the polymeric materials above T_g rapidly erases the anisotropy owing to the randomization of the molecular orientation. This type of photosensitive polymer doped with photochromic molecules has been applied to achieve "polarization holography" since laser light for holographic exposure is essentially polarized [48].

Attachment of photochromic units such as azobenzenes to the polymer backbone offers an alternative polymeric system demonstrating photoanisotropy. Polymeric liquid crystals substituted with azobenzenes in their side chains are interesting materials which show marked photobirefringence upon exposure to linearly polarized light [49 – 52][2]. The chromophores reorient in polymeric matrices so that the molecular axis is perpendicular to the electric vector of the actinic light. This molecular reorientation as a result of cumulative photoisomerization brings about the rearrangement of non-photochromic mesogenic groups surrounding the photosensitive groups to enhance the photoinduced birefringence. Photoinduced birefringence has been also observed in amorphous polymers with pendent azobenzene groups [53] and in amorphous films of liquid crystal polymers [54].

Recent works have revealed that the action of linearly polarized light brings about molecularly amplified reorientation of liquid crystal molecules which are in contact with substrate surfaces modified with photoactive molecules, as will be mentioned in the next section. The new-born liquid crystal cells have so far been fabricated by modifying a substrate surface with thin films of polymeric materials which are doped with photoreactive dyes, with thin films of photocrosslinkable polymers and with chemical adsorption methods to bind photochromic molecules.

[2]The photoinduced phenomena in photochromic side chain polymers are discussed in Chapter 2 of the book.

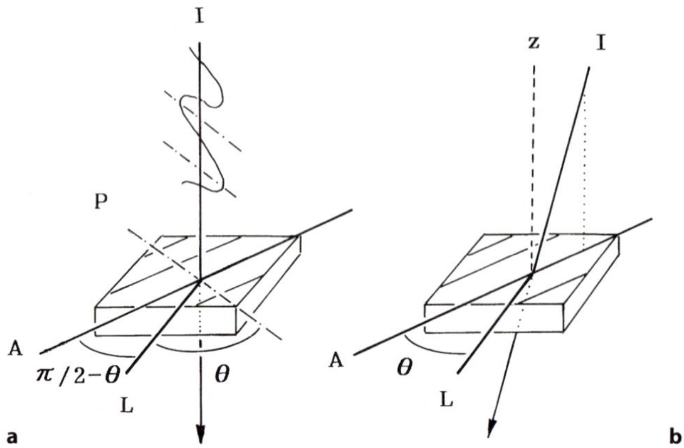

Fig. 4.6a,b. Alignment controls of nematic liquid crystals (**a**) upon irradiation with linearly polarized UV light and (**b**) upon slantwise irradiation with non-polarized UV light (*A*: alignment direction, *L*: tentative cell axis, *I*: direction of actinic light, *P*: electric vector of polarized light)

4.4.2
Polymer Thin Films Doped with Dichroic Dyes

Gibbons and his coworkers observed that a cell filled with a nematic liquid crystal doped with a dichroic azo dye is converted into a homogeneous alignment, the optical axis of which is perpendicular to the polarization plane after prolonged exposure to a polarized Ar-laser beam, as illustrated in Fig. 4.6 [55]. This alignment is stable for a long time, indicating that a polyimide thin layer covering the substrate plates memorizes the aligning ability permanently since the photoinduced reorientation of dichroic molecules in a mesophasic layer has to be subjected to thermal randomization to destroy the photoinduced uniaxial orientation. This observation suggests that an interface between the liquid crystal layer and the aligning polymer surface plays an essential role in the photoinduced homogeneous alignment and led them to examine the photoresponsiveness of a cell which is assembled by use of a substrate plate that is covered with a polyimide film doped with the dichroic dye in a relatively high concentration (33 wt %) to localize the dye molecules at the surface. Irradiation of the cell with an Ar laser beam brings about again the reorientation of liquid crystal molecules which become perpendicular to the polarization plane. The optically induced homogeneous alignment is also performed by illumination of the dye doped polymer thin film with polarized light prior to the cell assembly. It has been suggested that the photoinduced alignment is affected by various factors including dye structure, dye concentration, liquid crystal type, incident energy density and the nature of polymers to improve the photoresponsiveness.

This surface-assisted alignment regulation has been extended to a PVA system containing dissolved methyl orange which shows *trans-cis* photoisomerization and the relatively rapid thermal reversion [56]. This cell was fabricated by spin-coating a substrate surface with a PVA containing 14 wt % of the dye. In accordance with previous reports on polarization holography using the same photosensitive polymer [48], photoinduced birefringence emerges upon exposure of the photosensitive PVA film to an Ar-laser beam. A major part of the photoanisotropy is restored even after the photoirradiation is stopped, suggesting that the emergence of the photobirefringence arises not only from the photoisomerization of the doped dye itself, but also from the distortion of macromolecular chain conformation caused by the photoinduced structural alteration of the dye molecules. These observations imply that photoinduced homogeneous alignment may be attributed at least to the rearrangement of polymer chains.

4.4.3
Photocrosslinkable Polymers

Irradiation of a thin film of poly(vinyl cinnamate) derivatives (18) with linearly polarized UV light affords photobirefringence owing to an anisotropic cyclobutane crosslinking reaction [57, 58].

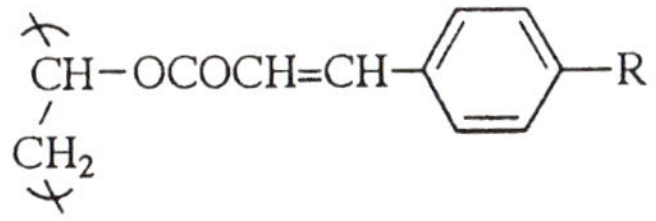

(XVIII)

The fact that the photoinduced anisotropy is diminished by embedding a plasticizer in the polymer matrix suggests that the birefringence arises from the molecular reorientation triggered by the action of polarized light. Azimuthal photocontrol of nematic liquid crystals can be achieved by fabricating a liquid crystal cell with use of a substrate plate which is spin-coated with the photocrosslinkable poly(vinyl *p*-methoxycinnamate) (Formula **XVIII** where R = OCH_3) [57]. Exposure of the film to non-polarized UV light before irradiation with linearly polarized light favorably improves the stability of the aligning ability of the film possibly because of the reduction of the thermally induced molecular mobility owing to the photocrosslinkage.

The emergence of the photobirefringence is interpreted as a result of angularly preferred consumption of cinnamate groups with a transition moment parallel to the electric vector of polarized light to cause an anisotropic distribution of the long axis of unreacted cinnamate residues and to create an anisotropic distribution of cyclobutane with their long axis also preferably aligned perpendicularly to the polarization plane. However, there are other possibilities to explain the formation of the photoanisotropy. One of these is based on the contribution of anisotropic *trans-cis* photoisomerization of the cinnamate groups as in the case of azobenzenes. The reorientation of the polymeric backbone chains may also contribute to the emergence of the optical anisotropy.

4.4.4
Azobenzene Molecular Layers Exposed to Polarized UV Light

During studies on photoalignment regulation to achieve a reversible alteration between a homeotropic state and a planar state which displays a schlieren or a marbled texture, irradiation with linearly polarized UV light to cause the *trans-cis* photoisomerization of monolayered azobenzenes was found to yield homogeneous alignment to improve optical quality with a uniform texture (Fig. 4.6a) [12, 59]. Surface modification of a substrate plate is performed by treating an aminated silica plate with an acrylated azobenzene (see Formula II, where $R = OCH_3, n = 0$) to bind the azo chromophore covalently through the Michael addition reaction (Fig. 4.3). A guest-host cell is assembled by sandwiching a nematic liquid crystal containing a dichroic dye between a couple of the surface modified plates and exposed to polarized UV light. The alignment axis of the photoinduced homogeneous state is examined by measuring a polarized absorption spectrum to reveal that liquid crystal molecules become reoriented perpendicular to the electric vector of the actinic light. By rotating the polarization plane, the alignment axis was controlled at will. The homogeneous orientation is converted again into the homeotropic alignment.

Without the dichroic dye in the liquid crystal, the axis of the photoinduced orientation of the mesophase becomes complicated and not exactly perpendicular to the electric vector of the actinic UV light. This different behavior arises from the absorption of the actinic light by the doped dichroic dye. It follows that the photoisomerization on the surface of the rear plate is more or less suppressed although the azobenzene units on the front surface photoisomerize effectively. This situation results in the surface-assisted homogeneous reorientation being triggered by the single photochromic layer on the front plate. On the other hand, UV light induces the photoisomerization on the rear plate when the LC does not contain the dye molecules. In this case, the anisotropic photochemistry is deformed since the polarization of the actinic light is distorted by passing through the birefringent mesophasic layer to influence the angular dependence of the alignment axis on the polarization plane as stated above. This was confirmed by replacing the azo-modified rear plate by a photochemically inert plate, which is modified with octadecyltriethoxysilane to give a homeotropic alignment, to obtain the orthogonal relationship of the homogeneous alignment.

A relaxation rate of the in-plane reorientation is monitored by measuring the transmittance intensity of linearly polarized probe light through the cell as a function of time after pulsed UV laser irradiation [59]. On the one hand, the liquid crystal reorientation was complete within 110 msec for homeotropic into randomly parallel mode upon exposure to nonpolarized laser light while it takes 90 msec to come to the end of the alignment alteration from a homeotropic into homogeneous mode with polarized laser light irradiation. The response time for the in-plane reorientation with polarized UV laser light can be followed by the change in the transmittance intensity of the probe light passed through the cell which is exposed to the polarized UV light in advance to yield a homogenous alignment. The reorientation takes place somewhat slower and is complete in 240 msec.

The azimuthal photocontrol of liquid crystal alignment by polarized photoisomerization arises from the reorientation of azobenzene units on a substrate surface in a way similar to the photoselection of azo chromophores embedded in polymeric systems, based on the angularly selective absorption of photons by the chromophores with molecular axis parallel to the polarization plane of the actinic UV light [60]. Although it remains disputable which isomer plays a major role in the azimuthal liquid crystal orientation, it has been suggested that the predominant *cis*-isomer units in a photostationary state let liquid crystal molecules lie parallel to the surface while a minor portion of the *trans*-isomer units induce the azimuthal orientation because the *cis*-azobenzene moiety destroys the liquid crystal phase because of its bent form. The photoinduced homogeneous alignment fades away on prolonged standing even in the dark because of the thermal reversion of the *cis*-isomer.

Furthermore, homogeneous alignment of the same guest-host cell is photochemically induced even though the excitation light is not linearly polarized. When the cell is subjected to slantwise UV exposure where the direction of the light propagation is tilted 10° to the normal to the cell surface, the mesophase displays an uniaxial orientation [61]. The orientation axis is parallel to the projection of the light path on the cell surface, as shown in Fig. 4.6b. The direction of homogeneous alignment is regulated at will by changing the incident direction of the actinic light, and subsequent irradiation with visible light reconverts the alignment into the original homeotropic mode. Conoscopic observation of the cell confirmed that the slantwise exposure makes the tilt angle of liquid crystal molecules tunable. The polarization of the actinic light due to reflection at the interfaces of the system is negligible upon consideration of the Fresnel's formulas. It is assumed that the photoisomerizability of the *trans*-isomer on the surface with random directions is angular-dependent since the electric vector of the incident UV light is perpendicular to the transition moment of the azobenzene units which possess the longer molecular axis parallel to the propagation direction of the light.

4.4.5 Surface Azobenzenes Exposed to Polarized Visible Light

It is well known that the photochemistry of azobenzenes is dependent on excitation wavelength; UV light for the π,π^* – transition affords the *cis*-isomer as a major component in a photostationary state whereas the *trans*-isomer is predominant over the *cis* counterpart upon excitation with light at longer wavelengths for the n,π^*–transition. This implies that persistent homogeneous orientation is photochemically achieved by photoirradiation of surface azobenzenes with visible (blue) light for the n,π^*-transition. In order to realize the in-plane reorientation of liquid crystal molecules by means of linearly polarized irradiation, a substrate surface covered with *trans*-azobenzene units should bring about a planar mode. As summarized in Table 4.1, liquid crystal alignment is affected by a substituent of the azobenzenes attached on surfaces; a hydrophobic substituent causes homeotropic alignment while a planar mode is attained by surface modification with azobenzenes bearing a polar substituent.

Introduction of *p*-cyanoazobenzene on silica surface through silylation with **XIX** affords a planar liquid crystal orientation with multi-domain structure.

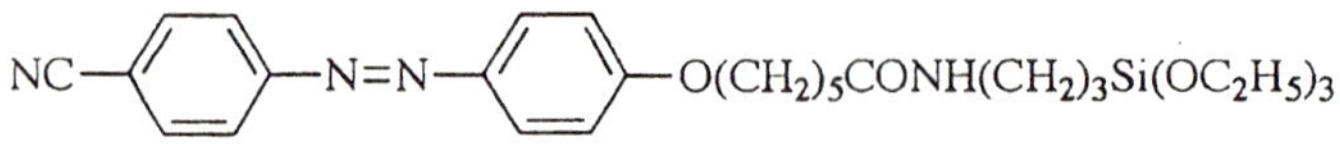

(**XIX**)

A hybrid liquid crystal cell was fabricated with an azo-modified plate and a plate treated with lecithin for homeotropic alignment and irradiated with linearly polarized light at >ca. 430 nm [62]. The alignment becomes homogeneous. It is noteworthy that the liquid crystal reorientation is markedly enhanced by heating the cell above $T_{N \to I}$ of the liquid crystal. Rotation of the polarization plane leads to the regulation of the alignment direction at will. Photoimages are formed by irradiating the whole area of the cell first with linearly polarized blue light to yield the homogeneous alignment, followed by subsequent imagewise illumination with polarized blue light after rotating the polarization plane at 45° with respect to the initial one. Interestingly, no change in the photoinduced optical anisotropy of the liquid crystal cells is observed upon prolonged storage at room temperature. Heating the cells at a higher temperature reduces the anisotropy, however.

A more sophisticated molecular design of the azobenzene moiety for modification of a silica surface has been undertaken to fulfill the following conditions to prepare an azobenzene silylating reagent **XX**.

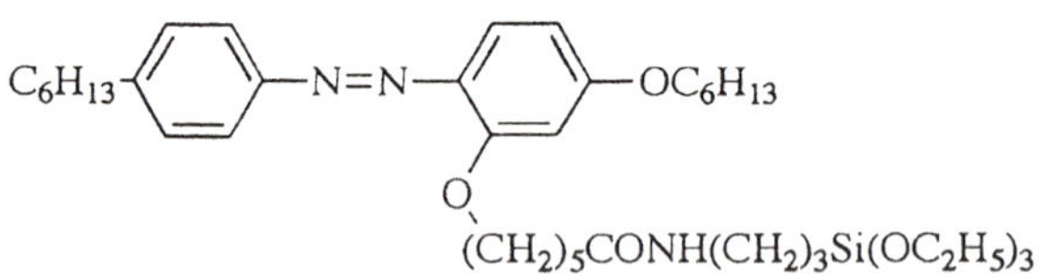

(**XX**)

The chromophore bears alkyl residues at both *p*-positions to mimic the size and shape of the nematic liquid crystal, and the linkage of the chromophore on substrate surface is made lateral to give a side-on type attachment so that the long molecular axis becomes oriented in parallel with the surface [63, 64]. The *trans/cis* ratios of the azobenzene in the photostationary states are approximately 9:1 and 1:9 upon irradiation with UV light at 365 nm and with visible light at >430 nm, respectively.

A hybrid cell is assembled with a plate surface-modified with the lateral azobenzene and a plate treated with lecithin and filled with a nematic liquid crystal (DON-103) of $T_{N \to I} = 74$ °C (Fig. 4.7). Enhancement of the emergence of optical anisotropy of a mesophase is observed again by elevating the cell temperature above $T_{N \to I}$. Homogeneous alignment was induced by irradiating with polarized blue light, for example, at 100 °C. When the cell was exposed at the same temperature to the polarized light after rotating the polarization plane at 45°, the alignment direction rotated along with the exposure time, and the azimuthal reorientation of liquid crystal is complete at an exposure energy of ca. 200 mJ·cm^{-2}. A prominent feature of this cell is an anomalous stability to heat treatment and exposure to nonpolarized actinic light. The photoinduced birefringence of the cell is not altered at all after heating at 100 °C for 1 h. The mesophase of the bulk layer is of course converted into an isotropic phase under

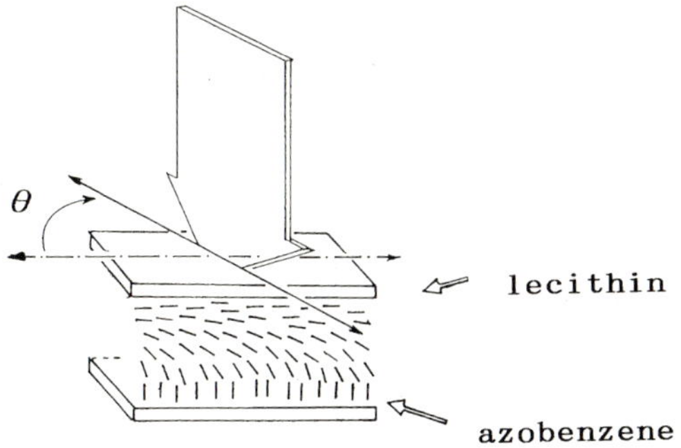

Fig. 4.7. Cell fabrication for in-plane alignment control

this condition. Even harsher storage conditions involving illumination with nonpolarized blue light leading to the photoisomerization and simultaneous heating at 100 °C for 1 h did not affect the homogeneous alignment. This kind of behavior contrasts dramatically with that of the cell which is surface-modified with *p*-cyanoazobenzene which displays far less stability toward heat treatment [62]. The extraordinary thermal stability of the alignment in the cell modified with azo-containing compound **XX** implied that the thermal as well as photochemical randomization of the photoanisotropic orientation of the surface azobenzene is markedly suppressed at the substrate surface. These data suggest that the boundary region between the substrate surface and the liquid crystal layer differs considerably from the bulk mesophase layer and is termed often as a "smectic-like layer" possessing a highly ordered structure [1]. In fact, the presence of a very thin birefringent layer above $T_{N \to I}$ is revealed experimentally as a result of wall-induced orientation of liquid crystal molecules on a surface [65, 66]. The thermal stability of the photoalignment most likely arises from the pretransitional ordered structure of an ultrathin mesophase layer, in which the randomization of the azobenzenes is highly suppressed due to the collective molecular properties of the aggregates with an ultrathin layer.

The emergence of the photoanisotropy can be interpreted as follows (Fig. 4.2). Continuous irradiation with the polarized light leads to selective excitation of the *trans*-isomer with the longer molecular axis with parallel to the polarization plane to bring about the photodichroism of the surface azobenzene. The photodichroism is indeed observed in a polarized absorption spectrum of an azo-modified plate after the polarized irradiation. As a result, the molecular axis of the azobenzenes rotates roughly perpendicular to the electric vector of the polarized actinic light. In this context, it has been proposed to call the azobenzene attached to the substrate surface a molecular rotor which rotates around the spacer unit as a molecular axle [63].

The effectiveness of the molecular rotor has been clarified by comparing the commanding ability of the head-on type with that of the side-on type of *p*-cyanoazobenzenes. For this purpose, a silylating residue is introduced at the *o*-position of a *p*-cyanoazobenzene derivative **XXI** to modify a silica surface. The rate of photoinduced reorientation of a liquid crystal, EXP-CIL, was evaluated by the following procedure; the cells were illuminated with polarized visible light for the homogeneous alignment in advance and subsequently exposed to the same light at a polarization plane angle at 45° with respect to the initial one to follow the rotated angle of reoriented alignment direction. Remarkable differences in photoresponsiveness between the head-on type cell and the side-on type cell can be observed [67]. The exposure energy for the reorientation is much reduced when the *p*-cyanoazobenzene is introduced in the side-on manner; the reorientation is complete at an exposure energy of less than 50 $mJ \cdot cm^{-2}$ for the side-on type azobenzene whereas about 200 $mJ \cdot cm^{-2}$ exposure energy is required for the head-on type. Furthermore, the rotated angles level off at the maximum angle of 45° for the side-on type cell while the head-on type cell exhibits a much smaller rotated angle of reorientation.

4.4.6
Spiropyran as a Command Molecule

The majority of studies on photoinduced molecular reorientation have been focused on azobenzene derivatives displaying geometrical photoisomerization. Exploration of other kinds of molecules commanding azimuthal liquid crystal alignment gives us an insight into a deeper understanding of the working mechanism and affords a clue to the molecular design of the command surfaces to lead to practical application. Spiropyrans are well-known photochromic compounds showing a reversible electrocyclic reaction through which the molecules become reoriented uniaxially by the action of linearly polarized light. Members of this photochromic family act as command surfaces to facilitate alignment photoregulation [68].

As mentioned above, when a silica surface is treated with a spiropyran **VIII** for silylation, alignment of nematic liquid crystals is planar and not altered upon exposure to UV light of the ring opening reaction [13]. This property is rather favorable for the azimuthal alignment control with polarized light. In fact, linearly polarized UV irradiation causes the emergence of homogeneous alignment of EXP-CIL owing to the in-plane orientation regulation. Again a cell temperature above $T_{N \rightarrow I}$ plays a critical role in the photocontrol, and UV exposure at an ambient temperature below $T_{N \rightarrow I}$ hardly gives rise to the in-plane alignment alteration. The photoinduced homogeneous alignment is stable on storage at ambient temperature. However, heat treatment of the cell at 120 °C for 5 min brings about the disappearance of the birefringence.

Because the photochromic unit is attached to a surface at the nitrogen atom of the indolenine moiety through a spacer, the long axis of the molecule containing two ring systems seems to be approximately parallel to the surface, demonstrating a molecular orientation similar to the side-on type azobenzene called a molecular rotor (Fig. 4.2). This conformation is suitable for the planar mode. Since the photoreaction is initiated by photon absorption of the 2*H*-nitrochromene moiety with the transition moment parallel to the molecular axis of the spiro skeleton. Therefore, polarized UV irradiation brings about the consumption of the spiropyran units with the molecular axis parallel to the electric vector of the polarized actinic light. Note that the UV light is simultaneously absorbed by the ring-opened form, photomerocyanine, to give a photostationary state which leads to the molecular reorientation of the photochromic units.

The stability of the photoinduced homogeneous alignment in the dark implies that the liquid crystal orientation is governed rather by the spiropyran itself since under these conditions the major part of the photomerocyanine retransforms into the ring-closed form. A closely related photochromic molecule, spirooxazine, can also act as a command molecule [69]. The chromophore was attached on silica surface in a similar fashion with use of a photochromic silylating reagent **XXII**.

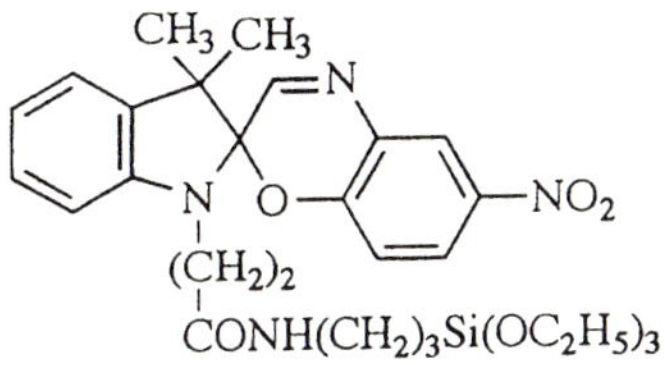

(**XXII**)

Linearly polarized UV light induces the homogeneous alignment with the axis perpendicular to the electric vector of the actinic light.

4.4.7 Photochromic Polymer Layer

From a practical viewpoint, azimuthal reorientation with polymer films is more favorable because of better availability of command surfaces. As mentioned previously, two approaches to photochemical alignment regulation have been explored by use of polymers doped with dichroic dyes [55, 56] and photocrosslinkable poly(vinyl cinnamate) derivatives [57, 58]. Since the photoalignment of liquid crystals is achieved by the photochemical events locally at substrate surfaces, monolayered as well as multilayered polymeric LB films afford significant information concerning the role of the polymer surfaces in commanding the azimuthal alignment of liquid crystals.

Linearly polarized UV irradiation of liquid crystal cells made up with a silica plate covered with LB films of amphiphilic azobenzene polymers (see Formula **XIV**, mAznPVA) yields the in-plane reorientation of liquid crystals from a homeotropic mode to afford homogeneous alignment with the axis perpendicular to the electric vector of the polarized light [23, 35–38] in a manner quite similar to the cases of cells fabricated with a plate surface-modified with an azobenzene covalently.

Table 4.6 shows the structural effect of LB films of the polymers on the photoregulativity for azimuthal reorientation upon exposure to linearly polarized UV light. The Table tells us that the photoresponsive behavior is also dependent on the number of accumulated layers. Azimuthal reorientation can be achieved specifically by monolayered films in general although there are some exceptions.

In contrast, multi-layered films cause homogeneous alignment perpendicular to the dipping direction, and no azimuthal reorientation is induced by irradiation with linearly polarized UV light. This means that the conformational orientation of the polymers probably wins over the photoinduced molecular orientation. The ability to command the in-plane reorientation is also a function of the occupying area of the azobenzene unit in a monolayered film. When the film is relatively closely packed, i.e., occupying areas of less that 0.33 nm^2 per azobenzene, the alignment is governed by the dipping direction and not allowed to be reoriented by the polarized light irradiation. On the other hand, the photoinduced azimuthal reorientation is achieved by LB monolayers occupying areas larger than 0.35 nm^2, but smaller than 1.02nm^2. These facts strongly suggest that the mobility of the azobenzene units at the polymer surface should be sufficiently secured to realize the commanding ability for the in-plane reorientation.

Quite a different type of LB film is also active for the in-plane alignment regulation. Polyimide LB films have been prepared by deposition of molecular layers of long alkylammonium salts of polyamic acids as precursors on a substrate which is baked for the cyclization to lead to the polyimide formation [70]. Using this technique, a monolayered polyimide LB film substituted with *p*-cyanoazobenzene units is deposited on a silica plate to assemble a liquid crystal cell exhibiting a planar mode [71]. Exposure of the cell to linearly polarized visible light results in the in-plane reorientation to form homogeneous alignment with excellent optical quality. It should be stressed here that the reorientation can be performed reversibly by the irradiation at room temperature below $T_{N \to I}$, in contrast to the command surfaces prepared by surface silylation to attach azobenzenes on silica surfaces.

Spin-coated films of polymers with pendent azobenzenes are also candidates for azimuthal alignment photocontrol [72]. When a homopolymer of 4-methacroyloxyazobenzene is (see Formula **XVI**, where $n = 0$, $R = H$) coated on a substrate plate and irradiated with linearly polarized visible light, molecular reorientation of the azochromophore is brought about to give persistent photodichroism. The polymeric film affords a homogeneous alignment of nematic liquid crystals when the cell fabrication is made after the photoirradiation, and no photoinduced birefringence is obtained when the polarized exposure is made after the cell assemblage. This is a great contrast to the fact that the photoinduced alignment of liquid crystals is a result of the irradiation of cells fabricated with an azo-modified silica plate, as described so far. Furthermore, the photoanisotropy of mesophases is so unstable that the homogeneous alignment fades away at room temperature even in the dark. This possibly arises from the permeation of liquid crystal molecules into the polymer film to relax the oriented state of the azobenzene units and is in line with the observation that no alignment is induced when the polarized irradiation is made after cell fabrication. The mobility of the azo-chromophore at the polymer surface of the cell in enhanced by contact with liquid crystals which act as plasticizers to inhibit the photoinduced molecular reorientation. This means unequivocally that the compatibility of liquid crystals with photoactive polymers plays a crucial role in the persistence of the photoinduced in-plane alignment of liquid crystals.

4.5
Future Aspects

Let us consider the degree of molecular amplification in photosensitive liquid crystal systems. Reversible transformations of mesophases are induced by photochromic dopants which are dissolved in host liquid crystals usually at least in a few mol %. In these cases, a single photochromic molecule causes the reorientation of approximately ten to one hundred liquid crystal molecules, and the degree of amplification is assumed to be 10 to 10^2. As mentioned in the introductory section, the degree of molecular amplification in the reversible alteration between homeotropic and planar modes amounts to roughly 10^4. Concerning the azimuthal alignment photocontrol by the command surfaces, the reorientational amplification is thought to consist of two processes; azimuthal reorientation at the surface region and the subsequent molecular rearrangement in bulk mesophases. Since the dichroic ratio of monolayered photochromic molecules on substrate surface is not far from 1:1 [64], the amplification degree of the azimuthal reorientation on surfaces is about 10 and may be multiplied by a factor of 10^4 for the bulk reorientation to yield 10^5 as the amplification degree for this system. This qualitative discussion reveals the uniqueness of the surface-mediated photocontrol of liquid crystal alignment.

The considerable amplification of molecular events performed by the command surfaces reminds one of the visual process which is initiated by the reversible change in the molecular structure of retinylidene unit embedded in opsin which is subjected to the deformation of the highly ordered network due to the geometrical photoisomerization of the conjugated polyene chromophore leading to the subsequent processes conjugated with visual sensory systems. In analogy to the concept of the command surfaces in which command molecules on solid surfaces trigger the macroscopic property change based on molecular rearrangement of liquid crystal systems through a "domino effect," the retinylidene unit in rhodopsin can be referred to as a command molecule which brings about the rearrangement of the biomolecules in the primary process of the visual system. We may call these photoresponsive systems "command molecular systems" in a broader sense. In this context, there is a common feature among the visual process and the photoresponsive liquid crystal systems.

The surface-assisted alignment regulation of liquid crystals is closely related with various interfacial phenomena including friction, lubrication, wettability, adhesion, epitaxial growth, bioadaptability and so forth. For instance, wettability is distinctly governed by molecular structures of monolayers covering solid surfaces [73]. In this respect, deeper understanding of the command surfaces for the liquid crystal alignment regulation affords essential information on and guidance for designing surfaces on molecular levels to improve the interfacial events and open a way to the functionalization of molecular films [74].

The technology of liquid crystal displays is growing rapidly along with remarkable advances in fundamental research into liquid crystal science. It has been predicted that the number of liquid crystal display devices rank with that of the conventional cathode ray tubes at the beginning of the next century. They all work by the application of an electric field despite the variety of their working principles. What kind of role can the alignment photocontrol techniques play in liquid crystal technologies? The alignment photocontrols can be reasonably applied in two ways; active and passive photofunctions. In the former, reversibility of the photoinduced alignment alteration is put to practical uses including erasable optical recordings of various types, laser-addressed display devices, spatial light modulators in optical computing systems, image processing and so on. The in-plane alignment photoregulation is in particular of great interest because the optical information is written-in and read-out simply by rotating the electric vector of polarized light from a single light source. An alternative application, though passive, is to afford aligning layers for fabricating liquid crystal display devices. Uniaxial orientation of liquid crystals in display devices is an inevitable technology and has been achieved conventionally by buffing the thin film surface of polymers like polyimides to get homogeneous alignment. The rubbing of the insulating polymer films brings about static electricity on the surface which attracts dust to cause severe defects in fine liquid crystal devices. Moreover, recent liquid crystal technologies require more sophisticated alignment to improve optical quality of display devices. The photoinduced azimuthal alignment control by means of photoactive polymers is one of the novel techniques for overcoming the problems encountered in the conventional rubbing treatment. Further systematic studies are needed to apply surface-mediated liquid crystal alignment to these practical applications.

4.6 References

1. Cognard J (1982) Mol Cryst Liq Cryst Suppl Ser 1: 1
2. Ichimura K (1990) In: Dürr H, Bouas-Laurent H (eds) Photochromism: Molecules and systems. Elsevier, Amsterdam
3. Sackmann E (1971) J Am Chem Soc 93: 7088
4. Haas WE, Nelson KF, Adams JE, Dir GA (1972) J Electrochem Soc 121: 1667
5. Tazuke S, Kurihara S, Ikeda T (1987) Chem Lett 911
6. Ichimura K (1991) In: Honda K (ed) Photochemical processes in organized molecular systems. North-Holland, Amsterdam, p 343
7. Ichimura K, Suzuki Y, Seki T, Hosoki A, Aoki K (1988) Lanmguir 4: 1214
8. Plueddemann EP (1982) Silane coupling agents, Plenum, New York
9. Ishida H (1984) Polym Composites 5: 101
10. a) Sagiv J (1980) J Am Chem Soc 112: 92 b) Ogawa K, Mino N, Nakajima K, Azuma Y, Ohmura T (1991) Langmuir 7: 1473
11. Aoki K, Seki T, Suzuki Y, Tamaki T, Hosoki A, Ichimura K (1992) Langmuir 8: 1007
12. Kawanishi Y, Tamaki T, Sakuragi M, Seki T, Suzuki Y, Ichimura K (1992) Langmuir 8: 2601
13. Aoki K, Ichimura K, Tamaki T, Seki T, Kawanishi Y (1990) Kobunshi Ronbunshu 47: 771
14. Aoki K, Seki T, Sakuragi M, Ichimura K (1992) Makromol Chem 193: 2163
15. Aoki K, Tamaki T, Seki T, Kawanishi Y, Ichimura K (1992) Langmuir 8: 1014
16. Aoki K, Kawanishi Y, Seki T, Tamaki T, Sakuragi M, Ichimura K (1992) Thin Solid Films 219: 226
17. Courtot P, Pichon R, Le Saint J (1976) Tetrahedron Lett 1177 and 1181
18. a) Yamamura S, Tamaki T, Seki T, Sakuragi M, Kawanishi Y, Ichimura K (1992) Chem Lett 543; b) Yamamura S, Tamaki T, Seki T, Sakuragi M, Kawanishi Y, Ichimura K (1993) Liq Cryst 13: 189
19. Nakahara H, Fukuda K, Shimomura M, Kunitake T (1988) Nippon Kagaku Kaishi 1001

20. Shimomura M, Ando R, Kunitake T (1983) Ber Bunsen-Ges Phys Chem 87: 1134
21. Seki T, Ichimura K (1989) Polym Commun 30: 109
22. Hiltrop K, Stegemeyer H (1978) Ber. Bunsen-Ges Phys Chem 82: 884
23. Seki T, Sakuragi M, Kawanishi Y, Suzuki Y, Tamaki T, Fukuda R, Ichimura K (1993) Langmuir 9: 211
24. Aoki K, Kawanishi Y, Seki T, Tamaki T, Ichimura K (unpublished results)
25. Ichimura K, Suzuki Y, Seki T, Kawanishi Y, Tamaki T, Aoki K, (1989) Jpn J Appl Phys 28: Suppl 28-3: 289
26. Suzuki Y, Sakuragi M, Tamaki T, Asano N, Aoki K, Ichimura K (unpublished results)
27. Ichimura K, Suzuki Y, Seki T, Kawanishi Y, Aoki K (1989) Makromol Chem, Rapid Commun 10: 5
28. Kawanishi Y, Seki T, Tamaki T, Ichimura K, Ikeda M, Aoki K (1990) Polym Adv Technol 1: 311
29. Kawanishi Y, Tamaki T, Seki T, Sakuragi M, Suzuki Y, Ichimura K (1991) Langmuir 7: 1314
30. Kawanishi Y, Tamaki T, Seki T, Sakuragi M, Ichimura K (1991) J Photopolym Sci Technol 4: 271
31. Seki T, Ichimura K (1989) Thin Solid Films 179: 77
32. Seki T, Tamaki T (1993) Chem Lett 1739
33. Collins-Gold L, Möbius D, Whitten DG (1986) Langmuir 2: 191
34. Nishiyama K, Fujihira M (1988) Chem Lett 1257
35. Sakuragi M, Seki T, Kawanishi Y, Tamaki T, Ichimura K, Fukuda R, Hiramatsu H, Fujiwara K (1991) J Photopolym Sci Technol 4: 279
36. Seki T, Sakuragi M, Kawanishi Y, Fukuda R, Hiramatsu H, Fujiwara K, Ichimura K (1992) J Photopolym Sci Technol 5: 569
37. Seki T, Sakuragi M, Kawanishi Y, Suzuki Y, Tamaki T, Ichimura K, Fukuda R, Hiramatsu H, Thin Solid Films 210/211: 836
38. Seki T, Fukuda R, Tamaki T, Ichimura K (1994) Thin Solid Films 243: 675
39. Sakuragi M, Tamaki T, Seki T, Suzuki Y, Kawanishi Y, Ichimura K (1992) Chem Lett 1763
40. Katayama N, Ozaki Y, Seki T, Tamaki T, Iriyama K. Langmuir (submitted)
41. Naruse H, Ohnishi K, Iwamoto M, Ichimura K, Seki T, Tamaki T, Iriyama K (1993) Jpn J Appl Phys Part 1 32: 2832
42. Naruse H, Yoshida H, Ohnishi K, Iwamoto M, Ichimura K (1994) Thin Solid Films 244: 1018
43. Michle J (1990) In: Dürr H, Bouas-Laurent H (eds) Photochromism: Molecules and systems. Elsevier, Amsterdam, p 919
44. Weigert F (1921) Naturwiss 29: 583
45. a) Barachevsky VA (1991) Proc SPIE 1559: 184; b) Barachevsky VA (1991) J Photopolym Sci Technol 4: 177
46. a) Jones P, Jones WJ, Williams G (1990) J Chem Soc, Faraday Trans 86: 1013; b) Jones P, Darcy DP, Attard GS, Jones WJ, Williams G (1989) Mol Phys 67: 1053
47. Tran-Cong Q, Kumazawa T, Yano O, Soen T (1990) Macromolecules 23: 3002
48. a) Todorov T, Nokolova L, Tomova N (1984) Appl Opt 23: 4309; b) Calixto S, Lessard RA (1984) Appl Opt 23: 4313
49. a) Eich M, Wendorff JH, Ringsdorf H, Reck B (1987) Makromol Chem, Rapid Commun 8: 59; b) Eich M, Wendorff JH (1987) Makromol Chem, Rapid Commun 8: 467; c) Anderle K, Birenheide R, Werner MJA, Wendorff JH (1991) Liq Cryst 9: 691
50. a) Stumpe J, Müller L, Kreysig D, Hauck G, Koswig HD, Ruhmann R, Rübner J (1991) Makromol Chem., Rapid Commun 12: 81; b) Ivanov S, Yakovlev I, Kostromin S, Shibaev V, Läsker L, Stumpe J, Kreysig D (1991) Makromol Chem., Rapid Commun 12: 709; c) Fischer T, Läsker L, Stumpe J, Kostromin SJ (1994) Photochem Photobiol 80: 453
51. Wiesner U, Antonietti M, Boeffel C, Spiess HW (1990) Makromol Chem 191: 2133; b) Wiesner U, Reynolds N, Boeffel C, Spiess HW (1992) Liq Cryst 11: 251
52. Ortler R, Brauchle C, Muller A, Riepl G (1989) Makromol Chem Rapid Commun 10: 189
53. a) Rochon P, Gosselin J, Natansohn A, Xie S (1992) Appl Phys Lett 60: 4; b) Xie S, Natansohn A, Rochon P (1992) Macromolecules, 25, 2268 and 5531; c) Xie S, Natansohn A, Rochon P (1993) Chem Mater 5: 403

54. a) Läsker L, Fischer T, Stumpe J, Kostromin S, Ivanov S, Shibaev V, Ruhmann R (1994) Mol Cryst Liq Cryst 246: 347
55. a) Gibbons WM, Shannon PJ, Sun ST, Swetlin BJ (1991) Nature 351: 49; b) Gibbons WM, Shannon PJ, Sun ST, Swetlin BJ (1992) Proc SPIE, 1665: 184; c) Gibbons WM, Shannon PJ (1992) Liq Cryst 12: 869
56. Iimura Y, Kusano J, Kobayashi S, Aoyagi Y, Sugano T (1993) Jpn J Appl Phys 32: L93
57. Schadt M, Schmitt K, Kozinkov V, Chigrinov V (1992) Jpn J Appl Phys Part 1 31: 2155
58. Dyadyusa AG, Marusii TY, Reznikov Y, Khizhnyak AI, JETP lett 56: 18
59. Kawanishi Y, Tamaki T, Seki T, Sakuragi M, Ichimura K (1992) Mol Cryst Liq Cryst 218: 153
60. Jones C, Day S (1991) Nature 351: 15
61. a) Kawanishi Y, Tamaki T, Ichimura K (1992) Polym Mater Sci Eng 66: 263; b) Kawanishi Y, Tamaki T, Ichimura K (1994) ACS Symp Ser 537: 434
62. Ichimura K, Hayashi Y, Kawanishi Y, Seki T, Tamaki T, Ishizuki N (1993) Langmuir 9: 857
63. Ichimura K, Hayashi Y, Ikeda Y, Ishizuki N (1993) Appl Phys Lett 63: 449
64. Ichimura K, Hayashi Y, Akiyama H, Ishizuki N (1993) Langmuir 9: 3298
65. a) Miyano K (1979) Chem Phys Lett 43: 51; b) Miyano K (1979) J Chem Phys 71: 4108
66. Yokoyama H, Kobayashi S, Kamei H (1984) J Appl Phys 56: 2645
67. a) Ichimura K (1994) Mol Cryst Liq Cryst 246: 331; b) Ishizuki N, Akiyama H, Ichimura K (unpublished results)
68. a) Ichimura K, Hayashi Y, Ishizuki N (1992) Chem Lett 1063; b) Ichimura K, Hayashi Y, Goto K, Ishizuki N (1993) Thin Solid Films 235: 101
69. Goto K, Akiyama H, Ichimura K (unpublished results)
70. Suzuki M, Kakimoto M, Konishi T, Imai Y, Iwamoto M, Hino T (1986) Chem Lett 395
71. Akiyama H, Kudo K, Ichimura K, Yokoyama S, Kakimoto M, Imai Y (1995) Langmuir 11: 1033
72. Ichimura K, Akiyama H, Ishizuki N, Kawanishi Y (1993) Makromol Chem Rapid Commun 14: 813
73. Whitesides GM, Laibinis PE (1990) Langmuir 6: 87
74. Swalen JD, Allara DL, Andrade JD, Chandross EA, Garoff S, Israelachvili J, McCarthy TJ, Murray R, Pease RF, Rabolt JF, Wynne KJ, Yu H (1987) Langmuir 3: 932

CHAPTER 5

Electric Field Poling of Nonlinear Optical Side Chain Polymers

C.P.J.M. van der Vorst and S.J. Picken

5.1 Introduction: Context and Scope

The role of optics in information technology is expected to grow considerably in the forthcoming years. This is due to the presently encountered limitations in purely electronic circuitry and transmission systems with respect to bandwidth, speed, power consumption and packing densities. The first areas in which optics has already obtained a strong position are to be found in long haul telecommunication networks. Fiber optic transmission lines have already widely been installed. Optical storage media markets are rapidly increasing. Optical switching and optical interconnections, ranging from intra-computer links to local area networks, will probably follow this trend to go optic very soon. To fulfill the needs for high speed and high capacity information technology, vast research programs have been started to develop materials and systems suitable for these applications. In the case of materials for optical switching and data storage, both organic and inorganic materials are being pursued. We confine ourselves here to Electro-Optic (EO) switching based on the linear EO or Pockels effect in materials with second-order Nonlinear Optical (NLO) properties.

In this area, organic materials might provide the systems developer with unique possibilities and a higher flexibility in comparison to inorganic materials [1–16]. Very high NLO coefficients appear to be possible in organic materials, which surpass the values of traditional inorganic materials such as lithium niobate. Moreover, the effects are predominantly electronic in nature in the organic NLO materials, which – in combination with a low dielectric constant – allows the very high operation speeds required for future communication systems.

Other advantages of organic materials are the possibility of chemical tailoring and the high radiation damage thresholds. The macroscopic second-order NLO properties of organic materials are based on the corresponding molecular properties (hyperpolarizabilities) of certain non-centrosymmetric building blocks. These NLO building blocks have to be arranged in a non-centrosymmetric way, either by growing crystals with a non-centrosymmetric structure [8, 9]; by incorporating them in Langmuir-Blodgett (LB) double layers [17, 18]; or by putting them into a polymer and orienting them with a strong electric field near the polymer's glass transition temperature [10–12, 14, 15, 19–21].

Of all organic NLO materials – crystals, LB layers and polymers – especially the last group looks promising [15]. The most important advantage of organic polymeric materials is the relatively easy processing of organic polymeric materials into practical devices. Especially the processing of polymers into integrated optic devices is very attractive, since no high temperature steps are required. For this reason, it can be anticipated, that polymeric optical devices can be integrated with electronic circuitry. The feasibility of this concept has already been demonstrated. Diemeer et al. 1989 integrated a polymeric waveguide structure with a laser diode on a GaAs substrate [22].

Incorporation of the hyperpolarizable (NLO) building blocks into a polymer is possible by doping [5, 6, 11, 12, 23, 24] or by chemical attachment. Chemical attachment allows higher concentrations (> 40 wt %) than doping (< 30 wt %) without problems of seggregation. Also, the thermal stability of the side chain orientation is improved by chemical anchoring. The NLO groups can be attached as side chains to the polymer backbone (side chain polymers) [14, 15, 19, 20] or they can be incorporated into the backbone itself (main chain polymers) [25, 26]. In this chapter, we will deal mainly with the side chain polymers. Special attention will be paid to the influence of axial order – either induced by the poling field or spontaneous liquid-crystalline axial order – on the poling of the NLO groups. Predictions of theoretical models will be compared with each other and verified – if possible – by experimental results.

5.2 The Electrooptical Effect in Optically Nonlinear Materials

5.2.1 Organic Materials

In organic materials, the macroscopic second-order NLO susceptibility $\chi^{(2)}$ is based on the corresponding molecular property β (hyperpolarizability) of certain NLO moieties [4–7, 10–12, 15]. The macroscopic property $\chi^{(2)}$ is defined by the general relationship between the components of electric field(s) E – at angular frequencies ω_1, ω_2 – and the components of the induced polarization density P at angular frequency ω:

$$P(\omega)_I = \chi^{(1)}(-\omega;\omega_1)_{IJ}\, E(\omega_1)_J + \chi^{(2)}(-\omega;\omega_1,\omega_2)_{IJK}\, E(\omega_1)_J\, E(\omega_2)_K + \dots \quad (5.1)$$

The molecular property β is defined similarly by the relationship between the components of the *local* electric field(s) E′ and those of the induced molecular dipole moment p:

$$p(\omega)_i = \alpha(-\omega;\omega_1)_{ij}\, E'(\omega_1)_j + \beta(-\omega;\omega_1,\omega_2)_{ijk}\, E'(\omega_1)_j\, E'(\omega_2)_k + \dots \quad (5.2)$$

In these equations, $\chi^{(1)}$ is the linear susceptibility and α is the linear polarizability. Summation over identical indices is assumed. Capital indices I, J, K = X, Y, Z are cartesian coordinates in the macroscopic frame of reference, whereas i, j, k = x, y, z are in the molecular frame. The accent in E′ symbolizes that it is a local field, related to the corresponding applied field E by a local field factor

$f(E' = f. E)$. Exact expressions can only be derived for local field factors in simple model cases, otherwise approximations are necessary. Usually, one uses the Lorentz-Lorenz expression $f = (n^2 + 2)/3$, with n the refractive index, at optical frequencies and the more complicated Onsager expression for lower frequencies. For a discussion of the local field problem, see e.g. Böttcher 1973 [27].

The macroscopic susceptibilities are composed as a sum of all corresponding molecular contributions per unit volume, each molecular component mapped onto the appropriate macroscopic unit vectors. For the second order terms, the relation can be written as follows:

$$\chi^{(2)}(-\omega;\omega_1,\omega_2)_{IJK} = Nf(\omega)f(\omega_1)f(\omega_2)\beta(-\omega;\omega_1,\omega_2)_{ijk}.\langle O_{Ii}O_{Jj}O_{Kk}\rangle \tag{5.3}$$

Here, N is the number density of hyperpolarizable groups. The O's are projections of the molecular axes onto the macroscopic frame and the brackets indicate an averaging over all molecular orientations weighted by the orientational distribution function G (see below). The averages are called polar order parameters. The $f(\omega_i)$ terms are frequency-dependent local field correction factors, which have been taken out of the averaging brackets for simplicity. A more rigourous treatment of the local field problem in this context is given by Meredith [28].

The linear susceptibility $\chi^{(1)}$ is related to optical refraction and absorption. The most common effects due to $\chi^{(2)}$ are Frequency Doubling or Second Harmonic Generation $(\chi^{(2)}(-2\omega;\omega,\omega))$ and the Pockels effect or linear EO effect $(\chi^{(2)}(-\omega;\omega,0))$. These are both special cases of three-wave mixing. Choosing the principle axes of the refractive index n as laboratory frame, the Pockels effect can be (re)defined in terms of a change of the refractive index in a modulating field E as:

$$\delta n_I = -\tfrac{1}{2}n_I^3 \cdot r_{IIK}.E_K \tag{5.4}$$

The Pockels coefficient r and the susceptibility $\chi^{(2)}$ are directly related to each other by the following equation:

$$\chi^{(2)}_{IIK} = -\tfrac{1}{2}n_I^4 \cdot r_{IIK} \tag{5.5}$$

Because the Pockels coefficient r_{IJK} is invariant under permutation of the first two indices, often a condensed notation is used in which r has only two indices. The first index L represents the combination IJ and may have the values 1 = XX, 2 = YY, 3 = ZZ, 4 = YZ = ZY, 5 = ZX = XZ, 6 = XY = YX. The second index is K, which may have the values 1 = X, 2 = Y and 3 = Z.

5.2.2 Poled Polymers

Absence of centrosymmetry is strictly required for second-order nonlinearity. This requirement holds both on a molecular and on a macroscopic level in organic materials with second-order NLO properties.

Well-known hyperpolarizable molecules consist of an extended conjugated Π electron system, noncentrosymmetrically substituted with a Π electron acceptor A at one end and donor D at the other. Examples of such non-centrosymmetric

AΠD groups are 4-dimethyl-amino 4′-nitrostilbene (DANS) and *p*-methoxy-nitro-stilbene (MONS).

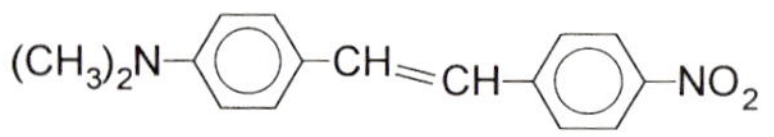

(DANS)

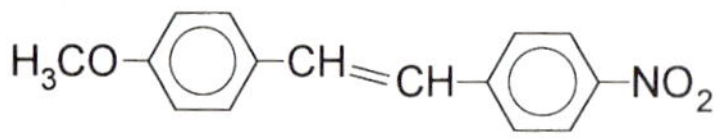

(MONS)

These molecules have their permanent dipole moment (μ) and one dominant hyperpolarizability component (β) along their long axis. Neglecting off-axis contributions, the molecules are quasi-rotationally symmetric, or rod-like.

To induce second-order nonlinearity on a macroscopic level, the orientational distribution function of the AΠD groups has to become non-centrosymmetric. With the AΠD groups dispersed in, or chemically attached to a polymer, this is accomplished by aligning the permanent dipole moments of the AΠD moieties with a strong electric field in the vicinity of the polymer's glass transition temperature T_g ("poling"). The obtained order can be frozen in (in a kinetic sense) by cooling well below T_g.

In a uniformly poled polymer sample, the refractive index is uniaxial with a long axis n_e in the poling direction (3 direction) and short axes n_o perpendicular to the poling direction (1 and 2 directions). Applying a modulation field E_{mod} in the poling direction, the Pockels effect is described by two Pockels coefficients, r_{33} and r_{13}, which are related to the susceptibilities $\chi^{(2)}_{333}$ and $\chi^{(2)}_{113}$, respectively. Each of these can be written in molecular terms. The relationships between all relevant quantities are described by the following two equations, expressing the Pockels effect in terms of r, $\chi^{(2)}$ or β [4–7, 10–12, 15]:

$$\delta n_e = -\tfrac{1}{2} n_e^3 r_{33} \cdot E_{mod} = n_e^{-1} \chi^{(2)}_{333} \cdot E_{mod} = n_e^{-1} NF\beta \langle \cos^3\theta \rangle \cdot E_{mod} \quad (5.6)$$

$$\delta n_o = -\tfrac{1}{2} n_o^3 r_{13} \cdot E_{mod} = n_o^{-1} \chi^{(2)}_{113} \cdot E_{mod} = n_o^{-1} NF\beta \langle \tfrac{1}{2} \cos\theta \sin^2\theta \rangle \cdot E_{mod} \quad (5.7)$$

The factor F contains corrections for local field effects. The expressions in brackets are the "polar order parameters". For quasi-rotationally symmetric AΠD molecules, only the angle θ between the permanent molecular dipole moment μ and the poling field E_{pol} appears in between the averaging brackets.

Since generally $\langle \cos^3\theta \rangle$ is larger than $\langle \tfrac{1}{2}\cos\theta \sin^2\theta \rangle$, the most efficient EO modulation is obtained by using the Pockels coefficient r_{33}, not r_{13}. If indeed r_{33} is chosen for modulation, $\langle \cos^3\theta \rangle$ is the relevant polar order parameter that must be optimized.

Polar order of AΠD molecules is the degree of alignment of the AΠD's dipole moment – a vector or first-rank tensor – along the direction of the poling field – another vector. Axial order is the degree of alignment of the AΠD molecule's long axis (second-rank tensor) along another axis, the director $\tilde{n}$. Here, this axis is parallel to the poling field direction. In the case of polar order, the orientational distribution function is non-centrosymmetric and the "up-down sym metry is broken", viz. more molecules are directed "upward" (in the field direction) than "downward" (opposite to the field direction). In the case of axial order, the up-down symmetry need not be broken and the orientational distribution function may remain centrosymmetric. The orientational distribution function $G(\theta)$ is defined in such a way that $G(\theta)\sin\theta d\theta = -G(\theta)d(\cos\theta)$ represents the number of molecules with a polar angle θ between θ and $\theta + d\theta$. $G(\theta)$ can be developed in a series expansion of Legendre polynomials $P_l(\cos\theta)$ of order l with coefficients $\langle P_l(\cos\theta)\rangle$. Odd-order $\langle P_l\rangle$ represent polar order, even-order l $\langle P_1\rangle$ represent axial order. "Our" polar order parameters $\langle\cos^3\theta\rangle$ and $\langle 1/2\cos\theta\sin^2\theta\rangle$ are linear combinations of the more general order parameters $\langle P_1\rangle$ and $\langle P_3\rangle$, e.g. $\langle\cos^3\theta\rangle = [3\langle P_1\rangle + 2\langle P_3\rangle]/5$. The most important axial order parameters are $\langle P_2\rangle$ and $\langle P_4\rangle$.

In this Chapter, we discuss the influence of axial order, in particular of liquid crystalline axial order, on the field-induced polar order parameters, especially on $\langle\cos^3\theta\rangle$. We will do this from a theoretical viewpoint using different models (Sect. 5.3) and compare the theoretical predictions with experimental results (Sect. 5.5).

5.3
Theoretical Models for the Calculation of Order Parameters

5.3.1
General Consideration

A number of simple theoretical models exist that can be used for the calculation of the field-induced order parameters such as $\langle\cos^3\theta\rangle$. These are the following:

- the isotropic model [5, 6];
- the Ising model [5, 6];
- the linearized model of Singer, Kuzyk and Sohn [12];
- our extension of the Maier-Saupe model [29, 30].

All these models are molecular statistical mean-field models, which implies that they do not describe the time-dependence of poling, omit excluded volume and entropy terms and do not explicitly treat two-particle interaction terms like dipole-dipole interaction. Instead, all two-particle interactions are put in an effective single-particle mean-field potential. Averages of any angle dependent function $A(\theta)$ (e.g. $\cos^3\theta$) are calculated using the orientational distribution G as a weight factor:

$$\langle A\rangle = \int_{-1}^{+1} d(\cos\theta)A(\theta)G(\theta) \tag{5.8}$$

The orientational distribution function is written in the mean-field approximation as

$$G(\theta) = Z^{-1} \cdot \exp[-U(\theta)/kT] \tag{5.9}$$

where T is the absolute poling temperature, k is Boltzmann's constant and $U(\theta)$ is the mean-field potential in the presence of an electric poling field. Z is used for normalization (partition function).

Table 5.1. Energy expressions in four molecular statistical models

	isotropic*	Ising*	SKS**	MSVP***
$U_0(\theta)$	0	$\begin{cases} 0\ (\theta = 0, \pi) \\ \infty\ (\theta \neq 0, \pi) \end{cases}$	no analytical expression	$-\varepsilon\langle P_2\rangle P_2(\cos\theta)$
$U_1(\theta)$	$-\mu E\cos\theta$	$-\mu E\cos\theta$	$-\mu E\cos\theta$	$-\mu E\cos\theta$
$U_2(\theta)$	0	0	0	$-\Delta\alpha E^2 P_2(\cos\theta)/3$
$\langle\cos^3\theta\rangle$	a/5	$a \equiv \mu E/kT$	$a[7+20\langle P_2\rangle + 8\langle P_4\rangle]/35$	no analytical expression
$\langle\frac{1}{2}\cos\theta\sin^2\theta\rangle$	a/15	0	$a[7+5\langle P_2\rangle - 12\langle P_4\rangle]/105$	no analytical expression
$\langle P_2\rangle$	$a^2/15$	1	input	calculated selfconsistently

*Meredith GR, van Dusen JG, Williams DJ (1982) Macromolecules 15: 1385
**Linearized model: Singer KD, Kuzyk MG, Sohn JE (1987) J Opt Soc Am B4: 968 Model SKS
***Extented Maier-Saupe model: van der Vorst CPJM, Picken SJ (1990) J Opt Soc Am B7: 320 Model MSVP
The single-particle energy $U(\theta)$ is a sum of three terms $U_i(\theta)$ (i = 1–3), each proportional to the i-th power of the field strength E of the local poling field. μ is the permanent molecular dipole moment, $\Delta\alpha$ is the anisotropy of the molecular polarizability, ε is the strength of the Maier-Saupe potential, P_2 is the second-order Legendre polynomical of $cos\theta$. An analytical expression for $U_0(\theta)$, the energy term favouring spontaneous axial order, is not given in the SKS model. Instead, it is assumed that the order parameters of the liquid crystalline host (input parameters $\langle P_2\rangle$ and $\langle P_4\rangle$) are transferred to the AIID dopant molecules.

Hereafter, we will use simply E instead of E_{pol} as symbol for the poling field, whenever no confusion between poling field and modulation field exists. To be more precise, E will be used for the local poling field (thereby leaving out the accent) related to the applied poling field by a local field factor. Assuming that this local field factor remains constant during poling, the local field problem is decoupled from the molecular statistical models. We will not go into more detail in the local field problem.

The mean field potential may consist of a number of energy terms ($U = U_0 + U_1 + U_2$), which can be found in Table 5.1. The most important practical differences between the four models are in the number of energy terms taken into account and in computational approximations. One such approximation is to truncate the Taylor series expansions of e^{Y} after the linear term, as applied e.g. in the model of Singer et al. This results in a linearized dependence of calculated polar order parameters as a function of the field strength, not showing any saturation, and restricts the applicability of the model to only moderate field strengths. The dominant energy term, taken into account by all four models, is the energy of the permanent dipole moment μ in the (local) poling field E, viz. $U_1 = -\mu E\cos\theta$. Other energy terms used in specific models will be discussed later.

5.3.2
Isotropic Model Versus Ising Model: Enhancement of Polar Order by Axial Order

All the above-mentioned molecular statistical models, which neglect the explicit treatment of dipole-dipole interaction, predict an enhancement of the electric field induced *polar* order – in particular of the order parameter $\langle \cos^3(\theta) \rangle$ – by *axial* order [5, 6, 10, 12, 15, 29, 30].

The first indication for such an enhancement is obtained by a comparison of the two oldest models, viz. the isotropic model with no initial axial order and the Ising model with perfect axial order [5, 6]. This perfect axial order is forced by the energy term U_0 (Table 5.1), which only allows the molecules to stand either upward or downward. If we compare the linearized solutions for $\langle \cos^3\theta \rangle$ in both models with each other, viz.

$$\langle \cos^3\theta \rangle_{\text{Ising}} = a \equiv \mu E/kT \tag{5.10}$$

and

$$\langle \cos^3\theta \rangle_{\text{isotropic}} = a/5 \tag{5.11}$$

(exact solutions are given in Ref. [30]), an enhancement is predicted by a factor of 5 [5, 6].

This enhancement can be understood qualitatively by the strong reduction in the number of allowed states in the Ising case [15, 30]. This is illustrated in Fig. 5.1. The Ising model allows only two states: $\theta = 0$ or "up" and $\theta = \pi$ or "down". These two states have *extreme* values of opposite sign for the dipole energy term $-\mu E\cos\theta$ and also for $\cos^3\theta$, both of which play a role in the value of $\langle \cos^3\theta \rangle$. The *maximal* energy separation between the two allowed states causes a relatively large asymmetry in the population of these states. This, added to the fact that their contribution $\cos^3\theta$ to the average value is also extreme, gives relatively few cancellations and, consequently, a large $\langle \cos^3\theta \rangle$. The isotropic

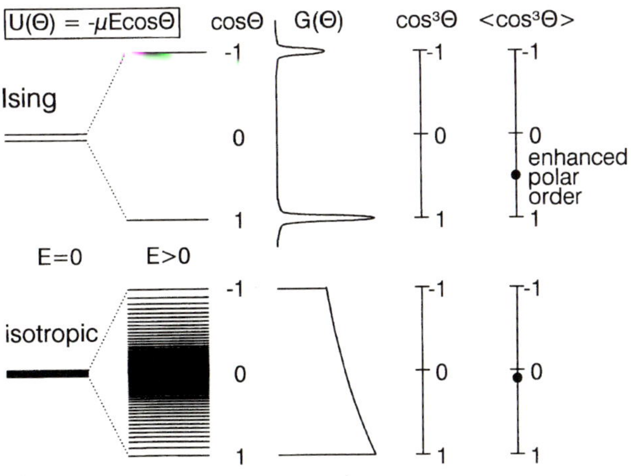

Fig. 5.1. Comparison of dipole energy in Ising and isotropic models. The enhancement of $\langle \cos^3\theta \rangle$ by axial order can be understood qualitatively from inspection of the dipole energy levels and the thermal population thereof in extreme cases of axial order

model, on the other hand, allows many states $\{\theta, \phi\}$, with $0 \leq \theta \leq \pi$ and $0 \leq \phi \leq 2\pi$. Here, ϕ is the azimuthal angle of μ in the macroscopic frame. All states can be grouped into pairs. The states within a pair, $\{\theta, \phi\}$ and $\{\pi - \theta, 2\pi - \phi\}$, are converted into each other by inversion. The two states of each pair have values of opposite sign for the dipole energy $-\mu E\cos\theta$ and for $\cos^3\theta$. Most pairs do *not* have *extreme* (absolute) values for $-\mu E\cos\theta$ and $\cos^3\theta$, as is the case in the Ising model. In fact, *small* values dominate. This is a result of a geometrical effect, which causes that states with θ in the vicinity of $\pi/2$ are most abundant ($\sin\theta$ maximal). A *small* energy separation between the two opposing states of a pair causes a rather symmetric population and therefore strong cancellation of the opposite contributions $\cos^3\theta$ of each state. This, added to the fact that the absolute value of these contributions is also small, results in a *small* $\langle\cos^3\theta\rangle$.

The enhancement of polar order by axial order only concerns the polar order parameter $\langle\cos^3\theta\rangle$. At first glance, the second polar order parameter $\langle\frac{1}{2}\cos\theta\sin^2\theta\rangle$ seems to be *negatively influenced* by axial order as the (linearized) solution in the isotropic model is a/15, whereas it is zero in the Ising model [5, 6]. Calculations with the MSVP model show that the relation between axial order and $\langle\frac{1}{2}\cos\theta\sin^2\theta\rangle$ is not so obvious (Sect. 5.3.5).

5.3.3
Liquid-Crystalline Axial Order

Since liquid crystals are characterized by considerable spontaneous axial order in a mesogenic phase below the clearing temperature T_{cl}, they have been considered as suitable poling media for AΠD molecules. If the liquid-crystalline (LC) axial order can be transferred to – or induced in – the AΠD molecules, this will possibly result in enhanced poling of the AΠD rods with respect to poling in an isotropic medium. Several research groups have tried to introduce LC properties into the polymeric poling medium in order to improve poling. Historically, the experimental work has been along two main lines – "mixed rod systems" and "single rod systems" [15] – for each of which specific theoretical models have been developed.

Mixed rod systems are the oldest ones considered. Starting from an existing LC system with mesogenic rod-like (but not hyperpolarizable) molecules A, hyperpolarizable molecule B are added in low concentrations. It is then hoped that the LC order of the mesogens A remains intact and is transferred to the likewise rod-like molecules B by alignment of the B molecules along the A molecules. The model of Singer et al. has been designed for this situation (Sect. 5.3.4). Examples of mixed rod system are e.g. guest-host systems, with the AΠD rodes doped in a LC side chain or main chain polymer, and side chain copolymers with different mesogenic and hyperpolarizable side groups (Sect. 5.5.2).

The recognition that the above mentioned A-B alignment inherently implies that the hyperpolarizable entities B themselves are also (slightly) mesogenic and therefore, in principle, capable of B-B alignment, forms the basis of the newer single-rod systems. Here, the mesogenic and hyperpolarizable properties are combined in a single type of rod-like molecule. The MSVP model applies to this situation (Sect. 5.3.5). Mixed rod systems are generally concentrated systems, such as side chain copolymers or homopolymers with one mesogenic, hyperpolarizable rod-like side group (Sect. 5.5.4).

Apart from the expected enhancement of $\langle\cos^3\theta\rangle$ *during poling* in liquid crystals, the poling induced alignment is also expected to show better thermal stability *after poling* [6]. This would be a second bonus from poling in a LC environment. As far as the authors know, this presumed second advantage has not yet been properly verified theoretically by any model describing the time-dependence of poling and relaxation in both isotropic and LC media. Time-dependent models do exist (cf. Sect. 5.3.10), but these have not been used to investigate the influence from LC axial order on the poling and relaxation kinetics. Experimental verification – to be discussed in Sect. 5.5 – is still lacking. Without performing any calculations, a slower thermal relaxation in an LC medium is expected from the mesogenic interactions forcing the AΠD molecules (in the extreme Ising case) to reside in either one of the two energy minima: "up" or "down". These two minima are separated from each other by a relatively high energy threshold – in a kinetic theory this corresponds to a high activation energy. This larger activation energy seems to be the important difference between a LC and an isotropic polymer environment. In addition to this, differences in viscosity may also be of importance. It should also be realized that not only the relaxation but also the poling will become slower in a LC polymer.

5.3.4 Singer-Kuzyk-Sohn (SKS) Model for Mixed Rod Systems

If the spontaneous axial order of rodlike liquid crystal molecules can be transferred by molecular alignment to the likewise rodlike AΠD molecules, this will possibly result in enhanced poling of the AΠD rods with respect to poling in an isotropic medium. The mixed rod systems have been the subject of both theoretical and experimental work. Singer, Kuzyk and Sohn developed a theoretical model for poling of AΠD dopants in a LC host (linear or SKS model). In their model [12], the axial order of the LC host is assumed to be independent of the field, not influenced by the dopant and transferred perfectly to the AΠD rods (without using an explicit energy expression for U_0). The axial order parameters $\langle P_2\rangle$ and $\langle P_4\rangle$ of the LC host enter explicitly as input parameters in the expression for the polar order parameters of the AΠD groups (Table 5.1). For $\langle\cos^3\theta\rangle$, the following relation is derived:

$$\langle\cos^3\theta\rangle_{SKS} = a\,.\,[1/5 + 4\langle P_2\rangle/7 + 8\langle P_4\rangle/35] \tag{5.12}$$

This again shows enhancement of $\langle\cos^3\theta\rangle$ by axial order. The degree of enhancement with respect to the solution of the isotropic model is smaller than the factor of 5 expected from the Ising model. Only in the limit of perfect axial order ($\langle P_2\rangle = 1 = \langle P_4\rangle$) is the enhancement by a factor of 5.

5.3.5
Maier-Saupe-van der Vorst-Picken (MSVP) Model for Single Rod Systems

The above mentioned transfer of axial order from the mesogenic rods A to the hyperpolarizable rods B by A-B alignment implicitly assumes the existence of *latent mesogenic properties* in the AΠD rods B themselves. This explains why also single rod systems with the hyperpolarizable and the mesogenic properties united in a single type of rod like AΠD molecule have gained interest, both theoretically and experimentally. For a practical application, the single rod system have the very important advantage over the mixed rod systems that much higher number densities of hyperpolarizable groups are possible, allowing – in principle – a larger Pockels coefficient.

In our model for poling in single rod systems [29, 30], the mesogenic properties of the AΠD rods are described by an effective single-particle potential U_0 (Table 5.1) that was used originally by Maier and Saupe [31–33] in their theory of the nematic phase including the nematic-to-isotropic (NI) phase transition at the clearing temperature T_{cl}. If the anisotropic mutual interactions between the AΠD rods are large enough to overcome thermal motion up to a certain temperature T_{cl}, an LC phase may result in the polymer for $T < T_{cl}$, just like in normal LC materials. The relationship between the "strength" ε of the potential U_0 and the clearing temperature T_{cl} is given by $kT_{cl} \approx 0.22\varepsilon$ without external fields present [31–33]. We use here the "zero-field" clearing temperature T_{cl} related to ε according to the above mentioned equation as an alternative model parameter instead of ε.

The interaction of the electric field with the permanent dipole moment μ of the AΠD groups is contained in the energy term U_1. The term U_2 contains the angle-dependent part of the interaction with the linearly induced dipole moment, with $\Delta\alpha$ the anisotropy of the linear polarizability α. U_2 is very small and by itself induces only axial order, whereas U_1 by itself induces both polar and axial order. As mentioned above, U_0 by itself may generate spontaneous axial order for $T < T_{cl}$. The interplay between the three energy terms U_0, U_1 and U_2 is complicated and, as no analytical equations can be derived, need numerical evaluation [15, 29, 30].

One of the effects, which can be calculated with our model, is the **shift of the clearing temperature in an electric field** E [30]. The field strength E represents in our model *a local* field strength experienced by the AΠD molecules. Local field corrections are left outside our model. The actual clearing temperature $T_{cl}(E)$ will be higher than the "zero-field" clearing temperature $T_{cl} \equiv T_{cl}(0)$. This is demonstrated by the results of calculations for a specific parameter set ($\mu = 7$ D; $\Delta\alpha = 47$ Å^3; $T_{cl} = 420$ K; $T_g = 380$ K $< T < 480$ K) shown in Fig. 5.2. Here, the axial order parameter $\langle P_2 \rangle$ is shown as a function of temperature T (which is assumed to be $\geq T_g$) for a number of local poling field strengths E. Without a field, the phase transition takes place at T_{cl} and is of first order: $\langle P_2 \rangle$ changes discontinuously at the transition and jumps between 0 and a value near 0.4. The phase transition is shifted towards higher temperatures in the presence of an electric field. Also, the discontinuous jump of $\langle P_2 \rangle$ at the transition decreases with increasing E, until it becomes zero at a critical end point T_{crit}. From thereon,

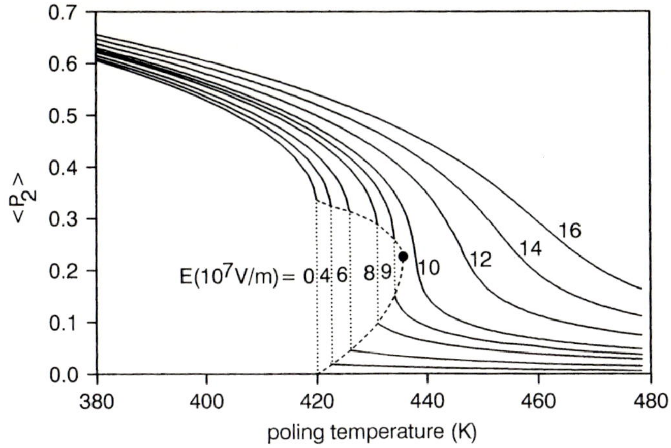

Fig. 5.2. Temperature and field strength dependence of axial order in a liquid crystal near the NI transition. $\langle P_2 \rangle$ has been calculated self-consistently using our model for μ=7 D, $\Delta\alpha$=47 A^3, $T_c \equiv T_c(E=0) = 420$ K, $T_g = 380K < T < 480K$, $0 < E < 160$ V/µm. A shift of the transition temperature $T_{cl}(E)$ with increasing local poling field strength E can be observed

the first-order phase transition does not become second-order, but is effectively lost: beyond T_{crit} no symmetry change occurs and all material properties change continuously. The field-dependent isotropic-to-nematic phase transition can also be treated using Landau theory (Sect. 5.3.11).

Since we are studying AΠD groups in a polymer, also the polymer properties must be taken into account. In particular, $T_{cl}(E)$ must be above the glass transition temperature T_g to provide sufficient mobility for an isotropic-to-nematic phase transition to occur at $T_{cl}(E)$ and, thus, for allowing a real LC phase between T_g and $T_{cl}(E) < T_g$. If such is the case, the LC phase can be "frozen in" below T_g. For $T_{cl}(E) < T_g$, a LC phase is not possible below T_{cl}, because mobility is insufficient to allow a real isotropic-to-nematic phase transition at $T_{cl}(E)$. If the zero-field clearing temperature is below the glass transition, T_{cl} only represents a **virtual phase transition** in the absence of an electric field. This virtual transition may turn into a real phase transition in the presence of an electric field, if the field-dependent clearing temperature $T_{cl}(E)$ can be raised above T_g. In this case, we have a **field-induced isotropic to nematic transition**. The field-induced transition is only a real phase transition, if it occurs below the critical end point T_{crit}.

The isotropic model takes only the U_1 term into account. This term by itself induces some axial order [30], being the result of a quadratic term in μE. Our model predicts much higher field-induced axial order, mainly because the latent mesogenic properties of the AΠD rods (U_0) manifest themselves more strongly in an electric field. The field-induced axial order becomes especially prominent if the system is "pulled" by the field through an isotropic-to-nematic transition. However, even on approaching such a transition, substantial field-induced axial order is already expected from our model. Figure 5.3a shows $\langle P_2 \rangle$ calculated for the parameters set $\mu = 7$ D; $\Delta\alpha = 47$ Å^3; $T_{cl} = 340$ K $< T_g = 380$ K [30]. For

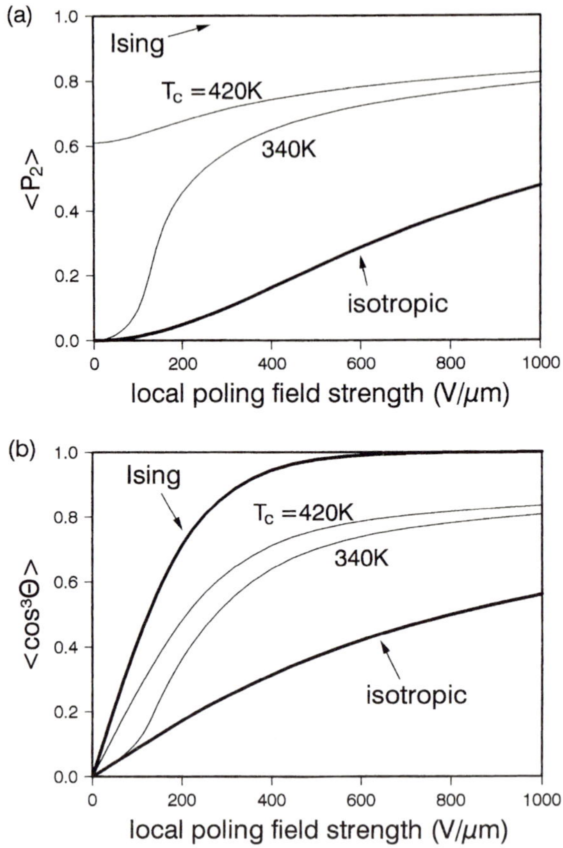

Fig. 5.3a,b. Field strength dependence of order parameters in initially LC and isotropic polymers. **a** axial order. The axial order parameter $\langle P_2 \rangle$ has been calculated self-consistently with our model for μ=7 D, $\Delta\alpha$=47 Å^3, T=T_g = 380 K and T_{cl} as indicated in the figure. Note the field-induced isotropic to nematic transition in the initially isotropic case ($T_{cl} < T_g$) taking place when $T_{cl}(E)$ ($> T_{cl}$) approaches T_g. For comparison, also the exact solutions for $\langle P_2 \rangle$ in the isotropic and Ising models are shown **b** polar order After $\langle P_2 \rangle$ was calculated self-consistently (Fig. 5.3a), $\langle \cos^3\theta \rangle$ was calculated for the same model parameters. Note the influence of the field-induced isotropic to nematic transition, leading to a superproportional dependence of $\langle \cos^3 \theta \rangle$ on E. For comparison, also the exact solutions in the isotropic and Ising models are shown

Fig. 5.4. Correlation between polar and axial order. For a wide range of input parameters, $\langle P_2 \rangle$ was calculated self-consistently with the MSVP model, after which $\langle \cos^3\theta \rangle$ and $\langle 1/2 \cos\theta \sin^2\theta \rangle$ were calculated. The correlation between polar and axial order was investigated by plotting *relative* values of the polar order parameters (either *relative* to the linearized solutions from the isotropic model, or *relative* to each other) versus $\langle P_2 \rangle$. All calculated points lie within bands of which the envelopes are shown. Fig. 5.4a shows the result for $\langle \cos^3 \theta \rangle$, Fig. 5.4b for $\langle \frac{1}{2} \cos\theta \sin^2\theta \rangle$ and Fig. 5.4c for their ratio. A strong positive correlation can be observed in Fig. 5.4a, demonstrating the enhancement of $\cos^3\theta$ with respect to the isotropic model by spontaneous or field-induced axial order

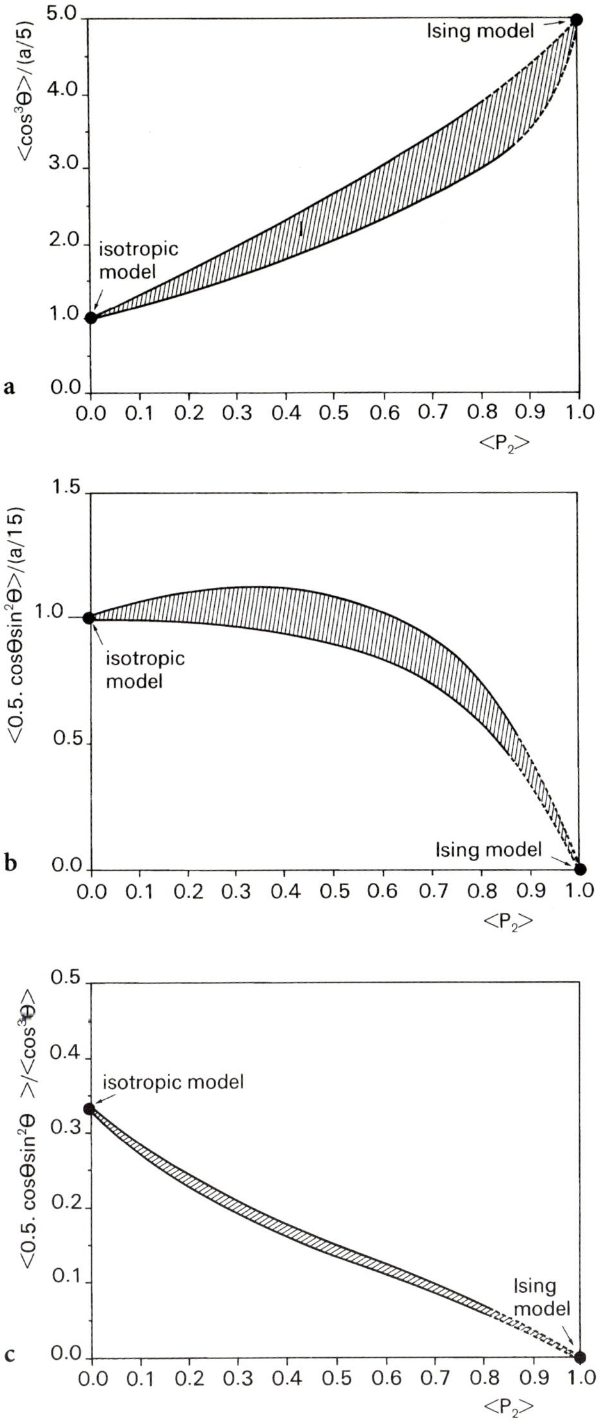

a
<cos³Θ>/(a/5)
5.0
4.0
3.0
2.0
1.0
0.0
Ising model
isotropic model
0.0 0.1 0.2 0.3 0.4 0.5 0.6 0.7 0.8 0.9 1.0
<P2>
b
<0.5. cosΘsin²Θ>/(a/15)
1.5
1.0
0.5
0.0
isotropic model
Ising model
0.0 0.1 0.2 0.3 0.4 0.5 0.6 0.7 0.8 0.9 1.0
<P2>
c
<0.5. cosΘsin²Θ >/<cos³Θ>
0.5
0.4
0.3
0.2
0.1
0.0
isotropic model
Ising model
0.0 0.1 0.2 0.3 0.4 0.5 0.6 0.7 0.8 0.9 1.0
<P2>

comparison, an initially LC model case with $T_{cl} = 420$ K > $T_g = 380$ K is also shown. After the field-induced transition, the difference between the initially isotropic and initially LC model cases is negligible.

Field-induced axial order is expected to enhance field-induced polar order $\langle\cos^3\theta\rangle$ just like the spontaneous axial order of liquid crystals is expected to do. Calculations with our model taking a wide range of input parameters into account show a general positive correlation between $\langle\cos^3\theta\rangle$ and $\langle P_2\rangle$, irrespective of whether the axial order is spontaneous and/or is induced by the electric field (Fig. 5.4a). Due to the enhancement of polar order by field-induced axial order, our model predicts a **super-proportional dependence** of $\langle\cos^3\theta\rangle$ as a function of E in isotropic polymers on approaching the above mentioned field-induced NI transition. Figure 5.3b – showing $\langle\cos^3\theta\rangle$ for the same model parameters as Fig. 5.3a – nicely demonstrates this [30]. Comparing this result with the poling result of the initially LC model case, the *net* advantage for poling from initial LC axial order is strongly reduced and nearly vanishes above the phase transition in the initially isotropic system (Fig. 5.3b).

Without performing calculations, a number of situations are conceivable which show a different dependence of $\langle\cos^3\theta\rangle$ as a function of E for poling above or in the vicinity of T_g. These situation differ mainly in the importance of the U_0-term relative to the U_1-term. With increasing T_{cl} (increasing ε), these are the following:

(1) For $T_{cl} \ll T_g$, a behavior is expected as described by the isotropic model. After a region with a linear dependence according to Eq. 5.6, saturation effects become manifest from a value of *a* of 1 and higher and lead to a region with sub-proportional behavior, described by the exact solution of the isotropic model. Any influence of field-induced axial order at extremely high field strengths is obscured by the continuous field dependence above T_{crit} and by saturation effects or cannot be observed experimentally because of dielectric breakdown;

(2) For a higher T_{cl}, field-induced axial order may lead to a region with a superproportional behaviour somewhere intermediate between the linear and the sub-proportional region. This situation is encountered in the above mentioned model case shown in Figs. 5.3a,b;

(3) For a slightly higher T_{cl}, the field-induced transition takes place somewhere in the middle of the original linear region. As a result, we may observe with increasing field strength: a first linear region (isotropic), a super-proportional region, which levels off to a second linear region (liquid crystalline) and finally a sub-proportional region. The straight lines of both linear regions intersect the origin;

(4) For T_{cl} approaching T_g, we may observe a field-induced phase transition at low field strength, after which the plot of $\langle\cos^3\theta\rangle$ vs E becomes a straight line intersecting the origin (LC linear region). The linear region is followed by the final sub-proportional region (saturation). With T_g below the critical end point T_{crit}, we might expect to see a discontinuity at the field-induced phase transition. In an experimental situation, the discontinuity may become smoothed because the glass transition itself is not a sharp transition at T_g

but has a certain width and, moreover, depends on the time scale of the experiment. A smoothed phase transition will then show a superproportional field dependence of $\langle \cos^3\theta \rangle$ followed by a leveling off towards a straight line.

(5) For $T_{cl} > T_g$, we immediately enter the LC linear region with a slope, intermediate between the ones of Eqs. 5.10 and 5.11 The linear region is followed by a subproportional region.

The enhancement of polar order by axial order only concerns the polar order parameter $\langle \cos^3\theta \rangle$. At first glance, the second polar order parameter $\langle \frac{1}{2}\cos\theta\sin^2\theta \rangle$ seems to be negatively influenced by axial order as the (linearized) solution in the isotropic model is a/15, whereas it is zero in the Ising model [5, 6]. Calculations with our model show that the relation between axial order and $\langle \frac{1}{2}\cos\theta\sin^2\theta \rangle$ is not so obvious (Fig. 5.4b). At low field strengths, where no significant axial order is induced yet, $\langle \frac{1}{2}\cos\theta\sin^2\theta \rangle$ is still linear in E and the solutions of initially isotropic and initially LC model cases *coincide*: $\langle \frac{1}{2}\cos\theta\sin^2\theta \rangle/(a/15) \approx 1$ (Fig. 5.4b). Here, spontaneous LC axial order apparently has no influence on this second polar order parameter! At higher field strength, where axial order is induced by the field – most prominently in initially isotropic model cases – and where also saturation effects start – most pronounced in initially LC model cases – the parameter $\langle \frac{1}{2}\cos\theta\sin^2\theta \rangle$ levels off. The levelling off is most prominent in the initially LC model cases. This result suggests that, although induced axial order and $\langle \frac{1}{2}\cos\theta\sin^2\theta \rangle$ are negatively correlated (Fig. 5.4b), saturation effects seem to play a bigger role in determining $\langle \frac{1}{2}\cos\theta\sin^2\theta \rangle$. After a first linear region, the second polar order parameter goes through a maximum and approaches zero (the Ising limit) only for extremely strong poling fields (not shown).

The behavior of the ratio of the two polar order parameters, $v \equiv \langle \cos^3\theta \rangle / \langle \frac{1}{2}\cos\theta\sin^2\theta \rangle$, can be found from the behavior calculated for the individual order parameters. For intermediate field strengths, the value of this ratio is around 3 for initially isotropic model cases and larger for initially LC cases, but certainly not ∞ as in the Ising limit. When appreciable axial order is induced, the model solutions of initially isotropic and initially LC cases approach each other, and both go only slowly to ∞ (Ising limit) for E approaching ∞ (not shown). Figure 5.4c shows the strong negative correlation between $v^{-1} = \langle \frac{1}{2}\cos\theta\sin^2\theta \rangle / \langle \cos^3\theta \rangle$ and $\langle P_2 \rangle$, corresponding to a strong positive correlation between v and $\langle P_2 \rangle$.

Based on theoretical calculations with our model – which also neglects dipole-dipole interaction – we questioned the presumed advantages of liquid crystallinity for improved poling in the concentrated single rod systems [15, 29, 30]. The first reason for our doubt was that real liquid crystals do not have the perfect axial order that is adopted in the Ising model. Thus, the enhancement with respect to the solution of the isotropic model will be smaller than the factor of 5. Using our model or the model of Singer et al. 1987 [12] and inserting realistic input parameters, an enhancement by a factor near the value 3 seems more likely. The second reason is that even initially isotropic polymers are predicted by our model to show enhanced poling with respect to the isotropic model solution. This makes the *net* effect of initial LC even smaller than the factor of 3 (See e.g. Fig. 5.3b).

5.3.6
Influence from Dipole-Dipole Interaction

The mutual interaction between the permanent dipole moments of the AΠD groups is not explicitly taken into account in any of the above mentioned molecular statistical models. Without performing elaborate calculations, some qualitative predictions can be made by taking into account geometrical effects in the statistics of averageing (sine-effects).

A first anticipation is that dipolar association (if significant) will have a negative influence on the order parameter $\langle \cos^3\theta \rangle$, since dipole-dipole association will be predominantly anti-parallel, both for energetic and for geometrical reasons.

Böttcher [27] considered the interaction energy between two dipoles μ_1 and μ_2 at a constant separation r as a function of the relative orientations. This calculation is of relevance for spherical molecules with a point dipole at their centers. Energy extremes are found for the two situations were the two dipoles are colinear – let us label these situations with "σ". The dipoles are parallel ("σp") in the absolute minimum of energy; they are anti-parallel ("σa") in the absolute maximum. Relative extremes occur when both dipoles are perpendicular to their separation vector r and (anti-) parallel to each other - let us label this with "π". The dipoles are anti-parallel ("πa") in the relative minimum, parallel ("πp") in the relative minimum (Fig. 5.5). If the molecules become elliptically deformed (or rod-like) with the long axis along their dipoles, the molecules can approach each other more closely in the π situation than in the σ situation, which may lead to a crossing of the energy levels. After the crossing, the anti-parallel orientation of type πa is energetically favored and corresponds to the absolute energy minimum (Fig. 5.5).

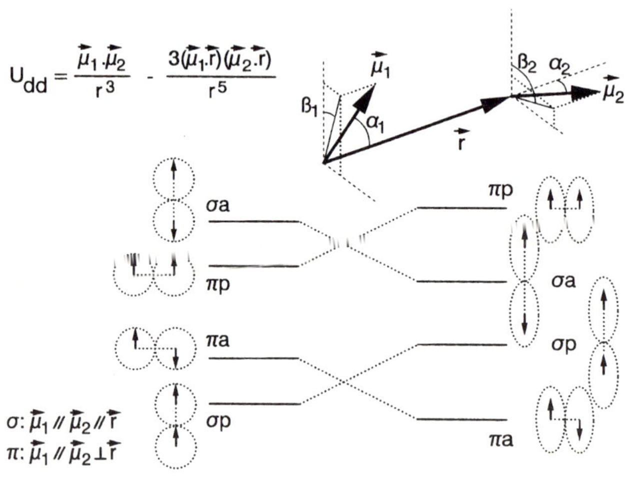

Fig. 5.5. Dipole-dipole energy as a function of mutual orientation. On the left, energy extrema are shown for spherical molecules with a central point dipole μ at closest proximity r (constant). On the right, the situation is shown for elongated molecules at closest proximity (r not constant). Note the crossing of the σ and π levels

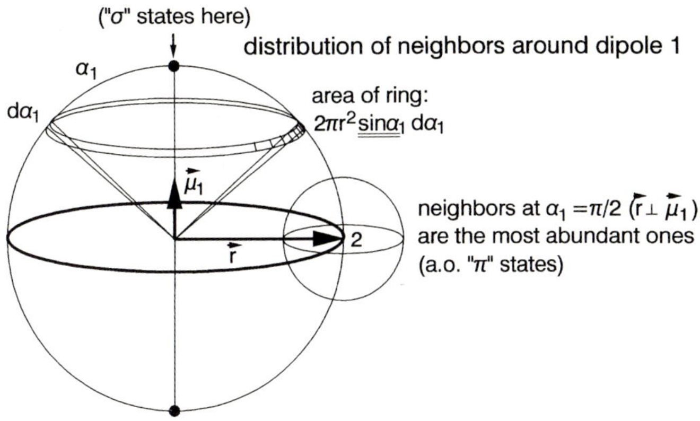

Fig. 5.6. Relative abundancy of different dipole pairs. Taking a simple isotropic distribution of the centers of gravity of dipoles 2 around a central 1, it is shown that the neighbors at right angle with respect to μ_1 are the most abundunt. The interaction between a pair of these most abundant neighbors in two extreme cases of axial order is shown in Fig. 5.7

To this energetic argument in favor of the πa association, a second, geometrical/statistical argument can be added. Let us consider the distribution of the centers of gravity of dipoles 2 around a central dipole 1 (Fig. 5.6). For simplicity, take spherical molecules again, keep the separation r constant, and assume an isotropic distribution. From inspection of Fig. 5.6, it is clear that the dipoles 2 with r approximately perpendicular to μ_1 ($\alpha_1 \approx 90°$: a.o. π pair states) are much more abundant than dipoles with r almost colinear with μ_1 ($\alpha_1 \approx 0°$ or $180°$: a.o. σ pair states) and therefore statistically much more important. Considering the dipole-dipole energy of μ_1 and μ_2 as a function of all orientations of μ_2, the energy minimum for the statistically most important dipoles 2 with $\alpha_1 \approx 90°$ is obtained in the anti-parallel situation πa, whereas the parallel situation πp corresponds to an energy maximum. The parallel situation σp may correspond to an absolute or relative energy minimum, but it is statistically unimportant. Hence, the tendency towards anti-parallel (πa) alignment is much larger than the tendency towards parallel (σp) alignment. This obviously diminishes $\langle \cos^3\theta \rangle$.

It can also be anticipated that the anti-parallel dipole-dipole association (πa) will be stronger in a case of strong axial order than without this axial order. This can be understood qualitatively using a scheme (Fig. 5.7) that bears some resemblance to Fig. 5.1 used in Sect. 5.3.2 to explain the enhancing effect of axial order on polar order. Let us start again in Fig. 5.6 with considering the statistically most abundant pair μ_1 and μ_2 with $\alpha_1 = 90°$ for different orientations of μ_2 with respect to its center of gravity. Consider now the dipole-dipole energy and the thermal population of pair states in two limiting cases of axial order: the Ising case with perfect axial order and the isotropic case with no axial order (Fig. 5.7). For $\alpha_1 = 90°$, the energy expression simplifies to $U_{dd} = \mu_1\mu_2\cos\Gamma_{12}/r^3$ with Γ_{12} the angle between μ_1 and μ_2.

In the Ising case, there are only two orientations possible for dipole 2, “up” or “down”. Since we started with dipole 1 “up”, these orientations correspond to

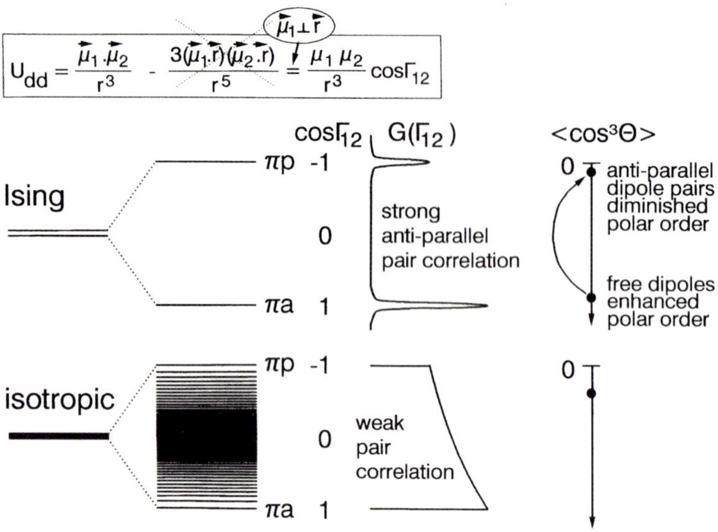

Fig. 5.7. Comparison of dipole-dipole energy in Ising and isotropic models. The interaction between the most abundant dipole pair of Fig. 5.6 is shown in the two models. The larger asymmetry in the thermal population of the maximally split-up levels of the Ising case causes stronger anti- parallel pair formation. This reduces polar order stronger in the Ising case than in the isotropic case, where dipole-dipole is relatively weak. Shown is the extreme case where the reduction is so strong in the Ising case, that polar order is even reduced with respect to the isotropic model solution

the pair states πp ($\Gamma_{12} = 0°$) and πa ($\Gamma_{12} = 180°$) respectively, each of which has an extreme value of the dipole-dipole energy. Because the energies of both pair states are (relatively) far apart, there will be a (relatively) large asymmetry in the population of both states, with the anti-parallel πa state situation more densely populated than the parallel situation. This corresponds to (relatively) strong anti-parallel dipolar association, which dimishes $\langle \cos^3\theta \rangle$.

In the isotropic case, all orientations of μ_2 are (initially) equally likely. In addition to the situations πp and πa with extreme energies, a lot of "intermediate" situations are also possible with non-extreme energies. A smaller energy separation between more parallel-like pair states and more anti-parallel-like pair states leads to less asymmetry in their thermal population and therefore to relatively weaker anti-parallel dipolar association. To this the geometrical/statistical argument may be added that, in fact, the most abundant pair states have angles Γ_{12} in the vicinity of 90°. Here, the energy separations are even smaller, resulting in even less asymmetry in the thermal population, therefore in relatively weak anti-parallel dipolar association.

Summarizing, there are two opposing tendencies in the effect of axial order on field-induced polar order. On the one hand, there is the enhancement by the reduced dimensionality. Assuming "free" dipoles without mutual interaction, this enhancement has maximum factor of 5 (Sect. 5.3.2). On the other hand, anti-parallel dipole-dipole association tends to reduce polar order, which tendency becomes stronger when axial order is higher. Thus, in addition to the enhancing effect on poling, axial order may also have an adverse effect through stronger anti-parallel pair correlation. It remains to be investigated which of the two effects "wins" under various conditions. For instance, pair correlation is expected to depend strongly on concentration. It might be conceivable, that $\langle\cos^3\theta\rangle$ as a function of concentration shows beyond the usual linear region a sublinear region and perhaps even a maximum, after which $\langle\cos^3\theta\rangle$ decreases with increasing concentration.

5.3.7
Ising Models Including Dipolar Interaction

Related to the phenomenon of antiparallel dipole association, we here describe variations in the Ising model that include dipole-dipole interaction. We start by examining just two point dipoles, for which we take the statistically most abundant pair of dipoles in Fig. 5.6, with $\alpha_1 = 90°$. We consider this pair of dipoles in the extreme case of perfect axial order (Ising limit) and in the absence of an electric field as already shown in Fig. 5.7. Only two situations are possible: parallel πp or anti parallel πa. The energy associated with these pair states relative to kT is + and –b, respectively, with $b = \mu^2/r^3kT$. The ratio μ^2/r^3 should be given a value representative for the material. The effect of an electric field on the πa pair state is zero and on the parallel pair state πp it gives a splitting in two levels + or –2a for orientation of the parallel pair anti-parallel or parallel to the field. Here, a is in the usual abbreviation of $\mu E/kT$. Now, similar to the standard Ising model it is easy to calculate the degree of field induced polarisation:

$$P/N\mu = \sinh(2a)/[\cosh(2a) + \exp(2b)] \tag{5.13}$$

In the limit of zero b, this expression reduces to the standard Ising model result for isolated dipoles, $P/N\mu = \tanh(a)$. For positive values of b and low a values (low E), this equation gives a reduced degree of polarisation compared to the standard Ising model, where the degree of polarisation is reduced in proportion with the initial fraction of anti-parallel dipole pairs. For larger a (larger E field strengths), the number of anti-parallel pairs is reduced and the degree of polarisation is enhanced. This gives rise to a superproportional behavior of P/Nμ with the field strength E. This is shown in Fig. 5.8a for μ = 7 D, r = 6 Å and T = 380 K (b = 4.3 and a/E = 0.00445 μm/V) compared to the standard Ising result Note that $P/N\mu$ is just the polar order parameter $\langle P_1\rangle = \langle\cos\theta\rangle$, ignoring local field effects. As all averages of odd powers of $\cos\theta$ have the same value in an Ising model, the above equation also can be applied to calculate $\langle\cos^3\theta\rangle$, appropriate for determining $\chi^{(2)}$.

A further extension of the model described above is to include two non-interacting dipoles (e.g. at large distance). We assume that the relative concentrations of paired and non-interacting dipoles is determined by a Boltzmann distribution of their energy levels. In this case, the two interacting dipoles have the same energy levels (relative to kT) as above (+/–b for the dipole-dipole interaction and +/–2a for the interaction of the parallel pair with the field). The single dipoles have the energy levels +/–a as in the standard Ising model. The

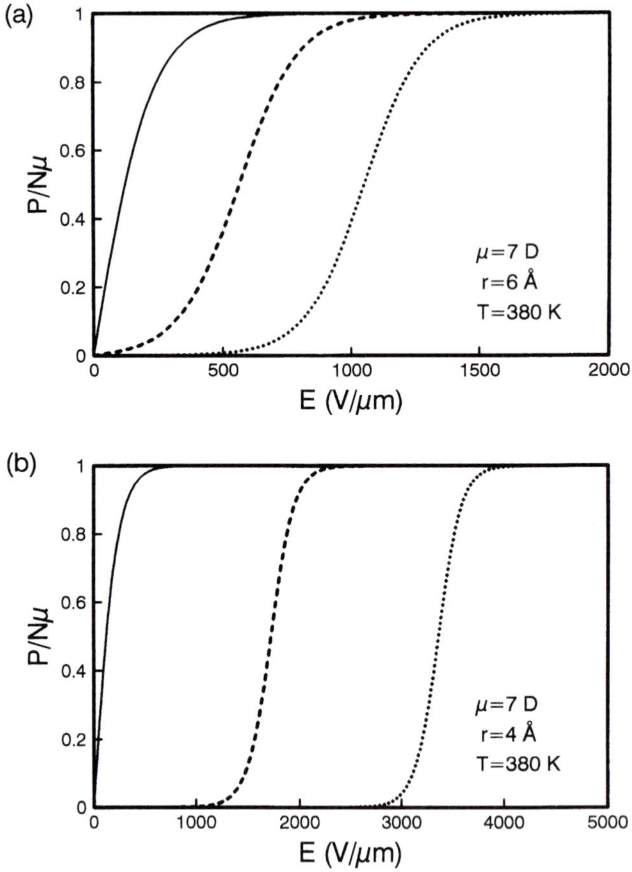

Figs. 5.8a,b. Ising models with dipolar interaction. The *solid lines* represents the field dependence of $P/N\mu = \langle\cos\theta\rangle$ in the original Ising model, viz. $\langle\cos\theta\rangle = \tanh(a)$ with $a \equiv \mu E/kT$. The *dotted lines* are from an extension of the Ising model, taking the interaction of a dipole pair with $b \equiv \mu^2/r^3kT$ into account. Another extension (*dashed lines*) includes free dipoles (see text). Input values are indicated in Figs. a and b

result of this calculation leads to a nearly identical expression as in Eq. 5.13, with the exception that 2b is replaced by b. Due to this similarity, it is not really necessary to describe the behavior of P/Nμ in further detail. Fig. 5.8a shows a result for μ = 7 D, r = 6 Å and T = 380 K.

In the above, we have obviously simplified the problem of dipole- dipole interaction enormously, so that only a qualitative agreement with experiments can be expected. The main importance of this simplified approach is the occurrence of a superproportional behavior for $P/N\mu$ analogous to the result from the MSVP model. In this case, however, the superproportional behavior is due to the decrease of the concentration of anti-parallel dipole pairs ("dimers") as a function of the applied field strength, rather than due to the rapid change in field induced axial order. In this case, we predict a superproportional behavior of $P/N\mu$ in a nematic phase with perfect axial order. The predicted change in dimerization will in principle also occur in initially isotropic systems, but the calculation would be much more complicated.

Whether or not the superproportional behavior can be observed experimentally remains to be investigated. For the input parameters used $\mu = 7$ D, $r = 6$ Å and $T = 380$, the superproportional behavior is predicted for field strengths smaller than 1000 V/µm – considered as the practical upper limit of local fields – and therefore observable at the strongest fields possible. It can be questioned whether the rather larger value of $r = 6$ Å is realistic. Substituting a lower value, e.g. $r = 4$ Å, shifts the superproportional behavior to higher fields beyond 1000 V/µm (Fig. 5.8b) and therefore outside the experimental range. It should be added that the parameter b is only an effective parameter, of which the dependence on μ and r is only given by $b = \mu^2/r^3$ in the case of point dipoles. Therefore, μ and r are also effective parameters, for which realistic values are not a priori clear.

5.3.8
Introduction of Pair Formation into MSVP Model

Anti-parallel dipolar association has been treated by Dalmolen, Picken, De Jong and De Jeu [34] in terms of a dimerization of individual dipoles. Instead of consisting of loosely associated dipoles, a material is described as a mixture of two species, which are in equilibrium with each other: monomers and dimers. The monomers are completely free (non-associated) dipoles, the dimers consist of two rigidly connected anti-parallel dipoles ("πa"). From polarized Raman measurements in the nematic and isotropic phases of a homologous series of *p*-alkyl-*p*′-cyanobiphenyls, they obtained values for $\langle P_2 \rangle$ and $\langle P_4 \rangle$. Information on dipolar association was obtained from permittivity measurements (Kirkwood-Fröhlich formalism). Using their dipole dimerization model, they were able to attribute low $\langle P_4 \rangle$ values to high dimer fractions, corresponding to large anti-parallel dipole association. They found an appreciable dimer fraction x_D in the isotropic state, which increased on cooling. In agreement with the above reasoning, they found a discontinuity at the phase transition, where the dimer fraction became larger in the nematic phase. The discontinuity was small, though, in the case of the cyanobiphenyls, where considerable pair formation was already found in the isotropic phase.

Dipolar association could be introduced semi-empirically into our model, along the line suggested by Dalmolen et al. [34]. Since the dimers (anti-parallel dipole pairs) have no net dipole moment, they do not contribute anymore to the poling and, effectively, lead to a reduction of the number density of polable (non-dimerized) dipoles by a factor of $(1-x_D)$. Theoretical values for $\chi^{(2)}$ can be calculated using our model for the order parameters and using empirical values for x_D from permittivity experiments to correct the number density.

5.3.9 Theoretical Model of Wang (1993) Including Pair Correlation

Wang [35] proposed a statistical mechanics method for the calculation of order parameters, including an angular dependent intermolecular interaction potential. It is shown that in addition to the alignment of dipoles to the electric field, the orientational pair correlation factor is also important in determining the orientational polar order parameters. With pair correlation included, the order parameter $\langle \cos^3\theta \rangle$ becomes a function of the NLO chromophore concentration. The concentration dependent part of this order parameter is shown to be proportional to the Kirkwood-Fröhlich parameter g associated with the orientational correlation of electric dipoles. In the weak field limit, the ratio of the two polar order parameters, $\langle \cos^3\theta \rangle / \langle \frac{1}{2}\cos\theta\sin^2\theta \rangle$ remains equal to 3, independent of the specific nature of the anisotropic intermolecular potential.

5.3.10 Model of Wu (1991) to Describe Transient Phenomena

With the above mentioned molecular statistical models, one can only calculate final values of order parameters that are obtained after infinite poling times; the models do not describe the time-dependence of poling or relaxation. Wu [36] made a theoretical model for the description of both steady-state and transient phenomena in poled polymers. The time-dependent orientational distribution function of the molecular dipoles was expanded in terms of Legendre polynomials with spherical modified Bessel functions as coefficients. An analytical relation was derived between the EO coefficient and the birefringence in the steady-state. The time-dependence of the onset and decay of the EO and optical properties was contained in a rotational diffusion equation with diffusion constant D. This equation was solved with the help of the recurrence relation for spherical modified Bessel functions. It was found that the onset of birefringence involved at least two time constants, with rise times of 1/2D and 1/6D, while the onset of the EO effect was dominated by the rise time of 1/2D. After removal of the DC poling field, the birefringence was found to relax with a time constant 1/6D, whereas the EO coefficient relaxed with a time constant 1/2D. This difference in time constant between the two relaxation processes was attributed to the difference in tensorial rank describing the birefringence and the EO effect. The author did not consider the influence of liquid crystallinity on either the steady-state values of the EO coefficients, nor on the dynamics of poling or relaxation.

5.3.11 Other Models

Apart from the models briefly described above, one of the most powerful methods to examine the phase diagrams is by using Landau theory. For low molecular weight nematics the best known theories are the Landau-De Gennes theory for the nematic phase [37] and the theories of De Gennes [38, 39] and

McMillan [40] for smectics. In the Landau-De Gennes theory, the free energy difference between the nematic and the isotropic phase is given by an expression of the form:

$$\Delta f = \tfrac{1}{2}C_1 Q_{\alpha\beta}Q_{\beta\alpha} - \tfrac{1}{3}C_2 Q_{\alpha\beta}Q_{\beta\gamma}Q_{\gamma\alpha} + \tfrac{1}{4}C_3 \left(Q_{\alpha\beta}Q_{\beta\alpha}\right)^2 \tag{5.14}$$

where $Q_{\alpha\beta}$ is the tensor order parameter $Q_{\alpha\beta} = S(N_{\alpha\beta} - 1/3\delta_{\alpha\beta})$, with $S = \langle P_2 \rangle$. In essence, the above equation is a straightforward expansion of the free energy near the transition in a Taylor series. Of the three coefficients C_1, C_2, and C_3 usually only the temperature dependence of C_1 is taken into account using an expression of the form $C_1 = h(T - T^*_{cl})$. This procedure then leads to the prediction of a first-order ($C_2 > 0$) or second-order ($C_2 = 0$) phase transition. By comparing the prediction of the Landau-De Gennes theory with experiments, values for h, T^*_{cl}, C_2 and C_3 can be found.

Apart from the NI phase transition and associated latent heat, the Landau-De Gennes theory allows the calculation of "nematic" fluctuations in the isotropic phase above T^*_{cl} and can obviously be extended to the case of externally applied fields e.g. by using terms of the form:

$$f_M = -\tfrac{1}{2}\mu_0^{-1}\Delta\chi B_\alpha B_\beta \tag{5.15}$$

or

$$f_E = -\tfrac{1}{2}\Delta\varepsilon E_\alpha E_\beta \tag{5.16}$$

respectively for applied magnetic and electric fields. In a review by Gramsbergen et al. [41], the application of Landau theory to the nematic phase is reviewed in more detail. Also the book by Vertogen and De Jeu [42] provides an excellent overview of this type of theory and compares it with other methods e.g. the Maier-Saupe and Onsager theories [43].

In addition, the work of Warner et al. should be mentioned, where a variety of interesting polymeric liquid crystals structures are treated in some detail, both main chain polymers [44–46] and side chain polymers [47–49]. Wang and Warner [45] examined the effect of external electric or magnetic fields on polarizable dipoles along the backbone of nematic main chain polymers. In close analogy to conventional nematics, a line of first-order transitions from nematic to paranematic was found as the field was increased, culminating in a critical end point as the critical field was attained. This critical point was related to those found by other authors. The equivalence of their results to the Landau approach in a field was demonstrated. Shortcomings in previous work on nematic polymers in field were analyzed. Gunn and Warner [46] examined the orientation of dipoles in the backbone by an external E field leading to an unusual dielectric response. Warner and Liu [49] studied the dielectric response for a side chain polymer theoretically. Assuming a stereoregular isotactic polymer structure – perhaps not very realistic for most materials studied sofar in experiments – they derived an unusual exponential dependence for the polarization.

Finally, the work of Khokhlov and Semenov and of Odijk should be mentioned. They studied the Onsager approach – based on excluded volume – for the main chain polymers in tremendous detail. For reviews of their work see Odijk

[50]. As regards the effect of an external electric field on such models (see Khokhlov and Semenov [51]), where a field induced phase transition is predicted with a critical end point and various polymer flexibility mechanisms are compared (stiff, freely jointed, persistent semi-flexible).

5.4 Measurement of the Electrooptical Effect in Polymer Samples

5.4.1 Measurements During Poling

Electrooptical experiments can be performed during poling, using the so-called DC-induced Pockels effect technique [15], which bears some resemblance with DC-induced Second Harmonic Generation (DCSHG or EFISH) in solutions of AΠD molecules [52]. Macroscopic second-order nonlinearity is induced in both cases by a DC poling field. The induced nonlinearity is characterized by the frequency doubling of a laser beam in DCSHG, whereas an additional AC modulation field is applied in the EO counterpart to measure the induced Pockels coefficients with phase sensitive detection. Usually, the DC and AC voltages are applied to the same electrode set.

Poling is also possible using the corona discharge from a sharp object. In this case, ionic species are deposited on top of the polymer and form an ionic top electrode. Because the ions have a low lateral mobility, the whole top electrode does not discharge during breakdown at a weak spot in the polymer, which reduces the damage to a minimum. Therefore, ultra-high poling field strengths (approaching the dielectric strength) are possible with corona poling. Because of the absence of a conductive top electrode, EO experiments are difficult to perform during corona poling, but SHG is possible [53, 54].

Another type of measurement that can be applied during poling is measuring the electric current through the sample during a temperature ramp (TSC) or at constant temperature as a function of time. The TSC technique (Thermally Stimulated Current, also called Thermally Stimulated Poling – TSP) can be applied to study the structure dependence of poling kinetics and thermodynamic [55]. A variant of this method is measuring the current in the poled samples due to relaxation while temperature is ramped up (Thermally Stimulated Discharge – TSD).

5.4.2 Electro-Optic Measuring Methods

Electrooptical measurements may differ in many respects, e.g. in the shape of the studied test samples (bulk polymer or thin layer; electrodes free standing or attached; co-planar or plane-parallel electrode configuration); in the propagation mode of the laser beam (transmission, reflection, guided); in the noise-reduction techniques applied (averaging, lock-in amplifier); or in whether measurements are performed after or during poling. We will not go into all these details and only briefly discuss another aspect separating the different EO measurements, viz. the basic principle that is used to convert the EO modulation of the refractive index into a measurable intensity modulation of a laser beam.

Several sample types allow EO measurements between crossed polarizers [56, 57]. The polarization state of the laser beam is linear after passing the first polarizer, but it becomes elliptical after passing the sample if the polarization direction makes an angle with the optical axis of the birefringent sample (not 0° nor 90°; often 45°). The birefringence is modulated by the Pockels effect, because the modulation of n_e (via r_{33}) and n_o (via r_{13}) is different. As a result, the polarization state of the transmitted light is modulated. This modulation is converted into an intensity modulation by the second polarizer, an external device. Measurement of this intensity modulation provides the value of $r_{33} - r_{13}$ (if birefringence is small).

If the incident light is polarized along or perpendicular to the optical axis, only the phase of the transmitted beam is modulated by the Pockels effect. This phase modulation can be converted into an intensity modulation by interference. For instance, with the sample placed in one arm of a Mach-Zehnder interferometer, the transmitted beam interferes with the beam through the empty arm [56, 58]. The polymer test samples in between plane-parallel, semi-transparent metal electrodes do not need an external device for interference, as they have an internal mechanism. Because of the high reflectivity of the metal layers (mirrors), the metal-polymer-metal sandwich forms a Fabry-Perot resonator cavity, where rays that have been internally reflected a different number of times interfere with each other. In the Fabry-Perot interferometric EO measurements, the transmission or reflection are measured as a function of either wavelength [59–61] or angle of incidence together with their EO modulation [57, 62–64]. With the incident beam s-polarized, r_{13} is measured. The coefficient r_{33} can be determined e.g. by combined crossed polarizers and Fabry-Perot experiments with s-polarization in a common set-up [57]. Both EO coefficients can also be obtained from angle-dependent Fabry-Perot experiments with only p-polarized light or from combined s and p measurement. Still another Fabry-Perot interferometric technique has been described by Meyrueix et al. [65–67].

Results of SHG measurements and several different EO measurements have been compared recently by Morichère et al. 1993 [68].

5.5 Comparison of Experimental Results with Theory

5.5.1 General Consideration

A comparison between absolute values of experimental and theoretical values of Pockels coefficients is complicated by many factors [15]. Theoretical values can be obtained from Eqs. (5.6–5.7) and the specific theoretical poling model by inserting the necessary input parameters. These input parameters, however, may not be known precisely. Since they are the result of other experiments, they contain experimental errors. The theoretical calculation of Pockels coefficients is complicated furthermore by the local field problem. Four local field factors, which only can be approximated, enter the calculations, three from the relation between macroscopic and molecular NLO properties (cf. Eqs. (5.3–5.7) and one for the poling field used in the poling models (Sect. 5.3.1). Usually, Lorentz local field factors (n^2+2/3) are used at optical frequencies, whereas Onsager local field factors are more appropriate at lower frequencies. For a general discussion of the local field problem, the reader is referred to the text books of Bötcher [27]; for a more specific treatment of the local field problem in poled polymers see Meredith [28]. The measured Pockels coefficients too are hampered by experimental errors, both experimental scatter and systematic errors e.g. due to piezoelectric contributions to apparent Pockels coefficients. Theoretical and experimental values must be compared under conditions that are "matched" to each other [15]. For instance, possible resonance enhancement of experimental values must be taken care of, either by performing the measurements far from resonance, or by applying a correction for resonance enhancement [69, 70]. Also, any decay of the EO coefficients during cooling and during or after switching the field off must be prevented, since these effects – though in themselves interesting – are not covered by the theoretical models. This is possible by measuring the EO coefficients during poling at T_g, although this may increase the piezoelectric contribution. Part of the "real" discrepancy between experiment and theory can be attributed to sterically hindered rotatability of the side chains [71–73], but probably also to significant anti-parallel dipole-dipole association.

A comparison between theory and experiment is easier for relative instead of absolute values of Pockels coefficients, for instance the value of r_{33} relative to that of r_{13} allows a comparison with predictions from poling models without many complications. The ratio r_{33}/r_{13} is approximately equal to the ratio of the corresponding polar order parameters viz. $v \equiv \langle\cos^3\theta\rangle / \langle\frac{1}{2}\cos\theta\sin^2\theta\rangle$. Another example is the (relative) behavior of a Pockels coefficient as a function of the poling field strength or poling temperature.

Of particular importance would be a comparison between experiment and theory for the value of $r_{33}(\langle\cos^3\rangle)$ measured in an LC phase relative to the value measured in the isotropic phase. To suppress temperature effects, the measurements or calculations preferably have to be performed at the same, or nearly the same temperature. Three experimental ways seem possible.

One is to perform the measurements during poling slightly above T_{cl}- in the isotropic phase – and slightly below T_{cl} – in the oriented LC phase. Disadvantages are stronger sample degradation and larger piezo-electric contributions because the polymer can be rather soft above T_g.

A second way, avoiding these problems, would be to perform the poling experiments below T_g in the frozen-in isotropic and the frozen-in single-domain LC phase. It remains to be investigated whether a poling temperature below T_g exists where poling in the frozen-in isotropic phase is already possible, whereas formation of an ordered LC phase is still too slow. This second route might also be used to investigate the influence of LC axial order on relaxation at a temperature below T_g.

A slightly different, third route might be to perform the measurements in the single-domain LC phase and in the polydomain LC phase, since the latter is expected to behave macroscopically as an isotropic medium in some respects - not so far as the influence of dipole-dipole interaction on the static values of the EO coefficients is considered, nor considering the kinetics of poling and relaxation. Here, we must prevent the poly-domain sample becoming single-domain while the AΠD moieties become oriented by the electric field. This third route has been applied successfully by Stupp et al. [23, 24] in a guest-host system. They used a strong magnetic field for domain pre-alignment and a weak electric field for dopant alignment. The electric field had to be weak to prevent electrical domain alignment in the polydomain sample.

5.5.2
Dilute Liquid-Crystalline Systems

Liquid-crystalline systems described in the literature with a low hyperpolarizability density N are either guest-LC host systems or LC copolymers. Both categories are examples of mixed rod systems. In the guest-LC host systems, the hyperpolarizable molecules are doped in a liquid-crystalline host polymer, which is either a side chain polymer [5, 6] or a main chain polymer [23, 24]. In the LC side chain copolymers [20, 74], separate hyperpolarizable and mesogenic groups are attached as side chains to the backbone.

The first mention in the literature of enhanced poling by liquid crystallinity, was by Meredith, Van Dusen and Williams [5, 6]. Their suggestion of enhanced poling, based on a comparison between the isotropic and Ising models, was accompanied by experimental results, which seemed to support the theoretical predictions. An LC polymer was doped with 2 wt/wt % DANS and poled with a field strength of 1.4 V/μm. $\chi^{(2)}_{333}$ was determined from SHG experiments and compared with a theoretical prediction using the Ising model. The result, $\chi^{(2)}_{\text{exp}} \approx 0.5^{*}\chi^{(2)}_{\text{Ising}} = 2.5^{*}\chi^{(2)}_{\text{isotropic}}$, was claimed to be consistent with an axial order parameter determined from dichroism measurements $\langle P_2(\cos\theta)\rangle \approx 0.3$ or $\langle\cos^2\theta\rangle \approx 0.5$), and was regarded as experimental proof for the enhancement [5, 6]. A direct comparison between experimental and theoretically calculated NLO coefficients is, however, a very dangerous thing to draw conclusions from, as we remarked in Sect. 5.5.1. This first experimental indication for enhanced poling should, therefore, be treated with caution.

Stupp et al. [23, 24] studied dopes of the well known chromophor Disperse Red 1 (4-[ethyl(2-hydoxyethyl)amino]-4′-nitroazobenzene) in a solidified LC main chain polymer. The host contained three different structural units, all aperiodically arranged and connected by ester bonds. The semiflexible chains formed a nematic phase in the temperature range between the melting point of 185 °C and 220 °C. The nematic phase may extend below 185 °C. Above 220 °C, a nematic-isotropic biphasic fluid was observed. Solvation of the dye was possible for concentrations up to 15 wt %. The samples were poled by an electric field below the melting point of the host using moderate field strengths which were too weak to align the domains and form a single-domain phase. Single-domain formation was possible before poling by aligning the main chains with a strong magnetic field. The magnetic (pre-)orientation only introduces macroscopic axial order, not polar order.

The second-order NLO properties induced by the electric field were studied in magnetically aligned (single-domain) and non-aligned (poly-domain) samples using SHG measurements. In the samples loaded with 5 wt % dye, it was found that magnetic alignment increased the SHG intensity by a factor of 6–9. Taking into account optical scattering effects, this increase could be ascribed to an enhancement of the field induced polar order in the monodomain nematic phase by a factor of about 3 with respect to the polydomain phase.

In addition, the authors found that poling proceeded more rapidly in the macroscopically aligned samples than in the non-aligned samples; thermal relaxation of polar order, on the other hand, was slower in the magnetically ordered system.

The enhancement of field induced polar order which is found is qualitatively in agreement with the poling models, considering that the multi-domain phase can be treated as isotropic in a molecular statistical model calculation taking into account only the energy of the permanent dipoles in the electric field (isotropic model). When also considering dipole-dipole interaction, or considering the kinetics of poling and relaxation, the polydomain nematic phase is expected to show more resemblance to the monodomain nematic phase than to a real isotropic phase. It is therefore difficult to draw conclusions about the influence of liquid crystallinity on the kinetics of poling and relaxation from the above-mentioned observations.

Le Barny et al. synthesized mixed-rod side chain copolymers [20], which were studied by dielectric relaxation spectroscopy. No positive experimental evidence for enhanced poling in the LC phase has been reported yet.

5.5.3
Concentrated Isotropic Systems

The mesogenic properties of the AΠD rods will only give rise to a LC phase in the single-rod systems, if their concentration is high enough, viz. if they are not diluted too much by the polymer. To achieve high concentrations without recrystallization problems, the AΠD rods have to be attached chemically e.g. as side chains to a polymer backbone. Experimentally, the important single-rod systems are mainly side chain homopolymers and also some copolymers, some of which indeed show a LC phase [19, 21]. In this subsection, the isotropic side chain homo- and copolymers will be reviewed, which do not show spontaneously a LC phase, but which do show behavior related to strong field-induced axial order.

A **field-induced phase transition** from an isotropic to a LC phase – as predicted by our model – has been reported recently by Gonin et al. [75, 76] in an isotropic copolyether and also in an isotropic homopolymethacrylate, both containing OCB (oxycyanobiphenyl) in the side chain.

A number of copolyethers were prepared by chemical modification of atactic polyepichlorohydrin with the sodium salt of 4-4′-hydroxycyanobiphenyl [75, 76]. The CH_2Cl side chains were partly substituted by the classical mesogenic group oxycyanobiphenyl (OCB), which is also slightly hyperpolarizable. The OCB group is attached to the backbone via a $-CH_2-$ spacer group. Although these polymers are copolymers and not homopolymers, they can be classified as single-rod systems, since the OCB moiety is the only rod-like group having hyperpolarizable or mesogenic properties. Substitution percentages larger than approximately 65 % resulted in formation of a LC (nematic) phase between T_g and $T_{cl} > T_g$. For instance, 80 % substitution gave a polymer "A" with a nematic phase between a T_g of 79.5 °C and a T_{cl} of 146.5 °C. A second copolyester "B" with a substitution percentage of 55 % had a T_g of 56 °C and was isotropic without a field [75, 76]. In addition to the OCB grafted copolyethers, commercially available OCB side chain homopolymers were also studied with OCB connected via a $-(CH_2)_3-$ spacer to the backbone (Merck). Polymer "C" was a polyacrylate with a smectic S_A phase between a T_g of 51.5 °C and a T_{cl} of 97 °C. The analogous polymethacrylate ("D") had a T_g of 97 °C and was isotropic at rest. Here, only results in the isotropic polymers B and D are reviewed. Results in the LC polymers A and C are mentioned in the next subsection.

The location of the *virtual* zero-field clearing temperature T_{cl} of polymer B was estimated in the region 40–56 °C by extrapolation of the measured T_{cl} values of A–B mixtures, using the mutual miscibility method [77]. Samples were poled with corona poling, which permits extremely high poling field strengths approaching the dielectric strength, usually a few hundred V/μm in polymers. The value of the axial order parameter $\langle P_2 \rangle$ in the poled polymer was determined from dichroitic absorption measurements and was found to be large, about 0.45. Also, the value of the ratio of second-harmonic coefficients (about equal to the ratio of the Pockels coefficients) was found to be much larger than the value of 3 in the isotropic model: $d_{33}/d_{13} = 11$. These results strongly suggest a transition to the nematic phase during poling. It was concluded that the enhancement of the NLO performance by the field-induced LC character is larger by a factor (slightly) more than 3 in comparison to the case of the isotropic model. Similar results were also obtained in the isotropic side chain homopolymethacrylate D.

A **superproportional behavior** of the EO effect as a function of poling field strength – theoretically predicted in isotropic concentrated polymers – has been observed on several occasions.

A first indication of superproportional behavior can be seen in the poling results of Hoechst Celanese polymers in the plenary papers by Stamatoff [78] and Haas et al. [79]. Figure 1 in the paper by Stamatoff [78] shows the Pockels coefficient as a function of the poling voltage. After a linear region (isotropic), one can observe a superproportional region (phase transition) which is followed again by a second linear region with a steeper slope (liquid crystalline). This resembles situation (3) (section 5.3.5) Figure 4 in the paper by Haas et al. [79] shows similar, though less detailed, poling results in a polymer HCC-1232. The authors do not give any experimental details and do not comment on the behavior shown.

Superproportional behavior has also been observed in our group in a DANS side chain polymer with T_g 140 °C [71, 72]. Test samples consisting of a spuncoat single layer of the polymer in between semitransparent, plane-parallel metal electrodes were studied during poling at 130 °C with the DC-induced Pockels effect technique, where a superposition of a DC poling voltage and an AC modulation voltage is applied to the sample electrodes. The EO effect was measured between crossed polarizers in transmission and in reflection. The applicable poling field strength is limited in the single layer test samples by dielectric breakdown at weak spots in the sample. Electric discharge at a weak spot is self-healing and only results in elimination of the weak spots (sparkling). In principle, this does not influence the measurements very much, unless the measure spot – where the laser hits the sample – itself is affected. Then, a new measure spot has to be selected, which may lead to a discontinuity (shift) in the field dependent measurements due to the nonuniformity of the layer thickness. Despite the increasing number of dielectric breakdowns at higher field strengths and occasional repositioning, superproportional behavior could be observed for field strengths in excess of 50 V/μm (Fig. 5.9) [72, 73].

Of great interest are the recent observations of Meyrueix et al. [67] concerning a polyurethane ("PU11") with the DANS-like azo chromophore DR17 as active side group. Not only did these authors find a value for r_{33}/r_{13} larger than 3, which increases with increasing poling field strength, but also a superproportional dependence of r_{33} and a subproportional dependence for r_{13}, all as predicted by our model.

Buckley [80] reports superproportional behavior in a PMMA sidechain polymer containing about 75 wt % DANS as active group.

Kelderman [81] mentions a superproportional field dependence during poling at 105 °C of a 75 wt % film of tetrapropoxytetranitrocalix[4]arene. Disregarding

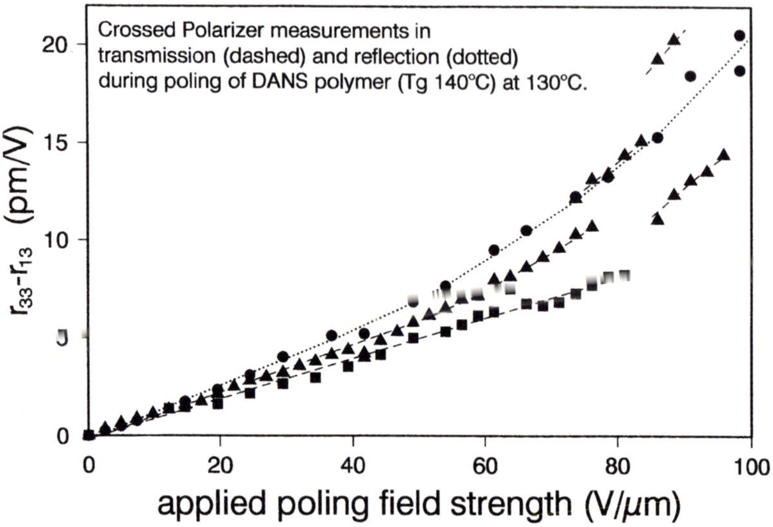

Fig. 5.9. Dependence of EO coefficients on applied poling field strength. This Figure shows the field dependence of a DANS side chain polymer as measured between crossed polarizers. Note the superproportional field dependence, tentatively associated with a field-induced isotropic-to-nematic transition. Discontinuities are due to dielectric breakdown at the measure spot. A new measure spot was selected, when the breakdown was "self-healing" [71]

the parabolic (?) curve fitted through the data points in Fig. 6.5 of his thesis, one can recognize a linear regime up to 20 V/μm, a superproportional regime between 20 and 30–40 V/μm, leveling off to a second almost linear regime between 30–40 and 50 V/μm (highest applied field strength). This resembles case (3) of Sect. 5.3.5.

At field strengths lower than in the superproportional region (of r_{33}), the predictions from our model coincide with those of the isotropic model. Both models then predict a linear field dependence of the polar order parameters $\langle \cos^3\theta \rangle$ and $\langle \frac{1}{2} \cos\theta \sin^2\theta \rangle$, and a value of their ratio $\langle \cos^3\theta \rangle / \langle \frac{1}{2} \cos\theta \sin^2\theta \rangle$ equal to 3. Measurements in several of our polymers indeed showed a linear field dependence of the Pockels coefficients at moderate field strength and a ratio r_{33}/r_{13} near the value of 3 [56, 57, 71, 72]. More recent results of ratios *smaller* than 3 [73] must probably be attributed to piezoelectric contributions to our *apparent* Pockels coefficients. Meyrueix et al. reported a method for correcting the apparent Pockels coefficients for the piezoelectric effect [67].

From the above, it can be inferred that the theoretical models which neglect dipole-dipole interaction apply even for concentrated (but isotropic) systems – at least in their qualitative predictions. However, looking more quantitatively, there are indications that anti-parallel dipole-dipole association does exist. For instance, experimental values of Pockels coefficients are often appreciably smaller than theoretically predicted values using Eqs. (5.6, 5.7) and the poling models for calculating the order parameters. Van der Vorst and Huyts [56] found a discrepancy by a factor of 5 between experiment and theory (isotropic model) in a MONS side chain polyacrylate, which was poled in the linear region. In a moderately poled DANS side chain polymer, a discrepancy by a factor 3 was found [57]. If resonance enhancement of the experimental Pockels coefficients is (better) taken into account, this discrepancy may become even larger. Experimentally, there are indications for steric hindrance during poling, which may account in part for the discrepancy between experiment and theory. It is, however, unlikely that sterical hindrance alone is responsible for such a large difference. Some degree of anti-parallel dipole-dipole association might well explain the remainder, perhaps the largest part, of the discrepancy between experiment and theory.

5.5.4
Concentrated Liquid-Crystalline Systems

Liquid crystalline side chain homopolymers with the mesogenic and hyperpolarizable properties united in a single type of rod-like side group (single-rod system) are expected to provide the largest Pockels coefficients as they combine a high number density of hyperpolarizable groups with liquid crystallinity, which is expected to enhance poling. Several groups have started synthesis along this route. LC side chain homopolymers with hyperpolarizable side chains have indeed been obtained in this way [19, 75, 76, 82].

The nematic OCB grafted side chain copolyether A and the smectic OCB homopolyacrylate C of Gonin et al. [75, 76] have already been mentioned in the previous subsection. Only minor experimental results were reported on polymer

C. The ration d_{33}/d_{13} in polymer A was determined to be 15±3 from SHG measurements. For $\langle P_2 \rangle$, a value was obtained of 0.55±0.02 from dichroism measurements. These experimental results were compared with the results of the theoretical models mentioned in Sect. 3.1. Reasonable agreement was obtained using either the SKS or the MSVP model. The ratio d_{33}/d_{13} significantly larger than 3 indirectly confirms that the polar order parameter $\langle \cos^3 \theta \rangle$ can be enhanced by liquid crystallinity, whereas $\langle \frac{1}{2} \cos\theta \sin^2 \theta \rangle$ is reduced.

Direct experimental evidence on enhanced poling in liquid crystals is scarce in the literature, which is rather strange considering that several groups ventured to study the LC option of poled polymers as early as the end of the 1980s. Of course, the enhancement of poling in a field-induced LC phase and also the superproportional behavior accompanying the field-induced transition, both discussed in the previous Sect. 5.5.3, can be regarded as positive evidence for enhanced poling in liquid crystals.

Results in our own laboratories from spontaneous LC side chain homopolymers suggested quite the contrary. Instead of an enhancing effect on poling from liquid crystallinity, we found indications of an adverse effect in two LC side chain homopolymers [82]. An adverse effect cannot, however, be concluded without ambiguity, because of certain experimental problems. Because of the importance of a possible adverse effect from liquid crystallinity on poling, we will deal with the experimental conditions in more detail.

One polymer was a MONS polyester with a nematic phase between T_g : 60 °C and T_{cl}: 85–90 °C, of which the structure of the repeating unit is shown in Fig. 5.10 [82]. The other was a MONS side chain polyurethane synthesized at University of Delft as part of a doctorate study, details of which will be published elsewhere [83]. This second polymer had a nematic to isotropic transition at about 58 °C; T_g was probably near T_{cl}.

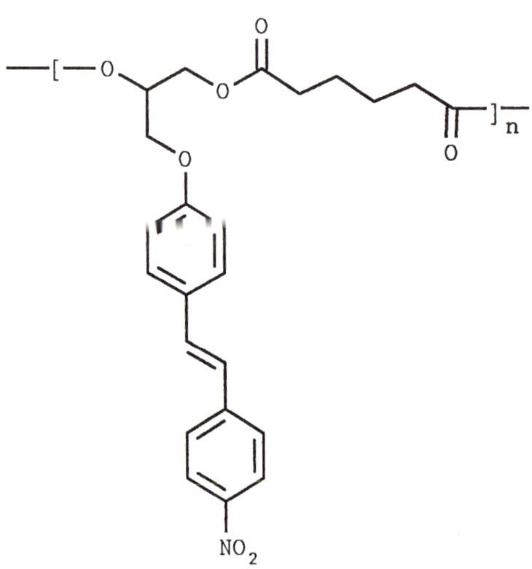

Fig. 5.10. Repeating unit of a nematic MONS side chain polyester. ($T_g = 60°$; $T_{cl} = 85-90°C$)

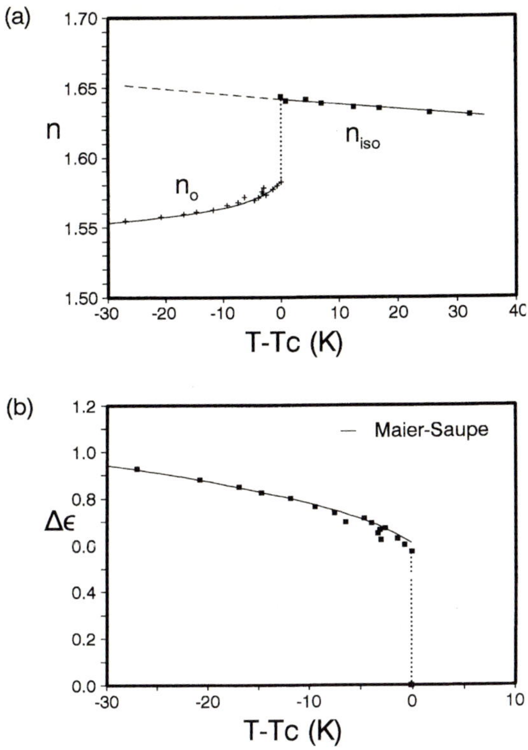

Fig. 5.11a,b. Temperature dependence of linear optical properties. **a** Result of temperature dependent measurements of the refractive index in the MONS polyester (see Fig. 5.10) using an Abbe refractometer; **b** Temperature dependence of permittivity anisotropy in the nematic phase fitted with the Maier-Saupe theory [78]

The refractive index of the MONS polyester in the isotropic phase slightly above T_{cl} was determined from prism-coupling experiments with a HeNe laser: $n_{iso} = 1.62$ (633 nm). With an Abbe refractometer using visible light effectively at the Na D line, a slightly higher value of about 1.65 was found. On cooling from the isotropic phase to the nematic phase, n suddenly became anisotropic at T_{cl}. Only the in-plane component of n (n_o) was measured in the nematic phase (Fig. 5.11a). Extrapolating the temperature dependence of n_{iso} to below T_{cl} and taking $n^2_{iso} = (n^2_e + 2n^2_o)/3$, the birefringence $\Delta n = n_e - n_o$ or the anisotropy of the permittivity $\Delta\varepsilon = \varepsilon_e - \varepsilon_o = n^2_e - n^2_o$ can be calculated. Figure 5.11b shows $\Delta\varepsilon$. Birefringence Δn and $\Delta\varepsilon$ grew further below T_{cl} with decreasing temperature. The observed temperature dependence can be understood from the linear relationship between $\Delta\varepsilon$ and the axial order parameter $\langle P_2(\cos\theta)\rangle$, viz. $\Delta\varepsilon = \Delta\varepsilon_{max} \cdot \langle P_2(\cos\theta)\rangle$, and the Maier-Saupe theory for $\langle P_2(\cos\theta)\rangle$. Agreement between experiment (data points) and theory (drawn curves) was obtained for $\Delta\varepsilon_{max} = 1.61$ (Fig. 5.11b).

Freshly prepared samples at ambient temperature showed a scattering texture indicating a frozen-in multi-domain LC phase. They became transparent on heating above T_{cl}. On cooling below T_{cl}, the scattering texture reappeared, if no electric field were applied. Scattering increased on cooling and became more intense when lower cooling rates were applied (larger domains). A transparent phase was frozen in at ambient temperature by using high cooling rates (from T_{cl} to ambient temperature in about a second). This transparent phase is either a real frozen-in isotropic phase, or just a multi-domain nematic phase with very small domains. The scattering multidomain texture of the nematic phase could be made transparent (single-domain) by shearing a polymer film between two microscope slides above T_g, or by applying an electric field of a few Volts per micrometer. The best results were obtained when either the shearing or the poling were performed during cooling through the isotropic to nematic phase transition at T_{cl}. Domain alignment was possible slightly below T_{cl} with DC or AC field strengths in excess of about 5 V/μm. It took a few minutes before the scattering texture disappeared and samples become transparent. Formation of a transparent single-domain sample by cooling through the phase transition was possible with DC or AC fields in excess of about 2 V/μm.

The EO effect was studied in two types of test samples, which differed in the electrode configuration: coplanar or plane-parallel samples. In both cases, electro-mechanical sample degradation occurred on poling in the vicinity of T_{cl}, where the polymer viscosity became too low to withstand the attractive forces between the electrodes. This complicated the comparative EO measurements in the nematic phase above T_{cl} and in the nematic phase below T_{cl}. Sample degradation was strongest in the plane-parallel samples with the top electrodes directly on top of the soft polymer layer, resulting in electrode crinkling and shortcircuiting. Sample degradation was less severe in the coplanar samples with both electrodes evaporated on the glass substate. Experiments in the coplanar samples had to be performed within half an hour; after that period, the electrode edges became too irregular. Results of crossed polarizer measurements obtained in two coplanar samples within 30 minutes are shown in Fig. 5.12 [82]. Sample degradation in the plane-parallel samples could only be prevented by applying polymer buffer layers in between the NLO layer and the electrodes. This allowed only relative measurements, since no equations were available for obtaining Pockels coefficients from multi-layer data.

Relative results from Fabry-Perot interferometric measurement with s polarized light (r_{13}) were brought into coincidence with the crossed polarizer results of the coplanar samples, showing the qualitative agreement between both types of measurements. Both suggest a significantly lower Pockels effect in the nematic phase than in the isotropic phase, quite the contrary of what was originally expected! In addition to a stepwise reduction of the EO effect at T_{cl} also a gradual decrease was observed on cooling. A similar result was obtained in the second LC polymer, the MONS polyurethane. These results cannot be ascribed to higher optical scattering in the LC phase, because scattering and depolarization effects had already been corrected for. The gradual temperature dependence, not the discontinuity at T_{cl} may, however, be attributed to piezo-electric contributions to the apparent Pockels coefficients. It was concluded that the smaller Pockels effect in the nematic phase was real and not an experimental artifact. The

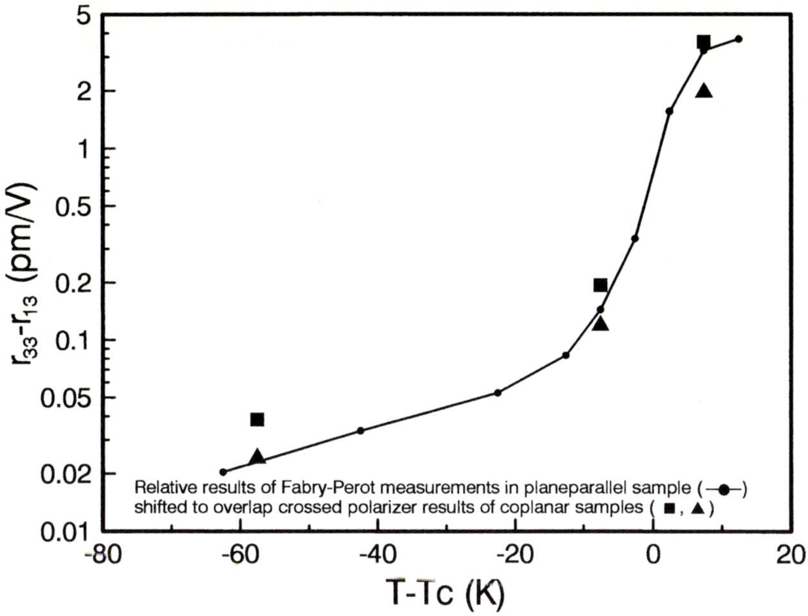

Fig. 5.12. Temperature dependence of NLO properties. Result of temperature dependent EO measurements in the nematic and isotropic phase of the MONS polyester (see Fig. 5.10). The T_{cl} has been taken as 87.5 °C. Apparently, the EO effect is smaller in the nematic phase, which was tentatively attributed to stronger anti-parallel dipolar association in the nematic phase in Ref. [82]. The dominant part of the apparent EO response in the planeparallel sample, however, seems to be piezo-electric in nature (see text)

smaller Pockels effect in the LC phase was attributed to larger anti-parallel dipole-dipole association in the nematic phase.

Since publication, new experiments have been performed in plane-parallel multi-layer samples above T_g, which strongly suggest that the apparent EO effect is predominantly piezo-electric. This piezo-electric "EO" effect grows with increasing temperature. A discontinuity at T_{cl} could, however, not be observed this time. Because of the dominant piezoelectric effect in the plane-parallel samples, not to mention interpretational problems due to differences in electrical conductivity between core and cladding (poling field strength unknown and perhaps temperature-dependent), it must be concluded that the multilayer plane-parallel samples are less suited for experiments far above T_g to be able to draw conclusions. Poling experiments in plane-parallel single layer samples below T_g in the frozen-in isotropic and frozen-in nematic phase did not show much difference, because of the formation of the nematic phase in the isotropic phase. Possibly, the frozen-in transparent "isotropic" phase was not really isotropic but poly-domain with small domain size. Of all the original experiments that suggest a smaller EO effect in the nematic phase, the above-mentioned crossed polarizer measurements in the coplanar samples close to T_{cl} remain.

5.5.5 Other Relevant Experiments

The experimental work of Dalmolen et al. in cyanobiphenyls [34] has already been mentioned in Sect. 5.3.8. From dielectric spectroscopy and employing Kirkwood-Fröhlich formalism, they found slightly larger anti-parallel pair formation in the nematic state than in the isotropic state.

Guan and Wang [84] performed SGH measurements on 2-methyl-4-nitroaniline/poly(methyl methacrylate) guest-host systems as a function of the NLO chromophore concentration and the poling field strength. The two polar order parameters were found to depend on the chromophore concentration, which was described in terms of the Kirkwood-Fröhlich parameter g using the model of Wang [35] (Sect. 5.3.9). For weak poling fields, the ratio of the two order parameters was independent of concentration and equal to 3. They state that this result is in agreement with a recent calculation. The Onsager local field factor, commonly used at low frequencies, but derived without considering pair correlations, was shown to be inadequate. The static dielectric constant of the guest-host system was extracted from the order parameter, using the Kirkwood-Fröhlich theory of dielectrics.

Amano and Kaino [85] studied the second-order NLO properties of a dye-attached polymer ("3R"). The dye has a conjugated diazo structure and a ß. μ that is larger than five times that of the monoazo dye ("2R"). For high dye concentrations, the attainable $\chi^{(2)}$ deviates from the theoretical linear relationship with the dye content. This indicates, according to the authors, that dipolar interaction increases as the dye content rises.

Toyshima et al. [86] examined the effect of an electric field on the temperature of the NI transition in dioxane solutions of poly-γ-benzyl-L-glutamate, ranging in concentration from 11 to 15 wt %, by means of polarization microscopy. The shift of the transition temperature $\Delta T_{cl}(E) = T_{cl}(E) - T_{cl}(0)$ was essentially proportional to E^2 for the very weak field strengths applied ($E < 5 \times 10^4$ V/m). A semiempirical equation with $\Delta T_{cl}(E)^2$ was derived, taking the structure of the solution into consideration, and compared with the experimental results.

5.6 Summary

It appears that the predictions from simple molecular statistical mean-field models, which neglect dipole-dipole interaction, compare well – at least qualitatively – in dilute systems with a low concentration of NLO groups and even in concentrated isotropic systems with a high NLO density. Special phenomena predicted in these concentrated systems – an electric field induced isotropic to nematic phase transition and a superproportional dependence of the EO effect as a function of the poling field strength – have indeed been observed experimentally. Apparently, the neglect of dipole-dipole interaction is not too serious an approximation invalidating the predictive power of the models. This does not, however, imply that the dipolar interaction is really negligible. Looking more quantitatively, the interaction might explain part of the discrepancy between theoretical and experimental values of EO coefficients. Concentrated LC polymers represent an even worse case, as far as the neglect of dipole-dipole interaction is concerned, since stronger anti-parallel dipolar association might be anticipated in the LC phase than in the isotropic phase. In fact, experiments in some concentrated LC side chain homopolymers seem to indicate a decrease of the EO coefficient in going from the isotropic to the LC phase. This would imply an effectively adverse instead of an enhancing effect from liquid crystallinity on field induced polar order. Examples of a positive influence from liquid crystallinity on poling in concentrated LC systems are rather scarce. However, the above-mentioned superproportional behavior of the EO effect during poling of initially isotropic concentrated polymers - tentatively attributed to a field-induced isotropic to nematic phase transition - can be regarded as an indirect experimental indication of the presence of such a positive influence.

Acknowledgements. The authors wish to thank Dr. C.T.J. Wreesmann, E.W.P. Erdhuisen and B.H.M. Hams for the organic syntheses, Prof. L.W. Jenneskens and Dr. J.W. Hofstraat for the DSC measurements, J.W. Mertens, Miss M. van Rheede and R. van Gassel for the preparation of the samples and several of the EO experiments. The authors are indebted to Prof. W. Mijs and Dr. Jin Shenggu for kindly supplying the MONS side chain polyurethane, which was synthesized as part of the doctorate study of Dr. Jin. Also, the authors thank D.F. van Embden for the artwork.

5.7 References

1. Williams DJ (ed) (1983) Nonlinear Optical Properties of Organic and Polymeric Materials. American Chemical Society, Washington (ACS Symposium Series, vol 233)
2. McArdle CB (ed) (1988) Side Chain Liquid Crystal Polymers. Blackie, Glasgow
3. Hornak LA (ed) (1992) Polymers for lightwave and integrated optics: Technology and applications. Marcel Dekker, New York
4. Williams DJ (1984) Angew Chem Int Ed Engl 23: 690
5. Meredith GR, van Dusen JG, Williams DJ (1982), Macromolecules 15: 1385
6. Meredith GR, van Dusen JG, Williams DJ (1983) In: Williams DJ (ed) Nonlinear optical properties of organic and polymeric materials. American Chemical Society, Washington, pp 109 (ACS Symposium Series, vol 233, chapter 5)
7. Garito AF, Singer KD, Teng CC (1993) In: Williams DJ (ed) Nonlinear optical properties of organic and polymeric materials. American Chemical Society, Washington, p 1 (ACS Symposium Series, vol 233, chapter 1)
8. Zyss J (1982) J Non-Cryst Sol 47: 211
9. Zyss J, Tsoucaris G (1986) Mol Cryst Liq Cryst 137: 303
10. le Grange JD, Kuzyk MG, Singer KD (1987) Mol Cryst Liq Cryst 150b: 567
11. Singer KD, Sohn JE, Lalama SJ (1986) Appl Phys Lett 49: 248
12. Singer KD, Kuzyk MG, Sohn JE (1987) J Opt Soc Am B4:968
13. Hill JR, Pantelis P, Davies GJ (1989) Institute of Physics Conference Series 103: 215
14. Möhlmann GR, van der Vorst CPJM (1988) In: McArdle CB (ed) Side Chain Liquid Crystal Polymers. Blackie, Glasgow, p 330 (chapter 12)
15. Van der Vorst CPJM, Horsthuis WHG, Möhlmann GR (1992) In: Hornak LA (ed) Polymers for lightwave and integrated optics: Technology and applications. Marcel Dekker, New York, p 365 (chapter 14)
16. Kaino T, Tomaru S (1993) Advanced Materials 5: 172

17. Neal DB, Petty MC, Roberts GG, Ahmad MM, Feast WJ, Girling IR, Cade NA, Kolinsky PV, Peterson IR (1986) Elec Lett 22: 460
18. Cresswell JP, Tsibouklis J, Petty MC, Feast WJ, Carr N, Goodwin M, Lvov YM (1990) SPIE Proceedings 1337: 358
19. Griffin AC, Bhatti AM, Hung RSL (1986) SPIE Proceedings 682: 65
20. le Barny P, Ravaux G, Dubois JC, Parneix JP, Njeumo R, Legrand C, Levelut AM (1986) SPIE Proceedings 682: 56
21. Stamatoff JB, Buckley A, Calundann G, Choe EW, DeMartino R, Khanarian G, Leslie T, Nelson G, Stuetz D, Teng CC, Yoon HN (1986) SPIE Proceedings 682: 85
22. Diemeer MBJ, Suyten FMM, Trommel ES, Möhlmann GR, Horsthuis WHG, van der Vorst CPJM, McDonach A, Copeland JM, Duchet C, Fabre P, Samso S, van Tomme E, van Daele P, Baets R (1989) Proceedings 15th European Conference on Optical Communications, p 425
23. Stupp SI, Lin HC, Wake DR (1992) Chem of Materials 4: 947
24. Stupp et al. (1992) Proceedings of the 25th Europhysics Conference on Macromolecular Physics, St. Petersburg, July 1992 (International Journal of Polymeric Materials, in print)
25. Willand CS, Williams DJ (1987) Ber Bunsenges Phys Chem 91: 1304
26. Ranon PM, Shi Y, Steier WH, Xu C, Wu B, Dalton LR (1993) Appl Phys Lett 62: 2605
27. Bottcher CFJ (1973) Theory of electric polarization. Elsevier, New York
28. Meredith GR (1987) SPIE Proceedings 824: 126
29. Van der Vorst CPJM, Picken SJ (1987) SPIE Proceedings 866: 99
30. Van der Vorst CPJM, Picken SJ (1990) J Opt Soc Am B7: 320
31. Maier W, Saupe A (1958) Z Naturforschung 13A: 564
32. Maier W, Saupe A (1959) Z Naturforschung 14A: 882
33. Maier W, Saupe A (1960) Z Naturforschung 15A: 287
34. Dalmolen LGP, Picken SJ, de Jong AF, de Jeu WH (1985) Journal Physique 46: 1443
35. Wang CH (1993) J Chem Phys 98: 3457
36. Wu JW (1991) J Opt Soc Am B8: 142
37. De Gennes PG (1971) Mol Cryst Liq Cryst 12: 193
38. De Gennes PG (1972) Sol State Comm 10: 753
39. De Gennes PG (1973) Mol Cryst Liq Cryst 21: 49
40. McMillan WL (1972) Phys Rev A6: 936
41. Gramsbergen EF, Longa L, de Jeu WH (1986) Phys Reports 135: 195
42. Vertogen G, De Jeu WH (1987) Thermotropic liquid crystals: Fundamentals Springer, Berlin Heidelberg New York
43. Onsager L (1949) Ann NY Acad Sci 51: 627
44. Warner M, Gunn JMF, Baumgartner AB (1985) J Phys A: Math Gen 18: 3007
45. Wang X-J, Warner M (1986) Phys Lett A119: 181
46. Gunn JMF, Warner M (1987) Phys Rev Lett 58: 393
47. Renz W, Warner M (1988) Proc Roy Soc A417: 213
48. Warner M (1988) Mol Cryst Liq Cryst 155: 433
49. Warner M, Liu H (1989) Liq Cryst 4: 325
50. Odijk T (1986) Macromolecules 19: 2313
51. Khokhlov AR, Semenov AN (1982) Macromolecules 15: 1272
52. Teng CC, Garito AF (1983) Phys Rev Lett 50: 350
53. Eich M, Sen A, Looser H, Yoon DY, Bjorklund GC, Twieg R, Swalen JD (1988) SPIE Proceedings 971: 128
54. Eich M, Ohl S, Zentel R, Baumann H, Goebel L, Kruger H, Frank W, Rschudi T (1992) SPIE Proceedings 1774: 27
55. Simon GP (1989) Pol Mag 30: 2227
56. Van der Vorst CPJM, Huyts RA (1989) SPIE Proceedings 1126: 6
57. Van der Vorst CPJM, van Weerdenburg CJM (1990) SPIE Proceedings 1337: 246
58. Sigelle M, Hierle R (1981) J Appl Phys 52: 4199
59. Uchiki H, Kobayashi T (1988) J Appl Phys 64: 2625
60. Kobayashi T, Uchiki H, Minoshima K (1988) SPIE Proceedings 971: 59
61. Kobayashi T, Minoshima K, Nomura S, Fukaya S, Ueki A (1989) SPIE Proceedings 1147: 182
62. Eldering CA, Knoesen A, Kowel ST (1990) SPIE Proceedings 1337: 348

63. Eldering CA, Knoesen A, Kowel ST (1991) J Appl Phys 69: 3676
64. Kowel ST, Eldering CA (1992) In: Hornak LA (ed) Polymers for lightwave and integrated optics: Technology and applications. Marcel Dekker, New York, p 473 (chapter 17)
65. Meyrueix R, Lecompte JP, Tapolsky G (1991) SPIE Proceedings 1560: 454
66. Meyrueix R, Lecompte JP, Tapolsky G (1991) Nonlinear Optics 1: 201
67. Meyrueix R (1993) SPIE Proceedings 2025 (in print)
68. Morichere D, Chollet PA, Fleming W, Jurich M, Smith BA, Swalen JD (1993) J Opt Soc Am B10: 1894
69. Bishop DM (1988) Phys Rev Lett B61: 322
70. Moylan CR, Swanson SA, Twieg RJ, Walsh CA, Thackara JI, Miller RD, Lee VY (1993) In: Organic Thin Films for Photonic Applications Topical Meeting, Toronto, October 5–7, 1993 (Optical Society of America, Technical Digest Series 17: 77)
71. Van der Vorst CPJM, van Rheede M (1992) SPIE Proceedings 1775: 186
72. Van der Vorst CPJM, van Rheede M, van Weerdenburg CJM (1992) Proceedings of the 25th Europhysics Conference on Macromolecular Physics, St. Petersburg, July 1992 (Int J of Pol Materials, in print)
73. Van der Vorst CPJM, van Rheede M, Hams BHM (1993) SPIE Proceedings 2025 (in print)
74. Bercovic G, Krongauz V, Yitzchaik S (1990) SPIE Proceedings 1442: 44
75. Gonin D, Noël C, le Borgne A, Gadret G, Kajzar F (1992) Makromol Chem., Rapid Commun 13: 537
76. Gonin D, Gadret G, Noël C, Kajzar F (1993) SPIE Proceedings 2025 (in print)
77. Saupe A (1974) In: Gray G, Winsor PA (ed) Liquid crystals and plastic crystals. Horwood, Chichester, vol 1, chapter 2
78. Stamatoff JB (1990) Int Photonics Research Topical Meeting, Hilton Head, S Carolina, March 1990 (Technical Digest Series 5, paper MA2)
79. Haas D, Teng CC, Yoon H, Man HT, Chiang K (1990) Int Photonics Research Topical Meeting, Hilton Head, S Carolina, March 1990 (Technical Digest Series 5, paper MF2)
80. Buckley A (1992) Adv Mater 4: 153
81. Kelderman E (1993) Molecules for second order non-linear optics Thesis, Twente University, Enschede
82. Van der Vorst CPJM, Picken SJ (1993) SPIE Proceedings 2025, in print
83. Jin S et al. (to be published)
84. Guan HW, Wang CH (1993) J Chem Phys 98: 3463
85. Amano M, Kaino T (1990) J Appl Phys 68: 6024
86. Toyoshima Y, Minami N, Sukigara M (1976) Mol Cryst Liq Cryst 35: 325

Printing: Saladruck, Berlin
Binding: Buchbinderei Lüderitz & Bauer, Berlin

Springer-Verlag and the Environment

We at Springer-Verlag firmly believe that an international science publisher has a special obligation to the environment, and our corporate policies consistently reflect this conviction.

We also expect our business partners – paper mills, printers, packaging manufacturers, etc. – to commit themselves to using environmentally friendly materials and production processes.

The paper in this book is made from low- or no-chlorine pulp and is acid free, in conformance with international standards for paper permanency.